Bio-Based Polymeric Films

Bio-Based Polymeric Films

Editors

Swarup Roy
Jong-Whan Rhim

MDPI • Basel • Beijing • Wuhan • Barcelona • Belgrade • Manchester • Tokyo • Cluj • Tianjin

Editors

Swarup Roy
School of Bioengineering and
Food Technology
Shoolini University
Solan
India

Jong-Whan Rhim
Department of Food and
Nutrition
Kyung Hee University
Seoul
Korea, South

Editorial Office
MDPI
St. Alban-Anlage 66
4052 Basel, Switzerland

This is a reprint of articles from the Special Issue published online in the open access journal *Polymers* (ISSN 2073-4360) (available at: www.mdpi.com/journal/polymers/special_issues/Biodegradable_Polymer_Films).

For citation purposes, cite each article independently as indicated on the article page online and as indicated below:

LastName, A.A.; LastName, B.B.; LastName, C.C. Article Title. *Journal Name* **Year**, *Volume Number*, Page Range.

ISBN 978-3-0365-5938-4 (Hbk)
ISBN 978-3-0365-5937-7 (PDF)

© 2022 by the authors. Articles in this book are Open Access and distributed under the Creative Commons Attribution (CC BY) license, which allows users to download, copy and build upon published articles, as long as the author and publisher are properly credited, which ensures maximum dissemination and a wider impact of our publications.

The book as a whole is distributed by MDPI under the terms and conditions of the Creative Commons license CC BY-NC-ND.

Contents

About the Editors vii

Swarup Roy and Jong-Whan Rhim
Advances and Challenges in Biopolymer-Based Films
Reprinted from: *Polymers* 2022, *14*, 3920, doi:10.3390/polym14183920 1

Nagaraj Basavegowda and Kwang-Hyun Baek
Advances in Functional Biopolymer-Based Nanocomposites for Active Food Packaging Applications
Reprinted from: *Polymers* 2021, *13*, 4198, doi:10.3390/polym13234198 3

Łukasz Łopusiewicz, Paweł Kwiatkowski, Emilia Drozłowska, Paulina Trocer, Mateusz Kostek and Mariusz Śliwiński et al.
Preparation and Characterization of Carboxymethyl Cellulose-Based Bioactive Composite Films Modified with Fungal Melanin and Carvacrol
Reprinted from: *Polymers* 2021, *13*, 499, doi:10.3390/polym13040499 27

Swarup Roy, Lindong Zhai, Hyun Chan Kim, Duc Hoa Pham, Hussein Alrobei and Jaehwan Kim
Tannic-Acid-Cross-Linked and TiO_2-Nanoparticle-Reinforced Chitosan-Based Nanocomposite Film
Reprinted from: *Polymers* 2021, *13*, 228, doi:10.3390/polym13020228 45

Shuo Li, Min Fan, Shanggui Deng and Ningping Tao
Characterization and Application in Packaging Grease of Gelatin–Sodium Alginate Edible Films Cross-Linked by Pullulan
Reprinted from: *Polymers* 2022, *14*, 3199, doi:10.3390/polym14153199 63

Daniela G. M. Pereira, Jorge M. Vieira, António A. Vicente and Rui M. S. Cruz
Development and Characterization of Pectin Films with *Salicornia ramosissima*: Biodegradation in Soil and Seawater
Reprinted from: *Polymers* 2021, *13*, 2632, doi:10.3390/polym13162632 81

Swarup Roy and Jong-Whan Rhim
Fabrication of Copper Sulfide Nanoparticles and Limonene Incorporated Pullulan/Carrageenan-Based Film with Improved Mechanical and Antibacterial Properties
Reprinted from: *Polymers* 2020, *12*, 2665, doi:10.3390/polym12112665 93

Katarina S. Postolović, Milan D. Antonijević, Biljana Ljujić, Slavko Radenković, Marina Miletić Kovačević and Zoltan Hiezl et al.
Curcumin and Diclofenac Therapeutic Efficacy Enhancement Applying Transdermal Hydrogel Polymer Films, Based on Carrageenan, Alginate and Poloxamer
Reprinted from: *Polymers* 2022, *14*, 4091, doi:10.3390/polym14194091 107

Enni Luoma, Marja Välimäki, Jyrki Ollila, Kyösti Heikkinen and Kirsi Immonen
Bio-Based Polymeric Substrates for Printed Hybrid Electronics
Reprinted from: *Polymers* 2022, *14*, 1863, doi:10.3390/polym14091863 135

Jake Richter, Moses Nnaji and Heungman Park
Solvent Effect to the Uniformity of Surfactant-Free Salmon-DNA Thin Films
Reprinted from: *Polymers* 2021, *13*, 1606, doi:10.3390/polym13101606 155

Anesh Manjaly Poulose, Arfat Anis, Hamid Shaikh, Abdullah Alhamidi, Nadavala Siva Kumar and Ahmed Yagoub Elnour et al.
Strontium Aluminate-Based Long Afterglow PP Composites: Phosphorescence, Thermal, and Mechanical Characteristics
Reprinted from: *Polymers* **2021**, *13*, 1373, doi:10.3390/polym13091373 . . . . **167**

About the Editors

Swarup Roy

Dr. Swarup Roy is currently working as an Assistant Professor at the School of Bioengineering and Food Technology, Shoolini University, India. Dr. Roy previously worked as a postdoctoral researcher in Kyung Hee University, and Inha University South Korea (2018–2021). He also worked as a research associate at IIT Indore (2016–2017). He received his Ph.D. in Biochemistry in 2016 from the University of Kalyani, India. His research work is based on the preparation and application of biopolymer-based composite material for food packaging.

Jong-Whan Rhim

Professor Jong-Whan Rhim received his Ph.D. in Food Engineering in 1988 from the Department of Food Science at North Carolina State University, Raleigh, North Carolina, USA. Currently, he is an emeritus professor in the Department of Food and Nutrition and serves as the director of the BioNanocomposite Research Center, at Kyung Hee University, Seoul, Korea. His current research interests are synthesizing and characterizing carbon and sulfur quantum dots and their composites for intelligent and active packaging applications.

Editorial

Advances and Challenges in Biopolymer-Based Films

Swarup Roy [1,*] and Jong-Whan Rhim [2,*]

1. School of Bioengineering and Food Technology, Shoolini University, Solan, Himachal Pradesh 173229, India
2. Department of Food and Nutrition, BioNanocomposite Research Institute, Kyung Hee University, 26 Kyungheedae-ro, Dongdaemun-gu, Seoul 02447, Korea
* Correspondence: swaruproy2013@gmail.com (S.R.); jwrhim@khu.ac.kr (J.-W.R.)

Today, biobased polymers derived from sustainable and renewable natural sources are of great interest as an alternative to control the severe damage already caused by petrochemical-based polymers. The extensive use of non-biodegradable plastics in the packaging sector produces an enormous amount of waste, ultimately ending up in landfills and the ocean. The scenario of packaging pollution has become more severe owing to the COVID-19 pandemic, and recently, many countries have to ban the use of single-use plastics, since worldwide yearly manufacturing of plastic materials is ~400 million tonnes; interestingly, about 40% of these materials are utilized for single-use packaging materials [1]. Thus, the present scenario urgently demands the replacement of synthetic plastics with biobased alternatives. The importance of biopolymers in packaging must be considered to provide a better and more sustainable future. Biopolymers are not a new concept and have been used since ancient times, but research on using biopolymers as a replacement for packaging materials began in the early 2000s. The use of biopolymers in developing packaging materials is a promising field of research as it comprehensively reduces plastic waste and decreases greenhouse gas emissions [2–4]. Varieties of biopolymers originating from renewable products and food waste, such as polysaccharides (cellulose, chitosan, pectin, carrageenan, agar, etc.), proteins (gelatin, soy protein isolate, zein, etc.), or their blends (gelatin/agar, chitosan/pullulan, pectin/agar, gelatin/zein, etc.), have been used in this regard for film production [5–8]. Biopolymers have great potential to replace conventional plastics due to their non-toxicity, biocompatibility, and fast degradability.

Moreover, biopolymers can make a good film with excellent physical properties. Furthermore, biobased polymers are good sources of the carriers of bioactive ingredients that can impart functionality to the packaging material to improve the food shelf-life of packed food [9,10]. Biopolymer-based film has been extensively used in fabricating various active and smart packaging films and coatings [11,12]. Current reports suggest that biobased-blend polymer-based packaging film showed comparable physical properties to convenient polymers-based film. Moreover, introducing active and intelligent packaging makes biobased polymers more popular in the packaging sector [13–15]. Even though the use of biopolymers is advantageous in many respects, especially to address plastic waste and food safety concerns, there are still many limitations compared to its counterpart, which need to be resolved to meet the requirement of synthetic plastics [16,17]. Synthetic plastics are easy to handle, cost-effective, highly flexible, and water-insoluble, which make them convenient for making a suitable product used in the packaging regime.

On the other hand, biobased polymers are costly and generally hydrophilic [18,19]. Therefore, improving hydrophobicity, cost minimization, and scale-up production of packaging film using biobased polymers could solve the drawbacks of biopolymers. As eco-friendly packaging materials, there are plenty of opportunities for biobased polymers in the food sector. The worldwide market price of biopolymers is increasing at a rate of ~6–7% [20]. Nevertheless, more research is still required to address the challenges related to biopolymers for their practicability as a potential material for food packaging films.

Citation: Roy, S.; Rhim, J.-W. Advances and Challenges in Biopolymer-Based Films. *Polymers* 2022, 14, 3920. https://doi.org/10.3390/polym14183920

Received: 26 August 2022
Accepted: 15 September 2022
Published: 19 September 2022

Publisher's Note: MDPI stays neutral with regard to jurisdictional claims in published maps and institutional affiliations.

Copyright: © 2022 by the authors. Licensee MDPI, Basel, Switzerland. This article is an open access article distributed under the terms and conditions of the Creative Commons Attribution (CC BY) license (https://creativecommons.org/licenses/by/4.0/).

Funding: This work was supported by the National Research Foundation of Korea (NRF) grant funded by the Korean government (MSIT) (No. 2022R1A2B02001422).

Conflicts of Interest: The authors declare no conflict of interest.

References

1. Chawla, S.; Varghese, B.S.; Chithra, A.; Hussain, C.G.; Keçili, R.; Hussain, C.M. Environmental Impacts of Post-Consumer Plastic Wastes: Treatment Technologies towards Eco-Sustainability and Circular Economy. *Chemosphere* **2022**, *308*, 135867. [CrossRef] [PubMed]
2. Bayer, I.S. Biopolymers in Multilayer Films for Long-Lasting Protective Food Packaging: A Review. *Sustain. Food Packag. Technol.* **2021**, *15*, 395–426. [CrossRef]
3. Díaz-Montes, E.; Castro-Muñoz, R. Edible Films and Coatings as Food-Quality Preservers: An Overview. *Foods* **2021**, *10*, 249. [CrossRef] [PubMed]
4. Roy, S.; Priyadarshi, R.; Ezati, P.; Rhim, J.-W. Curcumin and Its Uses in Active and Smart Food Packaging Applications—A Comprehensive Review. *Food Chem.* **2022**, *375*, 131885. [CrossRef] [PubMed]
5. Roy, S.; Rhim, J.-W. Starch/Agar-Based Functional Films Integrated with Enoki Mushroom-Mediated Silver Nanoparticles for Active Packaging Applications. *Food Biosci.* **2022**, *49*, 101867. [CrossRef]
6. Roy, S.; Priyadarshi, R.; Rhim, J.-W. Gelatin/Agar-Based Multifunctional Film Integrated with Copper-Doped Zinc Oxide Nanoparticles and Clove Essential Oil Pickering Emulsion for Enhancing the Shelf Life of Pork Meat. *Food Res. Int.* **2022**, *160*, 111690. [CrossRef] [PubMed]
7. Hadidi, M.; Jafarzadeh, S.; Forough, M.; Garavand, F.; Alizadeh, S.; Salehabadi, A.; Khaneghah, A.M.; Jafari, S.M. Plant Protein-Based Food Packaging Films; Recent Advances in Fabrication, Characterization, and Applications. *Trends Food Sci. Technol.* **2022**, *120*, 154–173. [CrossRef]
8. Umaraw, P.; Munekata, P.E.S.; Verma, A.K.; Barba, F.J.; Singh, V.P.; Kumar, P.; Lorenzo, J.M. Edible Films/Coating with Tailored Properties for Active Packaging of Meat, Fish and Derived Products. *Trends Food Sci. Technol.* **2020**, *98*, 10–24. [CrossRef]
9. Priyadarshi, R.; Roy, S.; Ghosh, T.; Biswas, D.; Rhim, J.-W. Antimicrobial Nanofillers Reinforced Biopolymer Composite Films for Active Food Packaging Applications—A Review. *Sustain. Mater. Technol.* **2022**, *32*, e00353. [CrossRef]
10. Souza, A.G.; Ferreira, R.R.; Paula, L.C.; Mitra, S.K.; Rosa, D.S. Starch-Based Films Enriched with Nanocellulose-Stabilized Pickering Emulsions Containing Different Essential Oils for Possible Applications in Food Packaging. *Food Packag. Shelf Life* **2021**, *27*, 100615. [CrossRef]
11. Roy, S.; Rhim, J.-W. Anthocyanin Food Colorant and Its Application in PH-Responsive Color Change Indicator Films. *Crit. Rev. Food Sci. Nutr.* **2021**, *61*, 2297–2325. [CrossRef] [PubMed]
12. Soltani Firouz, M.; Mohi-Alden, K.; Omid, M. A Critical Review on Intelligent and Active Packaging in the Food Industry: Research and Development. *Food Res. Int.* **2021**, *141*, 110113. [CrossRef] [PubMed]
13. de Abreu, D.A.P.; Cruz, J.M.; Losada, P.P. Active and Intelligent Packaging for the Food Industry. *Food Rev. Int.* **2012**, *28*, 146–187. [CrossRef]
14. Yam, K.L.; Takhistov, P.T.; Miltz, J. Intelligent Packaging: Concepts and Applications. *J. Food Sci.* **2005**, *70*, R1–R10. [CrossRef]
15. Yousefi, H.; Su, H.M.; Imani, S.M.; Alkhaldi, K.; Filipe, C.D.; Didar, T.F. Intelligent Food Packaging: A Review of Smart Sensing Technologies for Monitoring Food Quality. *ACS Sens.* **2019**, *4*, 808–821. [CrossRef] [PubMed]
16. Rosseto, M.; Rigueto, C.V.T.; Krein, D.D.C.; Balbé, N.P.; Massuda, L.A.; Dettmer, A. Biodegradable Polymers: Opportunities and Challenges. *Org. Polym.* **2019**, 110–119. [CrossRef]
17. Taherimehr, M.; YousefniaPasha, H.; Tabatabaeekoloor, R.; Pesaranhajiabbas, E. Trends and Challenges of Biopolymer-Based Nanocomposites in Food Packaging. *Compr. Rev. Food Sci. Food Saf.* **2021**, *20*, 5321–5344. [CrossRef] [PubMed]
18. Hoque, M.; Gupta, S.; Santhosh, R.; Syed, I.; Sarkar, P. Biopolymer-Based Edible Films and Coatings for Food Applications. *Food Med. Environ. Appl. Polysacch.* **2021**, 81–107. [CrossRef]
19. Kim, H.-J.; Roy, S.; Rhim, J.-W. Gelatin/Agar-Based Color-Indicator Film Integrated with Clitoria Ternatea Flower Anthocyanin and Zinc Oxide Nanoparticles for Monitoring Freshness of Shrimp. *Food Hydrocoll.* **2022**, *124*, 107294. [CrossRef]
20. Suhag, R.; Kumar, N.; Petkoska, A.T.; Upadhyay, A. Film Formation and Deposition Methods of Edible Coating on Food Products: A Review. *Food Res. Int.* **2020**, *136*, 109582. [CrossRef] [PubMed]

Review

Advances in Functional Biopolymer-Based Nanocomposites for Active Food Packaging Applications

Nagaraj Basavegowda and Kwang-Hyun Baek *

Department of Biotechnology, Yeungnam University, Gyeongsan 38541, Gyeongbuk, Korea; nagarajb2005@yahoo.co.in
* Correspondence: khbaek@ynu.ac.kr; Tel.: +82-52-810-3029

Abstract: Polymeric nanocomposites have received significant attention in both scientific and industrial research in recent years. The demand for new methods of food preservation to ensure high-quality, healthy foods with an extended shelf life has increased. Packaging, a crucial feature of the food industry, plays a vital role in satisfying this demand. Polymeric nanocomposites exhibit remarkably improved packaging properties, including barrier properties, oxygen impermeability, solvent resistance, moisture permeability, thermal stability, and antimicrobial characteristics. Bio-based polymers have drawn considerable interest to mitigate the influence and application of petroleum-derived polymeric materials and related environmental concerns. The integration of nanotechnology in food packaging systems has shown promise for enhancing the quality and shelf life of food. This article provides a general overview of bio-based polymeric nanocomposites comprising polymer matrices and inorganic nanoparticles, and describes their classification, fabrication, properties, and applications for active food packaging systems with future perspectives.

Keywords: biopolymer; nanocomposites; shelf life; food packaging; antimicrobial

Citation: Basavegowda, N.; Baek, K.-H. Advances in Functional Biopolymer-Based Nanocomposites for Active Food Packaging Applications. *Polymers* **2021**, *13*, 4198. https://doi.org/10.3390/polym13234198

Academic Editors: Swarup Roy and Jong-Whan Rhim

Received: 25 October 2021
Accepted: 27 November 2021
Published: 30 November 2021

Publisher's Note: MDPI stays neutral with regard to jurisdictional claims in published maps and institutional affiliations.

Copyright: © 2021 by the authors. Licensee MDPI, Basel, Switzerland. This article is an open access article distributed under the terms and conditions of the Creative Commons Attribution (CC BY) license (https://creativecommons.org/licenses/by/4.0/).

1. Introduction

Nanoscience and nanotechnology have matured into extremely active, vital, and expanding fields of research for developing small particles with multidimensional applications in the areas of nutrition, agriculture, cosmetics, paints and coatings, personal care products, catalysts, energy production, lubricants, security printing, molecular computing, structural materials, drug delivery, medical therapeutics, pharmaceuticals, and diagnostics [1]. The remarkably small size of these materials provides a large surface-area-to-volume ratio and, consequently, more surface atoms compared to their microscale counterparts. This improves the properties of materials with negligible defects on their surfaces [2]. Moreover, nanomaterials have been developed as nanocomposites, which are engineered solid materials that result when two or more different constituent materials with different physical and chemical properties are combined to create new substances [3]. Nanocomposites are hybrid materials consisting of mixtures of polymers and inorganic solids (such as clays and oxides) at the nanometer scale. The remarkably complicated structure of nanocomposites, in which one phase (such as nanoparticles (NPs) and nanotubes) has a nanoscale morphology, exhibits properties that are superior to those of microcomposites in an assembled structure [4].

Polymeric nanocomposites are produced by dispersing NPs or nanofillers into polymeric matrices. This reinforcement results in a matrix with unique, enhanced physical and mechanical properties [5]. The combination of these two materials produces a synergistic effect with special properties that are not exhibited by the individual components. Moreover, the preparation of nanocomposites with more than two components effectively assists in satisfying the design and strength requirements for specific applications [6]. Inorganic NPs have unique properties such as mechanical, magnetic, electrical, and catalytic characteristics. On the other hand, polymers are assemblies of different monomers with properties such as low weight, flexibility, and low-cost production. Combining these two substances yields novel, unique materials with high performance and unusual, incomparable properties [2]. Polymeric nanocomposites have attracted considerable attention because of their unique and enhanced mechanical, optical, thermal, diffusion barrier, magnetic, and electric properties compared with those of micro-, conventional, and individual components [7]. These unique and incomparable properties of polymeric nanocomposites and their synergistic multifunctions, achieved through the incorporation of multiple components into one compatible entity, have led to their broad application in different fields [8].

Biopolymers or biodegradable polymers are renewable natural resources generated from biological systems, such as plants, animals, and microorganisms, and/or chemically synthesized from the starting materials of natural fats or oils, sugars, and starch [9]. Natural biopolymers are alternatives to synthetic polymers obtained from non-renewable petroleum resources. Therefore, biodegradable polymers can be disintegrated or degraded by the enzymatic action of specific microorganisms, organic byproducts, methane, inorganic compounds, biomass, carbon dioxide, and water [10]. Biopolymers offer several advantages, such as low-cost extraction, biocompatibility, biodegradability, environmental friendliness, and lack of environmental toxicity. Therefore, biopolymers have been traditionally used in various industrial activities related to the biomedical [11], pharmaceutical [12], food [13], and environmental sectors [14]. Examples of biopolymers include protein isolates (soy, wheat, corn, gluten, whey, and gelatin), carbohydrates (pullulan and curdlan), polysaccharides (chitosan, alginates, starch, and cellulose derivatives), and lipids (bees, wax, and free fatty acids) [15]. Additionally, polylactic acid (PLA), polyvinyl alcohol (PVA), polycaprolactone (PCL), polyhydroxybutyrate (PHB), polybutylene succinate (PBS), and their blends are some examples of synthetic biopolymers [16].

Food packaging is a particularly critical step in the protection and preservation of food to ensure its safety and increase its shelf life. This factor plays a key role in the global food industry in terms of satisfying consumer demand for safe, fresh, high-quality, durable, and healthy food, along with addressing certain challenges such as cost efficiency, environmental issues, consumer convenience, and food safety regulations [17]. Food packaging has a crucial function in the modernized food industries, as package performs a sequence of tasks—primarily containment, safety, handiness, and communicating information. These functions must be evaluated and considered simultaneously during the packaging and development process, as they are all interconnected [18]. Therefore, the food industry is constantly in pursuit of novel and alternative technologies for improving critical parameters such as the quality, safety, and shelf life of food products. Food packaging is primarily used to protect food products against adverse external environmental elements, including microorganisms, heat, light, oxygen, moisture, enzymes, pressure, insects, dust, and dirt [19]. In addition, packaging provides tampering resistance, physical support, and chemical or biological requirements. It also prevents spoilage and contamination, and increases sensitivity by enabling the enzyme activity of food products in the process of storage, transport, and distribution [20]. The incorporation of functional nanomaterials into biopolymer matrices improves the physicochemical properties [21], mechanical and barrier properties, moisture stability, durability, and flexibility of food packaging materials. Moreover, this blending improves active and smart/intelligent packaging functions, such as antimicrobial, antioxidant, UV protective, and nanosensing characteristics for the detection of small organic molecules and gases [22].

Biopolymer-based nanocomposites are a new type of material that exhibit considerably enhanced properties such as barrier, mechanical, and thermal characteristics, and have been considered as novel and alternative packaging materials [23]. Biopolymer nanocomposites are bio-based multiphase materials composed of two or more constituents, in which the continuous phase (matrix) is a biopolymer, and the discontinuous phase (fillers) is composed of NPs. These packaging materials can interact with food by releasing certain active substances, such as antimicrobial and antioxidant agents, or by removing unfavorable elements such as water vapor and oxygen [24]. This review presents a systematic assessment of the most recent advances in research on the development of polymeric nanocomposites for food packaging systems based on their classification, fabrication methods, properties, and applications. In addition, the review highlights antimicrobial activities, the related mechanisms of action, and future perspectives for ensuring the safety of nanomaterials for active food-packaging systems. The development of smart (active/intelligent) food packaging materials using bio-based nanocomposite materials can play an important role not only in minimizing environmental issues but also in enhancing the functions of food packaging systems.

2. Polymer Nanocomposites

Composite materials are produced by combining two or more distinct constituents or phases with different physical or chemical properties [25]. Although the different materials do not blend into each other, they yield unique properties (mechanical, physical, thermal, and electrical) in a supplementary manner upon being composited and engineered into a complex architecture at the micro- or macro-scale levels [26]. Composite materials typically comprise two phases: a continuous phase (matrix) and a discontinuous phase (reinforcements). Matrix materials (ceramics, metals, or polymers) are responsible for maintaining the positions of the reinforcement materials, whereas the filler materials (fibers and particles) impart new properties to the matrix phase [26]. Polymers have gained significant attention owing to their unique properties such as low cost, high flexibility, low weight, high strength, specific stiffness, biocompatibility, and ease of production [27]. Nanocomposites or nanofillers such as organic and inorganic materials, clay, and carbon nanostructures are used as coatings. Organic nanofillers include polymer nanofibers, natural fibers, and natural clay [27], and inorganic nanofillers include metals (such as Au, Ag, and Fe), metal oxides (CuO, ZnO, FeO, and TiO) [28], and carbon (fullerenes, graphene, carbon nanotubes, and nanofibers) [29].

Polymeric nanocomposites are a combination of a polymer matrix (continuous phase) and inorganic NPs (discontinuous phase) with at least one dimension at the nanometer scale. These composites exhibit improved properties compared to those of polymers, such as high strength, thermal stability, electrical conductivity, chemical resistance, flame retardancy, and optical characteristics [30]. Commonly used polymers in the food packaging industry, as either matrices or substrates for coatings, include synthetic low-density polyethylene (LDPE), high-density polyethylene (HDPE), polyethylene terephthalate (PET), polypropylene (PP), polyamide (PA), ethylene vinyl alcohol (EVOH), and polystyrene (PS). Naturally occurring biodegradable polymers or biopolymers, such as polyhydroxyalkanoates (PHAs), PHB, PLA, poly(hydroxybutyrate-*co*-hydroxyvalerate) (PHVB), polyvinyl alcohol (PVOH), and PCL, which are polyesters produced by numerous microorganisms and the bacterial fermentation of sugars or lipids, are also similarly employed [31]. Various types of nanofillers have been used to fabricate nanocomposites. The most commonly used nanofillers are clay nanomaterials (such as montmorillonite, kaolinite, halloysite, saponite, hectorite, and laponite), silica NPs, carbon nanotubes (CNTs), nanosheets, graphene, silver, copper, zinc, titanium dioxide, copper oxide, zinc oxide, cellulose nanofibers, starch nanocrystals, chitosan, and chitin whiskers [32–34]. Achieving a uniform and homogenous dispersion of NPs in polymer matrices is the key challenge in obtaining nanocomposites with desirable properties [35]. The uniform dispersion of nanofillers (nanoscale dispersion) can lead to a large interfacial area in the composite matrix. This reinforcement depends on several factors, such as particle size, distribution, orientation, structure, and the properties and concentrations of the polymer matrix and filler [36].

2.1. Classification of Polymer Nanocomposites

Polymer nanocomposites can be categorized based on the dimensions of nanofillers (0D, 1D, 2D, and 3D), type of nanofiller (metal and metal oxide, metal sulfide, metal hydroxide, and silicate), type of polymer matrix (thermoplastic, thermoset, elastomer, natural, and biodegradable polymer matrix), and synthesis methods (ex situ, in situ, and simultaneous polymerization).

2.1.1. Classification Based on the Dimension of Nanofillers

Nanofillers can be categorized into zero- (0D), one- (1D), two- (2D), or three-dimensional (3D) nanoparticles according to their dimensions at the nanoscale (Figure 1).

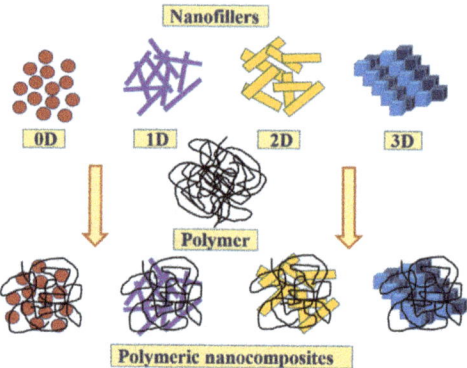

Figure 1. Classification of nanomaterials based on their dimensionality: zero- (0D), one- (1D), two- (2D), and three-dimensional (3D) nanocomposites.

Zero-dimensional (0D) nanofillers have dimensions that are all within the nano-range, such as NPs, that is, they do not have any dimension beyond 100 nm. These nanofillers may be amorphous, crystalline, metallic, ceramic, or polymeric in nature [37], and include materials such as gold, silver, and quantum dots (diameters in the range of 1–50 nm). One-dimensional (1D) nanofillers have a prominent dimension along one direction, and a nanostructure that is outside the nanometer range. These materials are long and have diameters of only a few nanometers [38]. These include nanotubes, nanorods, and nanowires of metals and metal oxides. Two-dimensional (2D) nanofillers have prominent dimensions in two directions and nanostructures that are outside the nanometer range. These large and extremely thin materials include nanofilms, nanosheets, nanowalls, nanofibers, and nanowhiskers; carbon nanotubes and montmorillonite are some of the noteworthy examples [39]. Three-dimensional (3D) nanofillers have dimensions along three directions, and nanostructures that are outside the nanometer range; however, the size of the individual blocks (structural units) is on the nanometer scale. These materials include nanoclays, nanogranules, and equiaxed NPs; zeolites are a notable example [40].

2.1.2. Classification Based on the Types of Nanofillers

Metal or metal oxide NPs are homogenously dispersed or spread onto the polymer matrix to form a homogenous nanophase-separated structure that imparts the nanocomposites with flame retardancy and thermal stability. Silica, zinc, magnesium, titanium oxide, zirconium oxide, aluminum oxide, and iron oxide are typical examples in this regard [41]. Metal-sulfide/polymer nanocomposites have attracted considerable attention owing to their thermal, optical, electrical, and mechanical properties. Nanocomposites containing metal sulfide NPs incorporated into a polymer matrix exhibit enhanced thermal stability compared to that of pure polymers [42]; CdS, ZnS, and HgS are prominent examples of this type of nanofiller. Metal hydroxide NPs, such as aluminum and magnesium hydroxide NPs, are known to exhibit remarkable flame retardancy and thermal stability. Magnesia and alumina powders or double hydroxides, such as zinc and alumina, have been deposited as fillers on the surfaces of polymer matrices to prepare flame-retardant polymer nanocomposites [43]. Silicate/polymer nanocomposites exhibit decent mechanical and thermal properties. The direct mixing of silica into polymers is the simplest method for preparing these silicate/polymer nanocomposites. This mixing can be achieved by solution and melt blending to prepare flame-resistant silicate/polymer nanocomposites [44].

2.1.3. Classification Based on Type of Polymer Matrix

As the name suggests, thermoplastic resins are polymers that can be molded/softened under pressure and heat, and hardened by cooling. Thermoplastic resins exhibit remarkable properties such, as high strength, high moldability, and chemical resistance, remolding, and recycling characteristics. For example, PS, PA, polyethylene (PE), and polyvinyl chloride (PVC) form thermoplastic networks [45]. Thermosetting polymers are generally liquid materials at room temperature and are the opposite of thermoplastics. They have three-dimensional covalent-bonded structures and, therefore, cannot be remolded by a heating–cooling process similar to that of thermoplastics. Moreover, these polymers harden irreversibly upon heating. Polyesters, vulcanized rubber, polyurethanes, and epoxy resins are examples of thermosetting polymers [46]. Elastomers, as their name suggests, exhibit elastic properties and can be stretched to a large extent without any damage. They are viscoelastic in nature because of their high viscosities and weak intermolecular forces. Polybutadiene, chloroprene, epichlorohydrin, and natural, silicone, and polyacrylic rubber materials are examples of elastomer matrices [47]. Biodegradable polymers or biopolymers are more soft and flexible than other polymeric materials. Starch, cellulose, chitosan, collagen, and proteins are the main sources of natural polymers, and PLA, PHB, and poly (3-hydroxybutyrate-co-3-hydroxyvalarate (PHBV) are typical examples of biopolymers.

2.2. Methods of Polymer Nanocomposite Preparation

Nanocomposites are prepared by incorporating nanofillers into polymer matrices to provide or improve the barrier, thermal, and mechanical properties of polymers. Polymer nanocomposites can be formulated by various approaches, including in situ polymerization, melt processing, and solution blending or casting methods.

2.2.1. In Situ Polymerization

This is an effective method that has been extensively used in the last few decades to prepare polymer nanocomposites. Typically, nanomaterials and monomers or multiple monomers are mixed in a suitable solvent, followed by polymerization with an appropriate reagent to yield polymer nanocomposites (Figure 2). This method enables the fabrication of well-defined multidimensional structures with distinct properties from the initial precursors. Homogenous dispersion in a polymer matrix can be achieved by this technique, which also assists in controlling the size, shape, and morphology of the nanomaterials [2].

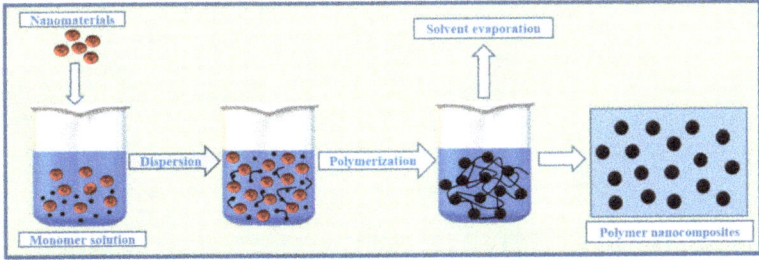

Figure 2. Schematic illustration of the in situ polymerization method.

2.2.2. Melt Processing

The melt processing technique is frequently used for thermoplastic polymers and is recognized as an economically viable, environmentally friendly, and green (solvent-free) technique. In this method, nanofillers or clay materials are incorporated into the polymer matrix by high-temperature annealing and rigorous mixing for a certain duration to encourage the intercalation and exfoliation of silicates, clay, or nanofillers, until a uniform distribution is achieved (Figure 3). The uniform distribution ensures a surface-modification-related compatibility with the host polymer and the processing conditions of the nanofiller [48]. Melt intercalation is a particularly attractive technique because of its durability and compatibility with current polymer processing techniques, such as extrusion and injection molding. This method is environmentally friendly because of the absence of solvents. Moreover, this method permits the use of polymers that cannot be prepared by the in situ polymerization and solution interaction methods [49].

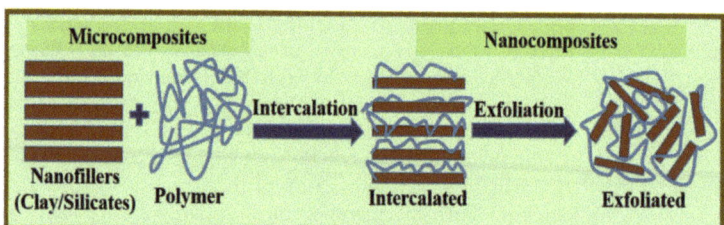

Figure 3. Schematic illustration of a nanofiller/polymer nanocomposite.

2.2.3. Solution Casting

This is the classical method for the production of polymer thin films with distinct thicknesses. In this technique, the polymer is dissolved in a specific solvent by continuous stirring, and the nanofillers are dispersed into the polymer solution to form a homogeneous mixture. This mixture is subsequently cast in a mold to evaporate the solvent and eventually yield thin films with polymer-oriented layers of intercalated clay [2]. Both organic solvents and water can be used to develop nanocomposites with either thermosets or thermoplastics (Figure 4). This method has been widely used to prepare nanocomposites containing water-soluble polymers, such as polyethylene oxide (PEO) and PVOH, and non-aqueous solvent-soluble polymers, including PCL and PLA in chloroform, and HDPE with xylene and benzonitrile solvent [49].

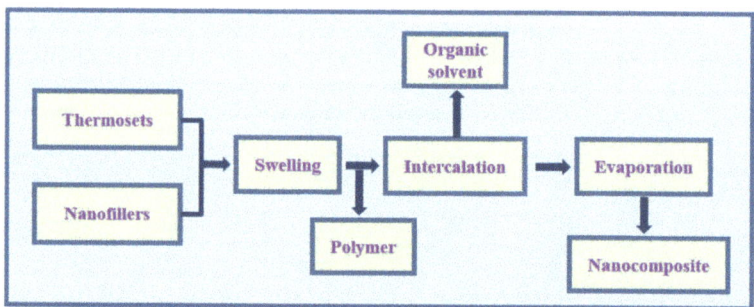

Figure 4. Schematic illustration of solution casting method.

2.3. Properties of Polymer Nanocomposites

The properties of polymer nanocomposites based on their microstructure, such as the degree of crystallinity, polar or non-polar groups, the presence of an amorphous phase, and the degree of crosslinking, are correlated with their high aspect ratios, the nature of nanofillers, the chemistry of polymer matrices, and the preparation method. The uniform dispersion of nanofillers in polymer matrices is essential for achieving the desired physical and mechanical characteristics. The barrier, mechanical, thermal, optical, and functional properties of nanocomposites are important parameters in food packaging systems.

2.3.1. Mechanical Properties

The primary intention of packaging is to preserve food from extrinsic deficiencies and influences, such as cracks and random breaks, in the packaging materials. The mechanical properties of polymers, such as strain at break, tenacity, and maximum stress, can be upgraded for use in food packaging systems [50]. Therefore, certain nanofillers are dispersed or reinforced in polymer matrices to improve properties such as strength and stiffness via a reinforcement mechanism [51]. This reinforcement primarily depends on the size, shape, concentration, orientation, surface area, dispersion state, and polydispersity of the nanofillers, possibly resulting in their grafting to the matrix polymers [52]. Nanocomposites prepared with small amounts of fillers exhibit a superior mechanical performance; in this regard, increasing the amount of filler diminishes all of the mechanical properties. These nanofillers can significantly improve the mechanical properties and firmness of the nanocomposites, which in turn alters their relaxation behavior and molecular mobility [53]. Appropriately distributed and aligned clay platelets are remarkably effective in improving the stiffness and mechanical properties of polymer materials, including Young's modulus (E), strain at break, and stress at break (σ_{max}) for polystyrene nanocomposites [54]. An adequate cohesion between the polymer and filler components increases the values of E, σ_{max}, and heat resistance, and improves the shear resistance, exfoliation, and corrosion resistance [55]. Different biodegradable polymers reinforced with chitin nanofiber showed higher tensile strength and lower elongation at break values [56].

Particle/polymer matrix interface plays the key role in determining the performance of advanced composite materials such as mechanical properties and dimensional stability. Interfacial adhesion occurs when two different materials such as particle and polymer matrix are blended or combined and create a better dispersion of materials into the matrices. However, the combination of materials must have the same properties, such as hydrophobic fillers and hydrophobic matrices or hydrophilic and hydrophilic materials to achieve a better interfacial adhesion and a strong bond between both materials [57]. Recent studies revealed that the effective mechanical reinforcement of polymeric nanocomposites containing spherical particle fillers is predicted based on a generalized analytical three-phase-series-parallel model [58]. The mechanical properties of particle–polymer matrix composites also depend strongly on the particle size and particle loading. A good adhesion between the fillers and the matrix is a prerequisite for high strength in the resulting composite.

2.3.2. Barrier Properties

Although polymeric materials offer several advantages in the packaging sector, a significant benefit is their intrinsic permeability to small molecules and other gases. Therefore, the loss of quality of packaged food products is due to either oxygen exposure or consistent changes in the movement of water vapor via the walls of the polymer packaging [35]. The permeability of gases and small molecules through the polymer matrices is controlled by various factors, such as the diffusivity, solubility, and morphology of the polymers. Therefore, the barrier properties of polymers are significantly correlated to their intrinsic ability to permit the exchange of low-molecular-weight substances. The structures of nanocomposites and the type and size of nanofillers can affect the degree of modification of the barrier properties of nanocomposites [59]. Well-dispersed nanofillers in the polymer matrix can influence the diffusivity and solubility of the penetrating molecules, particularly in interfacial domains, by increasing the diffusion length and the tortuous path of penetrating molecules to form an impermeable structure in the polymer matrix because of their high aspect ratio [60]. The barrier properties are also affected by the shape, polarity, and crystallinity of the diffusing molecule, the degree of crosslinking, and polymer chains [61]. Improved gas barrier properties and superior permeability have been exhibited by latex membranes and platelet-shaped fillers, respectively, compared to those of the neat membranes [62].

2.3.3. Thermal Properties

Thermal properties are crucial for the use of polymeric materials in a variety of applications, including packaging for consumer products. The low thermal conductivity of polymers or the mismatch between the thermal expansion characteristics of fillers and polymeric components is a major technological barrier [63]. Thermally stable neat polymers typically exhibit thermal conductivities in the range of 0.1–1.4 W/m·K; however, most nanofillers or nanomaterials exhibit high thermal conductivities in the range of 100–400 W/m·K [64]. The use of different types and concentrations of nanofillers plays a vital role in the thermal stability of polymeric nanocomposites, with the nanofillers exhibiting higher E (Young's modulus) values and lower thermal expansion coefficients than those of the polymer components [65]. Nanofillers act as barricades to heat and mass transfer, and reduce the diffusion of gaseous products and the molecular mobility of polymers, which prohibits heat-induced polymer degradation. The combined chemical and physical mechanisms also enhance the thermal stability of polymeric nanocomposites. These routes are the major mechanisms behind the thermal stability of polymeric nanocomposites [66].

2.3.4. Flame Retardancy

The propensity of materials to spread flame away from a fire source must be clearly understood, particularly for several thermoplastic materials, which tend to melt and produce flammable flow or drips, causing fire hazards. Therefore, the flame retardancy of polymeric materials must be improved by the incorporation of flame retardants [67]. Nanofillers such as CNTs and clay are attractive materials as flame retardants because they can concurrently improve the flammability and physical properties of polymeric nanocomposites. However, nanofillers do not exhibit noteworthy fire retardancy on their own, and are therefore combined with other fire retardants [68]. Nanofillers such as nanoclay particles or CNTs can decrease the flammability by prohibiting the vigorous bubbling effect during combustion-induced degradation. The addition of these nanofillers generally leads to the added benefit of improving the physical properties of nanocomposites compared to those of the polymer matrix [69]. Therefore, nanofiller-incorporated nanocomposites can form a constant protective solid layer consisting of clay particles and carbonaceous char (CNTs) on the burning surface [70].

2.3.5. Optical Properties

The incorporation of nanomaterials into polymer matrices provides the possibility of remarkable improvements in the optical properties of polymeric nanocomposites. The unique optical properties of nanomaterials are associated with the effects of dielectric restriction, dimensional quantization, and the excitation of local surface plasmons [71]. The spectral position and intensity of surface plasmon resonance (SPR) are extremely specific for various nanomaterials and strongly depend on both the spatial organization and properties of nanomaterials or nanofillers. The optical properties of nanocomposites can be regulated by altering the size, shape, and concentration of nanomaterials, as well as the dielectric constant of the polymer matrix [72]. The linear and non-linear optical properties of nanofiller-infused nanocomposites are influenced by the excitation of local SPR, especially the collective oscillations of the conduction electrons. The plasmon properties of nanocomposites are categorically related to their sub-micrometer-scale ordering [73]. Moreover, the addition of graphene and CNTs provides certain benefits to other useful optical properties.

3. Biodegradable Polymers (Biopolymers)

Petroleum-based polymers have the ability to satisfy all the packaging requirements of the food industry. However, they are non-biodegradable, non-renewable, or non-compostable, which can lead to serious issues related to disposal and waste generation worldwide, consequently leading to environmental damage. Therefore, the research on packaging must be focused on promoting and developing bio-based plastics that are alternatives or replacements for fossil fuels or synthetic polymers to effectively minimize waste disposal [74]. Biodegradable materials are capable of undergoing biological decomposition or degradation to yield water, methane, carbon dioxide, inorganic compounds, and biomass by enzymatic activities of microorganisms, depending on the environmental conditions of the process. Biopolymers are biodegradable polymers that comprise covalently bonded monomeric units, which construct chain-like molecules that can be degraded or metabolized by naturally occurring microbes [75]. Biodegradable polymers are typically derived from animal sources, agricultural feedstock, marine and food processing industrial wastes, or microbial sources, including starch, proteins, peptides, DNA, and RNA.

Biopolymers can be classified into the following groups based on the origin of the raw materials (renewable or non-renewable) and their manufacturing process (Figure 5): natural resources, microbial or renewable resources, and synthetic or fossil resources. Natural biopolymers are subdivided into polysaccharides (starch, wheat, cellulose, pectin, and chitosan) and proteins and lipids (gluten, soya, zein, peanut, casein, whey, gelatin, and collagen). Similarly, renewable polymers are categorized into microbial polymers, such as polyesters (PHB and PHBV), carbohydrates (pullulan and curdlan), and natural polymers, such as PLA. Synthetic polymers are further categorized into PVA and aliphatic and aromatic polymers (polyglycolic acid (PGA), PCL, polyester amides (PEAs), PVA, poly (L-lactide) (PLA), polybutylene adipate-*co*-terephthalate (PBAT), and polybutylene succinate-*co*-butylene adipate (PBSA)) [76,77].

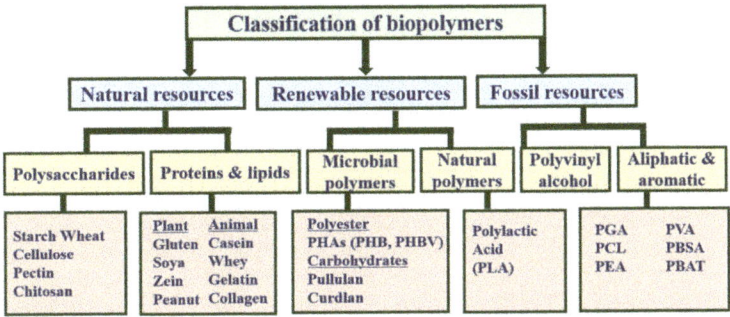

Figure 5. Classification of biopolymers for food packaging applications.

Biopolymers are regarded as the most promising materials for food packaging applications. However, they generally exhibit poor barrier and mechanical properties with regard to processing ability and end-use applications. In particular, the high gas and vapor permeability, brittleness, low heat-distortion temperature, and poor resistance to protracted processing operations of biopolymers significantly limit their industrial applications [78]. However, both natural and synthetic nanofillers can be used to improve their physical and mechanical properties. Fully biodegradable nanocomposites can be produced using polymer matrices and fillers derived from renewable resources [79]. Various examples of biopolymers used as packaging materials are listed in Table 1.

Table 1. Biopolymers with different properties used as packaging materials in the food industry.

Biopolymers	Source	Properties	Applications	Ref.
Cellulose	Agricultural waste	Highly crystalline, chemically and thermally stable, antimicrobial properties	Biodegradable packaging, microencapsulation	[80–82]
Starch	Potato, corn, wheat	Enhanced gas barrier and consistent with antioxidant and antimicrobial properties	Encapsulation and biodegradable packaging	[83,84]
Pectin	Apple pomace and citrus peels	Biodegradability, biocompatibility, edibility, and versatile physical and chemical properties	Biodegradable films for food packaging and microencapsulation	[85]
β-D-glucan	Oat and barley	Rheological, biocompatibility and biodegradable properties	Encapsulation matrix and for film-forming preparations	[86]
Chitosan	Crab, shrimp, crawfish	Moderate mechanical strength, low barrier properties, inherent antimicrobial properties	Biodegradable films, and microencapsulation	[83,87,88]
Gums	Acacia tree	Excellent adhesive strengths, enhanced structural, thermal and gas barrier properties	Adhesive packaging applications	[89]
Alginate	Marine brown algae	Low oxygen permeability, vapors, flexibility, and water solubility	Intelligent and green packaging technologies	[89]
Agar	Marine red algae	High transparency, permeability, thermal stability, or mechanical strength of the film	Food packaging applications	[90]
Carrageenan	Cell walls of seaweeds	Enhancing sensory properties, reducing moisture loss	Edible biodegradable films and coatings	[91]
Casein	Milk, yogurt and cheese	Biodegradability, high thermal stability, non-toxicity	Protein-based coatings and films in food packaging	[92]
Whey	Milk, yogurt and cheese	Excellent barrier characteristics for oxygen, oil, and aroma	Biodegradable films for food packaging	[93]
Gelatin	Cattle bones	Enhanced mechanical, and optical, barrier effect against gas flow	Gelatin-based coatings and films for food packaging	[94]
Zein	Corn protein	Good barrier properties, high compatibility	Bio-based packaging and edible coatings	[95]
Soy proteins	Soybeans	Remarkable gas barrier and weaker mechanical properties, better antimicrobial properties	Biodegradable films and microencapsulation	[96]
Collagen	Fish skin, bones, fins	Improved rheological properties, high-water absorption capacity	Smart and active packaging.	[97]
Wheat gluten	Wheat flour	Improved structural, surface, gas barrier, and water vapor properties	Paper coating and food packaging	[98]

The plastic industry promised to be a boon compared to other industries in its initial stages. Recent research and developments focused on the production and optimization of bio-derived products from various plant matters or biomass in a sustainable and economical way. Nevertheless, biopolymers occupy their own position and have the largest market share in plastic industries. Besides biodegradable plastics, natural biopolymers from polysaccharides and polypeptides have also been widely used for their biostability, sustainability, mechanical properties, biocompatibility, and minimum cytotoxicity in food packaging and other multifarious applications [99]. However, the future prospects of biopolymers seem to be the most promising way to enhance their mechanical and thermal properties. Biopolymers are not likely to replace all fossil fuels for packaging applications, where the cost of biopolymers need to be looked into objectively and addressed in light of environmental issues [100]. Most biopolymers are costly to produce, and since petroleum based polymers are cheaper, industries use them without considering the environmental factors. Furthermore, economic concerns must be addressed, as the future of all biopolymer products depends on their cost competitiveness, by-products, and social impact [101].

4. Applications of Bio-Based Nanocomposites for Food Packaging Systems

The incorporation of functional nanomaterials into polymer matrices can assist in the development of food packaging materials with improved mechanical and barrier properties. Moreover, the fundamental properties of packaging materials, such as flexibility, durability, resistance to temperature and humidity, and flame resistance, can be further altered by the addition and modification of different nanomaterials to improve the shelf life and quality of the food products [102]. Different natural and inorganic–organic nanofillers, including cellulose nanocrystals, zein NPs, and cellulose NPs, and inorganic nanomaterials such as clay and layered silicates (montmorillonite), mesoporous silica nanoparticles (MSNs), metal and metal oxide NPs (Ag, Au, Cu, ZnO_2, SiO_2, TiO_2, Fe_2O_3, and Al_2O_3), layered double hydroxides, nanotubes (CNTs), fullerenes, nanorods, and salts, are typically employed as nanoreinforcements [103]. Among the various types of food packaging materials, edible coatings or edible films in the form of films or thin layers are used to shield the food products and create a mass-transfer barrier. Edible coatings are more relevant for direct application to food products, whereas non-edible coatings are used as protective containers. The application of edible coatings can be found in the agriculture, bakery and cheese, and meat processing industries to furnish color, enzymes, flavors, antioxidants, and anti-browning compounds to food products [102].

Clay and silicates are natural inorganic compounds with variable chemical compositions, relative simplicity, and low cost, which have attracted research attention as potential nanomaterials owing to their availability and barrier, mechanical, and thermal properties. The combination of clays/silicates and polymers yields superior barrier properties and lengthens the diffusive path for infiltrating molecules. Nanoclays can be categorized into several subclasses, including montmorillonite (MMT), kaolinite, bentonite, halloysite, hectorite, sepiolite, and cloisite [103]. Prior to their incorporation into polymers, natural clays have been modified with organic compounds such as tetra-alkyl ammonium salts and alkyl amine to generate an intercalated and exfoliated mixed structure, which afforded superior properties to those of the original polymers [104]. Intercalated nanocomposites exhibit a multilayered structure with alternating nanofiller/polymer layers separated by a few nanometers; moreover, the exfoliated nanocomposites exhibit comprehensive polymer penetration with random dispersion of clay layers [105]. MMT clay has been extensively investigated for developing nanocomposites with a variety of polymers, such as nylon, PE, PVC, and starch. The amounts of incorporated nanoclays typically vary from 1% to 5% by weight and are one dimension smaller than 1 nm. The use of nanocellulose is considered an advanced approach for the preparation of sustainable food packaging in the form of both coatings and fillers. Additionally, nanocellulose fibers have been designed or modified to enhance their interaction with the matrix phase and improve the intrinsic properties of active and intelligent packaging systems [106].

CNTs, such as single-wall carbon nanotubes (SWCNTs) or multi-walled carbon nanotubes (MWCNTs), have also been incorporated into various polymers such as PVA, PA, and PET. CNTs also exhibit antibacterial properties, which are associated with their direct penetration of microbial cells and chemiresistive sensing [107]. Starch has been extensively studied as a substitute material for food packaging applications because of its biodegradability, availability, and non-toxicity, in addition to its stability in air. Moreover, starch enhances the tensile strength and modulus of pullulan films, with the positively charged ions present on the surfaces of these antimicrobial agents contributing to their antimicrobial action [108]. Chitosan is a natural cationic polymer known for its biocompatibility, biodegradability, non-toxicity, low cost, hydrophilicity, and antimicrobial activity, and is considered a potential polymer for food packaging, especially in the form of edible films and coatings [109]. Chitosan/polyethylene active antibacterial bags showed potential in inhibiting the activities of total mesophilic bacteria, molds, coliforms, and yeasts in chicken drumsticks, and in maintaining the color, pH, and hardness of samples [110]. Similarly, various metal and metal oxide NPs, such as Au, Ag, Cu, Zn, ZnO, TiO_2, SiO_2, and MgO, have been examined for diverse active-food-packaging applications. The properties of metal and metal oxide nanomaterials, such as mechanical strength, thermal and chemical stability, gas and water barrier properties, heat resistance, biodegradability, and active antimicrobial activities, have led to an improved performance in active food packaging applications [111]. Various types of nanocomposites used in food packaging applications are listed in Table 2.

Table 2. Examples of nanofillers and polymer matrices that have been applied as nanocomposites in food packaging systems.

Nanofillers	Polymer Matrix	Properties	Applications	Ref.
Cellulose nanocrystals	PLA	Oxygen barrier	Used as polar and non-polar simulants in food packaging materials	[112]
Cellulose nanocrystals	PLA	Mechanical and antimicrobial	Biocidal activity in food packaging industry	[113]
Organoclay	LDPE and HDPE	Rheological and barrier	Oxygen permeability of polymer decreasing slowly with increases in clay concentration	[114]
Starch nanocrystals	Potato starch	Mechanical and thermal	Biodegradable edible films for packaging	[115]
MMT	PCL	Mechanical	Biodegradable polymer nanocomposites for food packaging	[116]
Clay ZnO	PEA starch	Mechanical strength	Medical, agriculture, drug release, and packaging fields	[117]
Zein NPs	WPI (whey protein isolate)	Mechanical, water vapor barrier	Effective food packaging materials.	[118]
MMT	WPI	mechanical	WPI film for food packaging	[119]
Anionic sodium MMT	PET	Oxygen transmission rate decreased	Replacement of aluminum foil in food packaging systems	[120]
Cellulose whiskers	PEA starch	Tensile, thermomechanical	Biodegradable edible films for packaging	[121]
Cellulose nanocrystals	PLA	Mechanical and oxygen barrier	Biomaterial for food packaging systems.	[122]
MMT	Cellulose acetate	Mechanical	Replacing oil-based high performance plastics for food packaging	[123]
Starch nanocrystals	Polyurethane	Mechanical	Biomaterial for food packaging systems	[124]
Bacterial cellulose nanoribbons	Chitosan	Mechanical	New materials for the food packaging	[125]
Chitosan–tripolyphosphate NPs	Hydroxypropyl methylcellulose	Mechanical and barrier properties	Improved functionality to edible films for food packaging	[126]
Chitin whiskers	Starch	Mechanical, water vapor barrier	Improved properties to prolong the shelf life of packaged foods	[127]
Graphene	Poly(methyl methacrylate)	Heat resistant and barrier properties	Promising material for food packaging systems	[128]

Active and Intelligent Packaging Systems

Active and intelligent packaging are two forms of smart packaging that have been recently developed to enable the marketing of food products and provide passive protection against environmental conditions and contamination to extend the shelf life of food products. An active packaging material is a neat or modified substance that increases the shelf life of food products or enhances their safety or sensorial properties to maintain their quality [129]. Active packaging materials interact with packaged food and the environment in a certain manner and/or react to various stimuli, owing to their intrinsic properties or the incorporation of certain special additives in the packaging material to maintain the food quality [130]. Various categories of active food packaging exist, such as antimicrobial, antioxidant, oxygen scavenging, ethylene scavenging, liquid, moisture, odor, flavor absorbing, and ultraviolet barrier [131]. However, unlike intelligent (responsive) packaging, active packaging does not react to a specific trigger mechanism.

Intelligent packaging materials apply intelligent functions related to responsive packaging throughout the food supply chain, which include locating, registering, detecting, communicating, monitoring, and applying scientific logic. Therefore, these materials can trigger alerts for consumers by detecting microorganisms and food spoilage, extend shelf life, ease decision-making, improve the quality and safety of food, provide information, and warn of possible problems [132]. In addition, intelligent packaging can enable the release of antimicrobials, antioxidants, and other compounds upon detecting food spoilage under specific environmental changes to extend the shelf life of food products [133]. As discussed previously, the reinforcement of polymer matrices by nanofillers can improve the mechanical, barrier, and thermal properties of food packaging materials. Additionally, agents such as antimicrobials, antioxidants, nutraceuticals, coloring agents, flavors, and biosensors have been added to polymer matrices to enhance their smart functions, with regard to the quality, stability, and safety of food products [134]. Recently, Taherimehr et al. discussed the trends and challenges of biopolymer-based nanocomposites for food packaging applications [135]. However, the current review focused more on natural and inorganic–organic nanomaterials, clay, layered silicates, mesoporous silica nanoparticles, metal and metal oxide nanoparticles, layered double hydroxides, nanotubes, fullerenes, and nanorods employed as nanoreinforcements to develop nanocomposites for active and intelligent food packaging applications.

5. Antimicrobial Properties of Bio-Nanocomposites

Antimicrobial packaging is a robust technology that protects packaged food products from spoilage, which can occur via contamination by food-borne pathogens (bacteria, parasites, and viruses), leading to food-borne diseases [136]. This type of packaging can be prepared either by applying a coating layer (antimicrobial agent) within the packaging material or by incorporating an antimicrobial agent into the packaging material. Antimicrobial packaging of food products is a type of active packaging owing to the use of antimicrobial agents, growth inhibitors, and antimicrobial carriers [23]. Nanomaterials or nanocomposites, owing to their enhanced surface reactivity, high surface-to-volume ratio, and physicochemical and antimicrobial properties, impede the activity of microorganisms more efficiently than their micro- or macro-scale counterparts. The efficacy and performance of active food packaging have been enhanced through nanoencapsulation or the incorporation of natural antimicrobial-loaded nanocarriers [137]. The most commonly used antimicrobial nanocomposite materials include metal and metal oxide NPs (such as Au, Ag, Cu, ZnO, MgO, and TiO_2), natural biopolymers (chitosan), organic nanoclay (Ag-zeolite and MMT), enzymes (peroxidase and lysozyme), natural bioactive compounds (thymol, carvacrol, nisin, and isothiocyanate), and synthetic agents (EDTA, ammonium salts; benzoic, propionic, and sorbic acids) [23].

Different species of Gram-negative and Gram-positive bacteria, including *Salmonella* spp., *Staphylococcus aureus*, *Staphylococcus epidermis*, *Escherichia coli*, *Listeria monocytogenes*, *Pseudomonas aeruginosa*, *Enterococcus faecalis*, *Vibrio cholera*, and *Bacillus cereus*, are responsible for food spoilage. In addition, *Aspergillus* and *Rhizopus* (molds), and *Candida* and *Torulopsis* (yeasts) are also involved in foodborne infections [138]. A few recently reported antibacterial activities of metal and metal oxide nanocomposites are briefly described in the following. A carboxymethyl cellulose film coated with AgNPs exhibited antibacterial efficacy against *S. aureus* and *E. coli* [139]; cellulose acetate with AgNPs [140]. AuNPs with bacteriocin inhibited the activities of *E. coli*, *B. cereus*, *S. aureus*, and *Micrococcus luteus* [141]. Gellan gum-sodium carboxymethyl cellulose ((GC)-SiO_2) and GC-SiO_2-octadecyldimethyl-(3-trimethoxysilylpropyl)-ammonium chloride (ODDMAC) nanocomposites were effective against *B. cereus* and *S. aureus* [142]. Chitosan-ZnO coatings reduced the initial numbers of *E. coli* in white-brined cheese [143]. Table 3 provides a detailed list of select investigations that have examined the antimicrobial properties of different nanofillers.

Table 3. Examples of bio-based nanocomposites investigated for their antimicrobial properties.

Nanomaterials	Biopolymer	Pathogens	Applications	Ref.
Ag	Chitosan	*E. coli, Solmonella, S. aureus*	Active and intelligent food packaging	[144]
Ag	LDPE	*E. coli, E. faecalis, S. aureus*	Improved food quality and safety	[145]
Ag	Cellulose	*E. coli, S. aureus*	Potential bacterial barrier in food packaging	[146]
Au	PVA	*E. coli*	Active food packaging for banana fruits	[147]
CuS	Agar	*E. coli, L. monocytogenes*	Active food packaging	[148]
CuO	Agar, alginate, chitosan	*E. coli, L. monocytogenes*	UV-screening and food packaging	[149]
ZnO	Gelatin, cellulose	*E. coli, L. monocytogenes, S. aureus*	Active food packaging	[150]
TiO_2	Chitosan	*E. coli, S. aureus*	Active multifunctional food packaging	[151]
ZnO	Carboxymethyl cellulose	*E. coli, S. aureus*	Active food packaging	[152]
SiO_2	PHBV	*E. coli, S. aureus*	Eco-friendly, cost-effective food packaging materials.	[153]
SO_2	PA, PE	*E. coli, L. monocytogenes, S. aureus*	Active packaging for selected types of foods	[154]
ZnO	Soy protein isolate	*Aspergillus niger*	Ideal packaging matrix for food preservation	[155]
TiO_2	Zein, sodium alginate	*E. coli, S. aureus*	Improved shelf life and quality of food stuffs	[156]
MgO	PLA	*E. coli*	UV-screening and active food packaging	[157]
Carbon dots	Bacterial nanocellulose	*E. coli, L. monocytogenes*	UV-screening and forgery-proof packaging	[158]
SiO_2	Chitosan	*E. coli, S. aureus, S. typhimurium*	Active food packaging	[159]
CNTs	Allyl isothiocyanate	*Salmonella* spp.	Active packaging for shredded cooked chicken	[160]
MWCNTs	Chitosan, PLA	*E. coli, S. aureus, B. cinerea, Rhizopus*	Active packaging for fruits and vegetables	[161]
MSN	PHBV	*E. coli, S. aureus*	Interlayers or coatings for active food packaging	[162]
Cellulose	Agar	*E. coli, L. monocytogenes, S. aureus*	Active packaging for safety and shelf-life of food	[163]
Halloysite	Starch	*C. perfringenes, S. aureus, L. monocytogenes*	Active and useful barrier to control food contamination.	[164]
Chitosan	Fish gelatin	*S. aureus, L. monocytogenes, S. enteritidis, E. coli*	Greater flexible films, with decrease in water vapor permeability	[165]
MMT	Chitosan	*L. monocytogenes, E. coli, P. putida*	Antioxidant and antibacterial films for food preservation	[166]
Cinnamaldehyde nanoemulsions	Pectin, papaya puri	*E. coli, L. monocytogenes, S. aureus*	Environmentally friendly antimicrobial packaging material for food applications	[167]
Cellulose nanofiber	Starch	*B. subtilis, E. coli*	Biopolymer active food packaging	[168]
PLA nanofibers	PLA	*E. coli, S. aureus*	Effectively prolong the shelf-life of pork.	[169]

Mechanisms of Antimicrobial Action

Nanomaterials or nanocomposites can be employed as antimicrobial agents, growth inhibitors, antimicrobial carriers, and antimicrobial packaging films. Antimicrobial bio-nanocomposite films are applied as food packaging materials primarily for cheese, meat, bread, fish, poultry, vegetables, and fruits [170]. These nanocomposite materials have potent antibacterial effects through various mechanisms of action that interact precisely with microbial cells, including the disruption of cell walls, interruption of transmembrane electron transfer, oxidation of cell components, formation of reactive oxygen species (ROS), disruption of enzyme activity, destruction of internal cell organelles, prevention of DNA synthesis, and cellular death [171]. The probable mechanisms of antimicrobial action of the nanocomposites developed as active food packaging materials are illustrated in Figure 6.

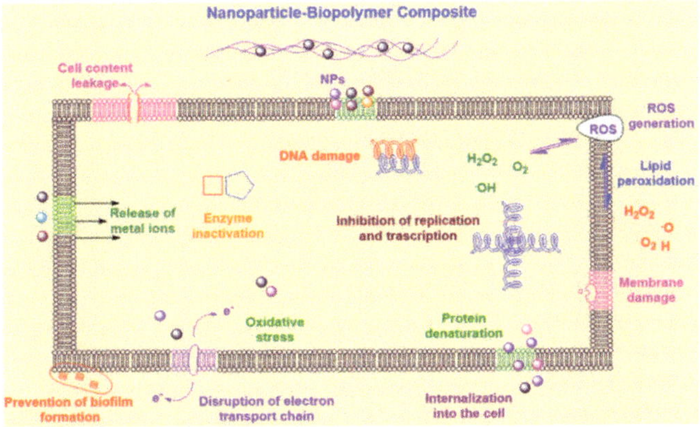

Figure 6. Schematic representation of antimicrobial mechanisms of action of nanocomposites designed for food packaging.

Nanocomposites have a remarkably positive zeta potential, which promotes their interaction with cell membranes by electrostatic binding to cell walls and releasing metal ions. The negatively charged bacterial membranes and positively charged nanocomposites induce electrostatic attraction and modify the permeability of the cell membrane. Therefore, disrupting the integrity of bacterial membranes is an efficient mechanism of action [172]. ROS production is an alternative mechanism that affects the physiological functions of cells and eventually damages DNA. Different types of ROS, such as superoxide anionic radicals, hydroxyl radicals, hydrogen peroxide, and the hydroxyl radical produced in mitochondria, exhibit varying levels of activity. Protein dysfunction is another mechanism of action by which nanocomposites bind to cytosolic proteins, such as DNA and enzymes, which leads to oxidative stress, damage of communication channels, peroxidation of cellular constituents, DNA strand breakage, lipid peroxidation, and modification of nucleic acids [15]. However, in the case of enzymes, carboxylation results in the loss of catalytic activity and accelerates protein degradation [173].

6. Conclusions and Future Trends

Food packaging plays a critical role in protecting food products from external contamination and maintaining their quality, integrity, and safety throughout their shelf life. Synthetic-polymer-based materials are predominantly used as packaging materials in the food industry because of their ease of production, versatility, affordability, functionality, and properties of low weight, flexibility, and low cost. However, these synthetic polymers are non-degradable, and most of the plastic waste and debris heavily pollute the environment. This necessitates the development and use of biodegradable polymer materials to resolve these environmental problems. Bio-based polymers or renewable-resource-based biopolymers, such as cellulosic plastics, starch, corn-derived plastics such as PHAs, and PLA are sustainable, high-performance materials with tremendous potential for replacing conventional petroleum-based food packaging materials. However, biopolymers have certain disadvantages compared to synthetic polymers, such as inferior thermal and mechanical properties (tensile strength and brittleness), moisture sensitivity, and water-vapor barrier performance.

The use of nanotechnology in the food sector can ensure food quality and safety by enhancing the potency of food packaging and the shelf life of food products. The application of nanotechnology to develop novel food packaging functions can enable enhancements in the properties of food, such as taste, healthiness, and nutritiousness via the packaging. The incorporation of nanoparticles or nanofillers in food packaging materials can improve various properties of biopolymers, such as mechanical properties, barrier properties against water and oxygen, protection against UV radiation, absorption of moisture, release of antimicrobials, and other environmental factors. The biodegradability of these nanocomposites can be modified by selecting appropriate polymers and nanomaterials to yield desired properties and enable their application in food packaging. In addition, biologically active substances, such as antimicrobials, growth inhibitors, and antimicrobial carriers, can be added to enable the desired functional properties. These nanocomposites can be modified by incorporating either organic or inorganic antimicrobial agents that exhibit excellent antibacterial activity against both Gram-positive and Gram-negative food-borne pathogens. The inclusion of antibacterial NPs, such as Ag, MgO, ZnO, TiO_2, graphene, and carbon dots, in bio-nanocomposite films enables their use as new active packaging materials to improve the quality and safety of food products over a longer period. Nanosensors for intelligent packaging can be designed to control the internal (detecting microorganisms and chemicals in the packaging) and external conditions of food products (detecting atmospheric influences).

The present review provides information related to the development of bio-based polymeric nanocomposites to improve the quality, safety, and shelf life of packaged food products. However, further research should focus on the effects of combinations of nanomaterials, such as bi-, tri-, and multi-metallic nanocomposites, to achieve better results. In addition, the molecular interactions of biopolymers with food matrices and the formulations of nanomaterials in polymer matrices to minimize organoleptic effects should be investigated. Furthermore, standard methods and probable toxicity evaluations of nanofillers and biopolymers should be established. Current toxicity tests have revealed that the toxicity of nanomaterials primarily depends on their size, shape, surface-to-volume ratio, doping concentration, and duration. Future research on active packaging must focus on improving the safety of nanomaterials, owing to the limited studies on the possible toxic influence of these packaging films. In addition, health and safety aspects, the management of environmental issues, and a hazard and risk assessment must be considered prior to their application as safe and effective food packaging materials. The use of these modern and alternative preservation techniques can significantly inhibit pathogens, extend the shelf life, and fulfill consumer demands such as the high quality, convenience, safety, freshness, taste, aroma, color, and texture of packaged food. These biopolymer-based nanocomposites exhibit tremendous potential for a wide range of applications in the food industry as sustainable, cost-effective, active, and intelligent packaging materials for food preservation.

Author Contributions: Data collection, N.B. and K.-H.B.; writing—original draft preparation, N.B.; writing—review and editing, K.-H.B. All authors have read and agreed to the published version of the manuscript.

Funding: This research was funded by NRF2019R1F1A1052625, NRF, Ministry of Education, Korea.

Institutional Review Board Statement: Not applicable.

Informed Consent Statement: Not applicable.

Data Availability Statement: Not applicable.

Acknowledgments: This work was carried out with the support of the Basic Science Research Program through the National Research Foundation of Korea (NRF) funded by the Ministry of Education (NRF2019R1F1A1052625).

Conflicts of Interest: The authors declare no conflict of interest.

References

1. Khan, I.; Saeed, K.; Khan, I. Nanoparticles: Properties, applications and toxicities. *Arab. J. Chem.* **2019**, *12*, 908–931. [CrossRef]
2. Shameem, M.M.; Sasikanth, S.M.; Annamalai, R.; Raman, R.G. A brief review on polymer nanocomposites and its applications. *Mater. Today Proc.* **2021**, *45*, 2536–2539. [CrossRef]
3. Sen, M. Nanocomposite materials. In *Nanotechnology and the Environment*; IntechOpen: London, UK, 2020.

4. Liu, X.; Antonietti, M. Molten salt activation for synthesis of porous carbon nanostructures and carbon sheets. *Carbon* **2014**, *69*, 460–466. [CrossRef]
5. Bustamante-Torres, M.; Romero-Fierro, D.; Arcentales-Vera, B.; Pardo, S.; Bucio, E. Interaction between Filler and Polymeric Matrix in Nanocomposites: Magnetic Approach and Applications. *Polymers* **2021**, *13*, 2998. [CrossRef]
6. Guerra, D.R. Influence of Nanoparticles on the Physical Properties of Fiber Reinforced Polymer Composites. Master Thesis, The Ohio State University, Columbus, OH, USA, 2009.
7. De Oliveira, A.D.; Beatrice, C.A.G. Polymer nanocomposites with different types of nanofiller. In *Nanocomposites-Recent Evolution*; Sivasankaran, S., Ed.; IntechOpen: London, UK, 2018; pp. 103–128.
8. Abdulkadir, A.; Sarker, T.; He, Q.; Guo, Z.; Wei, S. Mössbauer spectroscopy of polymer nanocomposites. In *Spectroscopy of Polymer Nanocomposites*; Elsevier: Amsterdam, The Netherlands, 2016; pp. 393–409.
9. Pawar, P.A.; Purwar, A.H. Biodegradable polymers in food packaging. *Am. J. Eng. Res.* **2013**, *2*, 151–164.
10. Folino, A.; Karageorgiou, A.; Calabrò, P.S.; Komilis, D. Biodegradation of wasted bioplastics in natural and industrial environments: A review. *Sustainability* **2020**, *12*, 6030. [CrossRef]
11. Park, S.-B.; Lih, E.; Park, K.-S.; Joung, Y.K.; Han, D.K. Biopolymer-based functional composites for medical applications. *Prog. Polym. Sci.* **2017**, *68*, 77–105. [CrossRef]
12. Singh, B.G.; Das, R.P.; Kunwar, A. Protein: A versatile biopolymer for the fabrication of smart materials for drug delivery. *J. Chem. Sci.* **2019**, *131*, 1–14. [CrossRef]
13. Jung, E.Y.; Jin, S.K.; Hur, S.J. Analysis of the effects of biopolymer encapsulation and sodium replacement combination technology on the quality characteristics and inhibition of sodium absorption from sausage in mice. *Food Chem.* **2018**, *250*, 197–203. [CrossRef] [PubMed]
14. Narayanan, N.; Gupta, S.; Gajbhiye, V.T.; Manjaiah, K.M. Optimization of isotherm models for pesticide sorption on biopolymer-nanoclay composite by error analysis. *Chemosphere* **2017**, *173*, 502–511. [CrossRef]
15. Chaudhary, P.; Fatima, F.; Kumar, A. Relevance of nanomaterials in food packaging and its advanced future prospects. *J. Inorg. Organomet. Polym. Mater.* **2020**, *30*, 5180–5192. [CrossRef]
16. Shankar, S.; Rhim, J.-W. Bionanocomposite films for food packaging applications. *Ref. Modul. Food Sci.* **2018**, *1*, 1–10.
17. Grumezescu, A.M. *Food Preservation*; Academic Press: Cambridge, MA, USA, 2016; ISBN 0128043741.
18. Soundararajan, N.; Borkotoky, S.S.; Katiyar, V. 7. Up-to-date advances of biobased and biodegradable polymers in food packaging. In *Sustainable Polymers for Food Packaging*; De Gruyter: Berlin, Germany, 2020; pp. 127–144, ISBN 3110648032.
19. Skinner, G.A. Smart labelling of foods and beverages. In *Advances in Food and Beverage Labelling: Information and Regulations*, 1st ed.; Berryman, P., Ed.; Woodhead Publishing: Souston, UK, 2015; pp. 191–205.
20. Varelis, P.; Melton, L.; Shahidi, F. *Encyclopedia of Food Chemistry*; Elsevier: Amsterdam, The Netherlands, 2018; ISBN 0128140453.
21. Jafarzadeh, S.; Salehabadi, A.; Nafchi, A.M.; Oladzadabbasabadi, N.; Jafari, S.M. Cheese packaging by edible coatings and biodegradable nanocomposites; improvement in shelf life, physicochemical and sensory properties. *Trends Food Sci. Technol.* **2021**, *116*, 218–231. [CrossRef]
22. Primožič, M.; Knez, Ž.; Leitgeb, M. (Bio) Nanotechnology in Food Science—Food Packaging. *Nanomaterials* **2021**, *11*, 292. [CrossRef] [PubMed]
23. Rhim, J.-W.; Park, H.-M.; Ha, C.-S. Bio-nanocomposites for food packaging applications. *Prog. Polym. Sci.* **2013**, *38*, 1629–1652. [CrossRef]
24. Othman, S.H. Bio-nanocomposite materials for food packaging applications: Types of biopolymer and nano-sized filler. *Agric. Agric. Sci. Procedia* **2014**, *2*, 296–303. [CrossRef]
25. Tsai, S.W.; Hahn, H.T. *Introduction to Composite Materials*; Routledge: Boca Raton, FL, USA, 2018; ISBN 0203750144.
26. Sharma, A.K.; Bhandari, R.; Aherwar, A.; Rimašauskienė, R. Matrix materials used in composites: A comprehensive study. *Mater. Today Proc.* **2020**, *21*, 1559–1562. [CrossRef]
27. Vera, M.; Mella, C.; Urbano, B.F. Smart polymer nanocomposites: Recent advances and perspectives. *J. Chil. Chem. Soc.* **2020**, *65*, 4973–4981. [CrossRef]
28. Sothornvit, R. Nanostructured materials for food packaging systems: New functional properties. *Curr. Opin. Food Sci.* **2019**, *25*, 82–87. [CrossRef]
29. Nasrollahzadeh, M.; Issaabadi, Z.; Sajjadi, M.; Sajadi, S.M.; Atarod, M. Types of nanostructures. *Interface Sci. Technol.* **2019**, *28*, 29–80.
30. Gacitua, W.; Ballerini, A.; Zhang, J. Polymer nanocomposites: Synthetic and natural fillers a review. *Maderas Cienc. Tecnol.* **2005**, *7*, 159–178. [CrossRef]
31. Vasile, C. Polymeric nanocomposites and nanocoatings for food packaging: A review. *Materials* **2018**, *11*, 1834. [CrossRef] [PubMed]
32. Olivera, N.; Rouf, T.B.; Bonilla, J.C.; Carriazo, J.G.; Dianda, N.; Kokini, J.L. Effect of LAPONITE®addition on the mechanical, barrier and surface properties of novel biodegradable kafirin nanocomposite films. *J. Food Eng.* **2019**, *245*, 24–32. [CrossRef]
33. Albdiry, M.T.; Yousif, B.F. Toughening of brittle polyester with functionalized halloysite nanocomposites. *Compos. Part B Eng.* **2019**, *160*, 94–109. [CrossRef]
34. Lin, Y.; Hu, S.; Wu, G. Structure, dynamics, and mechanical properties of polyimide-grafted silica nanocomposites. *J. Phys. Chem. C* **2019**, *123*, 6616–6626. [CrossRef]

35. Müller, K.; Bugnicourt, E.; Latorre, M.; Jorda, M.; Echegoyen Sanz, Y.; Lagaron, J.M.; Miesbauer, O.; Bianchin, A.; Hankin, S.; Bölz, U. Review on the processing and properties of polymer nanocomposites and nanocoatings and their applications in the packaging, automotive and solar energy fields. *Nanomaterials* **2017**, *7*, 74. [CrossRef] [PubMed]
36. Zaïri, F.; Gloaguen, J.-M.; Naït-Abdelaziz, M.; Mesbah, A.; Lefebvre, J.-M. Study of the effect of size and clay structural parameters on the yield and post-yield response of polymer/clay nanocomposites via a multiscale micromechanical modelling. *Acta Mater.* **2011**, *59*, 3851–3863. [CrossRef]
37. Sharma, A.K.; Kaith, B.S.; Panchal, S.; Bhatia, J.K.; Bajaj, S.; Tanwar, V.; Sharma, N. Response surface methodology directed synthesis of luminescent nanocomposite hydrogel for trapping anionic dyes. *J. Environ. Manag.* **2019**, *231*, 380–390. [CrossRef]
38. Lim, C.; Shin, Y.; Jung, J.; Kim, J.H.; Lee, S.; Kim, D.-H. Stretchable conductive nanocomposite based on alginate hydrogel and silver nanowires for wearable electronics. *APL Mater.* **2018**, *7*, 31502. [CrossRef]
39. Hu, X.; Ke, Y.; Zhao, Y.; Lu, S.; Deng, Q.; Yu, C.; Peng, F. Synthesis, characterization and solution properties of β-cyclodextrin-functionalized polyacrylamide/montmorillonite nanocomposites. *Colloids Surf. A Physicochem. Eng. Asp.* **2019**, *560*, 336–343. [CrossRef]
40. Chatterjee, T.N.; Das, D.; Roy, R.B.; Tudu, B.; Hazarika, A.K.; Sabhapondit, S.; Tamuly, P.; Bandyopadhyay, R. Development of a nickel hydroxide nanopetal decorated molecular imprinted polymer based electrode for sensitive detection of epigallocatechin-3-gallate in green tea. *Sens. Actuators B Chem.* **2019**, *283*, 69–78. [CrossRef]
41. Zhang, S.; Chen, S.; Yang, F.; Hu, F.; Yan, B.; Gu, Y.; Jiang, H.; Cao, Y.; Xiang, M. High-performance electrochromic device based on novel polyaniline nanofibers wrapped antimony-doped tin oxide/TiO2 nanorods. *Org. Electron.* **2019**, *65*, 341–348. [CrossRef]
42. Mbese, J.Z.; Ajibade, P.A. Preparation and characterization of ZnS, CdS and HgS/poly (methyl methacrylate) nanocomposites. *Polymers* **2014**, *6*, 2332–2344. [CrossRef]
43. Wang, D.-Y.; Leuteritz, A.; Wang, Y.-Z.; Wagenknecht, U.; Heinrich, G. Preparation and burning behaviors of flame retarding biodegradable poly (lactic acid) nanocomposite based on zinc aluminum layered double hydroxide. *Polym. Degrad. Stab.* **2010**, *95*, 2474–2480. [CrossRef]
44. Zou, H.; Wu, S.; Shen, J. Polymer/silica nanocomposites: Preparation, characterization, properties, and applications. *Chem. Rev.* **2008**, *108*, 3893–3957. [CrossRef] [PubMed]
45. Koo, J.H. *Polymer Nanocomposites: Processing, Characterization, and Applications*; McGraw-Hill Education: New York, NY, USA, 2019; ISBN 1260132315.
46. Lampman, S. *Characterization and Failure Analysis of Plastics*; ASM International: Geauga County, OH, USA, 2003; ISBN 1615030735.
47. De, S.K.; White, J.R. *Rubber Technologist's Handbook*; Smithers Rapra Publishing: Shropshire, UK, 2001; Volume 1, ISBN 1859572626.
48. Erceg, M.; Jozić, D.; Banovac, I.; Perinović, S.; Bernstorff, S. Preparation and characterization of melt intercalated poly (ethylene oxide)/lithium montmorillonite nanocomposites. *Thermochim. Acta* **2014**, *579*, 86–92. [CrossRef]
49. Rhim, J.-W. Potential use of biopolymer-based nanocomposite films in food packaging applications. *Food Sci. Biotechnol.* **2007**, *16*, 691–709.
50. Zeman, S. Mechanical properties to measure resistance of food packaging materials to external influences. *Tech. Sci.* **2007**, *10*, 35–40. [CrossRef]
51. Sen, S.; Thomin, J.D.; Kumar, S.K.; Keblinski, P. Molecular underpinnings of the mechanical reinforcement in polymer nanocomposites. *Macromolecules* **2007**, *40*, 4059–4067. [CrossRef]
52. Youssef, A.M. Polymer nanocomposites as a new trend for packaging applications. *Polym. Plast. Technol. Eng.* **2013**, *52*, 635–660. [CrossRef]
53. Azizi Samir, M.A.S.; Alloin, F.; Dufresne, A. Review of recent research into cellulosic whiskers, their properties and their application in nanocomposite field. *Biomacromolecules* **2005**, *6*, 612–626. [CrossRef] [PubMed]
54. Zheng, X.; Wilkie, C.A. Flame retardancy of polystyrene nanocomposites based on an oligomeric organically-modified clay containing phosphate. *Polym. Degrad. Stab.* **2003**, *81*, 539–550. [CrossRef]
55. Hussain, F.; Hojjati, M.; Okamoto, M.; Gorga, R.E. Polymer-matrix nanocomposites, processing, manufacturing, and application: An overview. *J. Compos. Mater.* **2006**, *40*, 1511–1575. [CrossRef]
56. Heidari, M.; Khomeiri, M.; Yousefi, H.; Rafieian, M.; Kashiri, M. Chitin nanofiber-based nanocomposites containing biodegradable polymers for food packaging applications. *J. Consum. Prot. Food Saf.* **2021**, *16*, 1–10. [CrossRef]
57. Taib, M.N.A.M.; Julkapli, N.M. Dimensional stability of natural fiber-based and hybrid composites. In *Mechanical and Physical Testing of Biocomposites, Fibre-Reinforced Composites and Hybrid Composites*; Elsevier: Amsterdam, The Netherlands, 2019; pp. 61–79.
58. Martinez-Garcia, J.C.; Serraïma-Ferrer, A.; Lopeandía-Fernández, A.; Lattuada, M.; Sapkota, J.; Rodríguez-Viejo, J. A Generalized Approach for Evaluating the Mechanical Properties of Polymer Nanocomposites Reinforced with Spherical Fillers. *Nanomaterials* **2021**, *11*, 830. [CrossRef]
59. Shankar, S.; Rhim, J. Polymer nanocomposites for food packaging applications. In *Functional and Physical Properties of Polymer Nanocomposites*; John Wiley & Sons, Ltd.: West Sussex, UK, 2016. [CrossRef]
60. Saritha, A.; Joseph, K. Barrier properties of nanocomposites. *Polym. Compos.* **2013**, *2*, 185–200.
61. Pillai, S.K.; Ray, S.S. Inorganic-organic hybrid polymers for food packaging. In *Functunal Polymers in Food Science: From Technology to Biology*; Scrivener Publishing: Beverly, MA, USA, 2015; pp. 281–322.
62. Duncan, T.V. Applications of nanotechnology in food packaging and food safety: Barrier materials, antimicrobials and sensors. *J. Colloid Interface Sci.* **2011**, *363*, 1–24. [CrossRef]

63. Irshad, H.M.; Hakeem, A.S.; Raza, K.; Baroud, T.N.; Ehsan, M.A.; Ali, S.; Tahir, M.S. Design, Development and Evaluation of Thermal Properties of Polysulphone–CNT/GNP Nanocomposites. *Nanomaterials* **2021**, *11*, 2080. [CrossRef]
64. Hussain, A.R.J.; Alahyari, A.A.; Eastman, S.A.; Thibaud-Erkey, C.; Johnston, S.; Sobkowicz, M.J. Review of polymers for heat exchanger applications: Factors concerning thermal conductivity. *Appl. Therm. Eng.* **2017**, *113*, 1118–1127. [CrossRef]
65. Yoon, P.J.; Fornes, T.D.; Paul, D.R. Thermal expansion behavior of nylon 6 nanocomposites. *Polymer* **2002**, *43*, 6727–6741. [CrossRef]
66. Jineesh, A.G.; Mohapatra, S. Thermal properties of polymer–carbon nanocomposites. In *Carbon-Containing Polymer Composites*; Springer: Berlin/Heidelberg, Germany, 2019; pp. 235–270.
67. Xue, Y.; Guo, Y.; Rafailovich, M.H. Flame retardant polymer nanocomposites and interfaces. In *Flame Retardants*; Zafar, F., Ed.; IntechOpen: London, UK, 2019; pp. 41–62.
68. Arao, Y. Flame retardancy of polymer nanocomposite. In *Flame Retardants*; Springer International Publishing: Cham, Switzerland, 2015; pp. 15–44.
69. Kashiwagi, T.; Du, F.; Douglas, J.F.; Winey, K.I.; Harris, R.H.; Shields, J.R. Nanoparticle networks reduce the flammability of polymer nanocomposites. *Nat. Mater.* **2005**, *4*, 928–933. [CrossRef] [PubMed]
70. Zanetti, M.; Kashiwagi, T.; Falqui, L.; Camino, G. Cone calorimeter combustion and gasification studies of polymer layered silicate nanocomposites. *Chem. Mater.* **2002**, *14*, 881–887. [CrossRef]
71. Kreibig, U.; Vollmer, M. *Optical Properties of Metal Clusters*; Springer Science & Business Media: Berlin/Heidelberg, Germany, 2013; Volume 25, ISBN 3662091097.
72. Sakhno, O.; Yezhov, P.; Hryn, V.; Rudenko, V.; Smirnova, T. Optical and nonlinear properties of photonic polymer nanocomposites and holographic gratings modified with noble metal nanoparticles. *Polymers* **2020**, *12*, 480. [CrossRef]
73. Ponyavina, A.N.; Kachan, S.M. Plasmonic spectroscopy of 2D densely packed and layered metallic nanostructures. In *Polarimetric Detection, Characterization and Remote Sensing*; Springer: Berlin/Heidelberg, Germany, 2011; pp. 383–408.
74. Ahmed, S. *Bio-Based Materials for Food Packaging: Green and Sustainable Advanced Packaging Materials*; Springer: Berlin/Heidelberg, Germany, 2018; ISBN 981131909X.
75. Sorrentino, A.; Gorrasi, G.; Vittoria, V. Potential perspectives of bio-nanocomposites for food packaging applications. *Trends Food Sci. Technol.* **2007**, *18*, 84–95. [CrossRef]
76. Bordes, P.; Pollet, E.; Avérous, L. Nano-biocomposites: Biodegradable polyester/nanoclay systems. *Prog. Polym. Sci.* **2009**, *34*, 125–155. [CrossRef]
77. Imran, M.; Revol-Junelles, A.-M.; Martyn, A.; Tehrany, E.A.; Jacquot, M.; Linder, M.; Desobry, S. Active food packaging evolution: Transformation from micro-to nanotechnology. *Crit. Rev. Food Sci. Nutr.* **2010**, *50*, 799–821. [CrossRef] [PubMed]
78. Koh, H.C.; Park, J.S.; Jeong, M.A.; Hwang, H.Y.; Hong, Y.T.; Ha, S.Y.; Nam, S.Y. Preparation and gas permeation properties of biodegradable polymer/layered silicate nanocomposite membranes. *Desalination* **2008**, *233*, 201–209. [CrossRef]
79. Petersson, L.; Kvien, I.; Oksman, K. Structure and thermal properties of poly (lactic acid)/cellulose whiskers nanocomposite materials. *Compos. Sci. Technol.* **2007**, *67*, 2535–2544. [CrossRef]
80. Wu, C.; Zhu, Y.; Wu, T.; Wang, L.; Yuan, Y.I.; Chen, J.; Hu, Y.; Pang, J. Enhanced functional properties of biopolymer film incorporated with curcurmin-loaded mesoporous silica nanoparticles for food packaging. *Food Chem.* **2019**, *288*, 139–145. [CrossRef]
81. Hassan, B.; Chatha, S.A.S.; Hussain, A.I.; Zia, K.M.; Akhtar, N. Recent advances on polysaccharides, lipids and protein based edible films and coatings: A review. *Int. J. Biol. Macromol.* **2018**, *109*, 1095–1107. [CrossRef] [PubMed]
82. Desai, K.G.H.; Jin Park, H. Recent developments in microencapsulation of food ingredients. *Dry. Technol.* **2005**, *23*, 1361–1394. [CrossRef]
83. Chakravartula, S.S.N.; Lourenço, R.V.; Balestra, F.; Bittante, A.M.Q.B.; do Amaral Sobral, P.J.; Dalla Rosa, M. Influence of pitanga (Eugenia uniflora L.) leaf extract and/or natamycin on properties of cassava starch/chitosan active films. *Food Packag. Shelf Life* **2020**, *24*, 100498. [CrossRef]
84. Lu, D.R.; Xiao, C.M.; Xu, S.J. Starch-based completely biodegradable polymer materials. *Express Polym. Lett.* **2009**, *3*, 366–375. [CrossRef]
85. Espitia, P.J.P.; Du, W.-X.; de Jesús Avena-Bustillos, R.; Soares, N.D.F.F.; McHugh, T.H. Edible films from pectin: Physical-mechanical and antimicrobial properties—A review. *Food Hydrocoll.* **2014**, *35*, 287–296. [CrossRef]
86. Peltzer, M.; Delgado, J.F.; Salvay, A.G.; Wagner, J.R. β-Glucan, a promising polysaccharide for bio-based films developments for food contact materials and medical applications. *Curr. Org. Chem.* **2018**, *22*, 1249–1254. [CrossRef]
87. Akhter, R.; Masoodi, F.A.; Wani, T.A.; Rather, S.A. Functional characterization of biopolymer based composite film: Incorporation of natural essential oils and antimicrobial agents. *Int. J. Biol. Macromol.* **2019**, *137*, 1245–1255. [CrossRef]
88. Ruiz-Navajas, Y.; Viuda-Martos, M.; Sendra, E.; Perez-Alvarez, J.A.; Fernández-López, J. In vitro antibacterial and antioxidant properties of chitosan edible films incorporated with Thymus moroderi or Thymus piperella essential oils. *Food Control* **2013**, *30*, 386–392. [CrossRef]
89. Katiyar, V.; Tripathi, N. Functionalizing gum arabic for adhesive and food packaging applications. *Plast. Res. Online* **2019**. [CrossRef]
90. Mostafavi, F.S.; Zaeim, D. Agar-based edible films for food packaging applications-A review. *Int. J. Biol. Macromol.* **2020**, *159*, 1165–1176. [CrossRef]

91. Sanchez-García, M.D. Carrageenan polysaccharides for food packaging. In *Multifunctional and Nanoreinforced Polymers for Food Packaging*; Elsevier: Amsterdam, The Netherlands, 2011; pp. 594–609.
92. Chen, H.; Wang, J.; Cheng, Y.; Wang, C.; Liu, H.; Bian, H.; Pan, Y.; Sun, J.; Han, W. Application of protein-based films and coatings for food packaging: A review. *Polymers* 2019, *11*, 2039. [CrossRef] [PubMed]
93. Cortés-Rodríguez, M.; Villegas-Yépez, C.; González, J.H.G.; Rodríguez, P.E.; Ortega-Toro, R. Development and evaluation of edible films based on cassava starch, whey protein, and bees wax. *Heliyon* 2020, *6*, e04884. [CrossRef] [PubMed]
94. Ramos, M.; Valdes, A.; Beltran, A.; Garrigós, M.C. Gelatin-based films and coatings for food packaging applications. *Coatings* 2016, *6*, 41. [CrossRef]
95. Bayer, I.S. Zein in Food Packaging. In *Sustainable Food Packaging Technolology*; Athanassiou, A., Ed.; WILEY-VCH Publishing: Weinheim, Germany, 2021; pp. 199–224.
96. Gautam, S.; Sharma, B.; Jain, P. Green Natural Protein Isolate based composites and nanocomposites: A review. *Polym. Test.* 2021, *99*, 106626. [CrossRef]
97. Lionetto, F.; Esposito Corcione, C. Recent applications of biopolymers derived from fish industry waste in food packaging. *Polymers* 2021, *13*, 2337. [CrossRef]
98. Guillaume, C.; Pinte, J.; Gontard, N.; Gastaldi, E. Wheat gluten-coated papers for bio-based food packaging: Structure, surface and transfer properties. *Food Res. Int.* 2010, *43*, 1395–1401. [CrossRef]
99. Wu, F.; Misra, M.; Mohanty, A.K. Challenges and new opportunities on barrier performance of biodegradable polymers for sustainable packaging. *Prog. Polym. Sci.* 2021, *117*, 101395. [CrossRef]
100. Guillard, V.; Gaucel, S.; Fornaciari, C.; Angellier-Coussy, H.; Buche, P.; Gontard, N. The next generation of sustainable food packaging to preserve our environment in a circular economy context. *Front. Nutr.* 2018, *5*, 121. [CrossRef]
101. Ibrahim, M.S.; Sani, N.; Adamu, M.; Abubakar, M.K. Biodegradable polymers for sustainable environmental and economic development. *MOJ Biorg. Org. Chem.* 2018, *2*, 192–194.
102. Nile, S.H.; Baskar, V.; Selvaraj, D.; Nile, A.; Xiao, J.; Kai, G. Nanotechnologies in food science: Applications, recent trends, and future perspectives. *Nano-Micro Lett.* 2020, *12*, 1–34. [CrossRef]
103. Alfei, S.; Marengo, B.; Zuccari, G. Nanotechnology application in food packaging: A plethora of opportunities versus pending risks assessment and public concerns. *Food Res. Int.* 2020, *137*, 109664. [CrossRef]
104. Guo, F.; Aryana, S.; Han, Y.; Jiao, Y. A review of the synthesis and applications of polymer–nanoclay composites. *Appl. Sci.* 2018, *8*, 1696. [CrossRef]
105. Luduena, L.N.; Alvarez, V.A.; Vazquez, A. Processing and microstructure of PCL/clay nanocomposites. *Mater. Sci. Eng. A* 2007, *460*, 121–129. [CrossRef]
106. Silva, F.A.G.S.; Dourado, F.; Gama, M.; Poças, F. Nanocellulose bio-based composites for food packaging. *Nanomaterials* 2020, *10*, 2041. [CrossRef] [PubMed]
107. Liu, S.F.; Petty, A.R.; Sazama, G.T.; Swager, T.M. Single-walled carbon nanotube/metalloporphyrin composites for the chemiresistive detection of amines and meat spoilage. *Angew. Chem. Int. Ed.* 2015, *54*, 6554–6557. [CrossRef] [PubMed]
108. Fahmy, H.M.; Eldin, R.E.S.; Serea, E.S.A.; Gomaa, N.M.; AboElmagd, G.M.; Salem, S.A.; Elsayed, Z.A.; Edrees, A.; Shams-Eldin, E.; Shalan, A.E. Advances in nanotechnology and antibacterial properties of biodegradable food packaging materials. *RSC Adv.* 2020, *10*, 20467–20484. [CrossRef]
109. Muzzarelli, R.A.A.; Muzzarelli, C. Chitosan chemistry: Relevance to the biomedical sciences. In *Polysaccharides I: Structure, Characterization and Use*; Springer: Berlin/Heidelberg, Germany, 2005; Volume 186, pp. 151–209.
110. Soysal, Ç.; Bozkurt, H.; Dirican, E.; Güçlü, M.; Bozhüyük, E.D.; Uslu, A.E.; Kaya, S. Effect of antimicrobial packaging on physicochemical and microbial quality of chicken drumsticks. *Food Control* 2015, *54*, 294–299. [CrossRef]
111. Jagadish, K.; Shiralgi, Y.; Chandrashekar, B.N.; Dhananjaya, B.L.; Srikantaswamy, S. Ecofriendly synthesis of metal/metal oxide nanoparticles and their application in food packaging and food preservation. In *Impact of Nanoscience in the Food Industry*; Elsevier: Amsterdam, The Netherlands, 2018; pp. 197–216.
112. Fortunati, E.; Peltzer, M.; Armentano, I.; Torre, L.; Jiménez, A.; Kenny, J.M. Effects of modified cellulose nanocrystals on the barrier and migration properties of PLA nano-biocomposites. *Carbohydr. Polym.* 2012, *90*, 948–956. [CrossRef] [PubMed]
113. Yang, W.; Fortunati, E.; Dominici, F.; Giovanale, G.; Mazzaglia, A.; Balestra, G.M.; Kenny, J.M.; Puglia, D. Synergic effect of cellulose and lignin nanostructures in PLA based systems for food antibacterial packaging. *Eur. Polym. J.* 2016, *79*, 1–12. [CrossRef]
114. Horst, M.F.; Quinzani, L.M.; Failla, M.D. Rheological and barrier properties of nanocomposites of HDPE and exfoliated montmorillonite. *J. Thermoplast. Compos. Mater.* 2014, *27*, 106–125. [CrossRef]
115. Sessini, V.; Arrieta, M.P.; Kenny, J.M.; Peponi, L. Processing of edible films based on nanoreinforced gelatinized starch. *Polym. Degrad. Stab.* 2016, *132*, 157–168. [CrossRef]
116. Marras, S.I.; Kladi, K.P.; Tsivintzelis, I.; Zuburtikudis, I.; Panayiotou, C. Biodegradable polymer nanocomposites: The role of nanoclays on the thermomechanical characteristics and the electrospun fibrous structure. *Acta Biomater.* 2008, *4*, 756–765. [CrossRef] [PubMed]
117. Yu, J.; Yang, J.; Liu, B.; Ma, X. Preparation and characterization of glycerol plasticized-pea starch/ZnO–carboxymethylcellulose sodium nanocomposites. *Bioresour. Technol.* 2009, *100*, 2832–2841. [CrossRef]
118. Oymaci, P.; Altinkaya, S.A. Improvement of barrier and mechanical properties of whey protein isolate based food packaging films by incorporation of zein nanoparticles as a novel bionanocomposite. *Food Hydrocoll.* 2016, *54*, 1–9. [CrossRef]

119. Wakai, M.; Almenar, E. Effect of the presence of montmorillonite on the solubility of whey protein isolate films in food model systems with different compositions and pH. *Food Hydrocoll.* **2015**, *43*, 612–621. [CrossRef]
120. Jang, W.-S.; Rawson, I.; Grunlan, J.C. Layer-by-layer assembly of thin film oxygen barrier. *Thin Solid Films* **2008**, *516*, 4819–4825. [CrossRef]
121. Svagan, A.J.; Hedenqvist, M.S.; Berglund, L. Reduced water vapour sorption in cellulose nanocomposites with starch matrix. *Compos. Sci. Technol.* **2009**, *69*, 500–506. [CrossRef]
122. Sung, S.H.; Chang, Y.; Han, J. Development of polylactic acid nanocomposite films reinforced with cellulose nanocrystals derived from coffee silverskin. *Carbohydr. Polym.* **2017**, *169*, 495–503. [CrossRef]
123. Hassan-Nejad, M.; Ganster, J.; Bohn, A.; Pinnow, M.; Volkert, B. Bio-based nanocomposites of cellulose acetate and nano-clay with superior mechanical properties. *Macromol. Symp.* **2009**, *280*, 123–129. [CrossRef]
124. Chen, G.; Wei, M.; Chen, J.; Huang, J.; Dufresne, A.; Chang, P.R. Simultaneous reinforcing and toughening: New nanocomposites of waterborne polyurethane filled with low loading level of starch nanocrystals. *Polymer* **2008**, *49*, 1860–1870. [CrossRef]
125. Velásquez-Cock, J.; Ramírez, E.; Betancourt, S.; Putaux, J.-L.; Osorio, M.; Castro, C.; Gañán, P.; Zuluaga, R. Influence of the acid type in the production of chitosan films reinforced with bacterial nanocellulose. *Int. J. Biol. Macromol.* **2014**, *69*, 208–213. [CrossRef]
126. De Moura, M.R.; Aouada, F.A.; Avena-Bustillos, R.J.; McHugh, T.H.; Krochta, J.M.; Mattoso, L.H.C. Improved barrier and mechanical properties of novel hydroxypropyl methylcellulose edible films with chitosan/tripolyphosphate nanoparticles. *J. Food Eng.* **2009**, *92*, 448–453. [CrossRef]
127. Qin, Y.; Zhang, S.; Yu, J.; Yang, J.; Xiong, L.; Sun, Q. Effects of chitin nano-whiskers on the antibacterial and physicochemical properties of maize starch films. *Carbohydr. Polym.* **2016**, *147*, 372–378. [CrossRef] [PubMed]
128. Ramanathan, T.; Abdala, A.A.; Stankovich, S.; Dikin, D.A.; Herrera-Alonso, M.; Piner, R.D.; Adamson, D.H.; Schniepp, H.C.; Chen, X.; Ruoff, R.S. Functionalized graphene sheets for polymer nanocomposites. *Nat. Nanotechnol.* **2008**, *3*, 327–331. [CrossRef] [PubMed]
129. Dobrucka, R.; Cierpiszewski, R. Active and intelligent packaging food-Research and development—A Review. *Pol. J. Food Nutr. Sci.* **2014**, *64*, 7–15. [CrossRef]
130. Fuertes, G.; Soto, I.; Carrasco, R.; Vargas, M.; Sabattin, J.; Lagos, C. Intelligent packaging systems: Sensors and nanosensors to monitor food quality and safety. *J. Sens.* **2016**, *2016*, 1–8. [CrossRef]
131. Biji, K.B.; Ravishankar, C.N.; Mohan, C.O.; Gopal, T.K.S. Smart packaging systems for food applications: A review. *J. Food Sci. Technol.* **2015**, *52*, 6125–6135. [CrossRef]
132. Yam, K.L.; Lee, D.S. *Emerging Food Packaging Technologies: Principles and Practice*; Elsevier: Amsterdam, The Netherlands, 2012; ISBN 0857095668.
133. Brockgreitens, J.; Abbas, A. Responsive food packaging: Recent progress and technological prospects. *Compr. Rev. Food Sci. Food Saf.* **2016**, *15*, 3–15. [CrossRef]
134. Ramos, Ó.L.; Pereira, R.N.; Cerqueira, M.A.; Martins, J.R.; Teixeira, J.A.; Malcata, F.X.; Vicente, A.A. Bio-based nanocomposites for food packaging and their effect in food quality and safety. In *Food Packaging and Preservation*; Elsevier: Amsterdam, The Netherlands, 2018; pp. 271–306.
135. Taherimehr, M.; YousefniaPasha, H.; Tabatabaeekoloor, R.; Pesaranhajiabbas, E. Trends and challenges of biopolymer-based nanocomposites in food packaging. *Compr. Rev. Food Sci. Food Saf.* **2021**, *20*, 5321–5344. [CrossRef]
136. Anvar, A.A.; Ahari, H.; Ataee, M. Antimicrobial properties of food nanopackaging: A new focus on foodborne pathogens. *Front. Microbiol.* **2021**, *12*, 690706. [CrossRef] [PubMed]
137. Damm, C.; Münstedt, H.; Rösch, A. The antimicrobial efficacy of polyamide 6/silver-nano-and microcomposites. *Mater. Chem. Phys.* **2008**, *108*, 61–66. [CrossRef]
138. Vilela, C.; Kurek, M.; Hayouka, Z.; Röcker, B.; Yildirim, S.; Antunes, M.D.C.; Nilsen-Nygaard, J.; Pettersen, M.K.; Freire, C.S.R. A concise guide to active agents for active food packaging. *Trends Food Sci. Technol.* **2018**, *80*, 212–222. [CrossRef]
139. He, Y.; Li, H.; Fei, X.; Peng, L. Carboxymethyl cellulose/cellulose nanocrystals immobilized silver nanoparticles as an effective coating to improve barrier and antibacterial properties of paper for food packaging applications. *Carbohydr. Polym.* **2021**, *252*, 117156. [CrossRef] [PubMed]
140. Marrez, D.A.; Abdelhamid, A.E.; Darwesh, O.M. Eco-friendly cellulose acetate green synthesized silver nano-composite as antibacterial packaging system for food safety. *Food Packag. Shelf Life* **2019**, *20*, 100302. [CrossRef]
141. Thirumurugan, A.; Ramachandran, S.; Shiamala Gowri, A. Combined effect of bacteriocin with gold nanoparticles against food spoiling bacteria-an approach for food packaging material preparation. *Int. Food Res. J.* **2013**, *20*, 1909–1912.
142. Rukmanikrishnan, B.; Jo, C.; Choi, S.; Ramalingam, S.; Lee, J. Flexible ternary combination of gellan gum, sodium carboxymethyl cellulose, and silicon dioxide nanocomposites fabricated by quaternary ammonium silane: Rheological, thermal, and antimicrobial properties. *ACS Omega* **2020**, *5*, 28767–28775. [CrossRef] [PubMed]
143. Al-Nabulsi, A.; Osaili, T.; Sawalha, A.; Olaimat, A.N.; Albiss, B.A.; Mehyar, G.; Ayyash, M.; Holley, R. Antimicrobial activity of chitosan coating containing ZnO nanoparticles against E. coli O157: H7 on the surface of white brined cheese. *Int. J. Food Microbiol.* **2020**, *334*, 108838. [CrossRef]
144. Qin, Y.; Liu, Y.; Yuan, L.; Yong, H.; Liu, J. Preparation and characterization of antioxidant, antimicrobial and pH-sensitive films based on chitosan, silver nanoparticles and purple corn extract. *Food Hydrocoll.* **2019**, *96*, 102–111. [CrossRef]

145. Da Costa Brito, S.; Bresolin, J.D.; Sivieri, K.; Ferreira, M.D. Low-density polyethylene films incorporated with silver nanoparticles to promote antimicrobial efficiency in food packaging. *Food Sci. Technol. Int.* **2020**, *26*, 353–366. [CrossRef]
146. Chen, Q.-Y.; Xiao, S.-L.; Shi, S.Q.; Cai, L.-P. A One-Pot Synthesis and Characterization of Antibacterial Silver Nanoparticle–Cellulose Film. *Polymers* **2020**, *12*, 440. [CrossRef]
147. Chowdhury, S.; Teoh, Y.L.; Ong, K.M.; Zaidi, N.S.R.; Mah, S.-K. Poly (vinyl) alcohol crosslinked composite packaging film containing gold nanoparticles on shelf life extension of banana. *Food Packag. Shelf Life* **2020**, *24*, 100463. [CrossRef]
148. Roy, S.; Rhim, J.-W. Effect of CuS reinforcement on the mechanical, water vapor barrier, UV-light barrier, and antibacterial properties of alginate-based composite films. *Int. J. Biol. Macromol.* **2020**, *164*, 37–44. [CrossRef] [PubMed]
149. Shankar, S.; Wang, L.-F.; Rhim, J.-W. Preparation and properties of carbohydrate-based composite films incorporated with CuO nanoparticles. *Carbohydr. Polym.* **2017**, *169*, 264–271. [CrossRef] [PubMed]
150. Ahmadi, A.; Ahmadi, P.; Sani, M.A.; Ehsani, A.; Ghanbarzadeh, B. Functional biocompatible nanocomposite films consisting of selenium and zinc oxide nanoparticles embedded in gelatin/cellulose nanofiber matrices. *Int. J. Biol. Macromol.* **2021**, *175*, 87–97. [CrossRef]
151. Lan, W.; Wang, S.; Zhang, Z.; Liang, X.; Liu, X.; Zhang, J. Development of red apple pomace extract/chitosan-based films reinforced by TiO2 nanoparticles as a multifunctional packaging material. *Int. J. Biol. Macromol.* **2021**, *168*, 105–115. [CrossRef]
152. Li, X.; Ren, Z.; Wang, R.; Liu, L.; Zhang, J.; Ma, F.; Khan, M.Z.H.; Zhao, D.; Liu, X. Characterization and antibacterial activity of edible films based on carboxymethyl cellulose, Dioscorea opposita mucilage, glycerol and ZnO nanoparticles. *Food Chem.* **2021**, *349*, 129208. [CrossRef] [PubMed]
153. Ojha, N.; Das, N. Fabrication and characterization of biodegradable PHBV/SiO2 nanocomposite for thermo-mechanical and antibacterial applications in food packaging. *IET Nanobiotechnol.* **2020**, *14*, 785–795. [CrossRef]
154. Maćkiw, E.; Mąka, Ł.; Ścieżyńska, H.; Pawlicka, M.; Dziadczyk, P.; Rżanek-Boroch, Z. The Impact of Plasma-modified Films with Sulfur Dioxide, Sodium Oxide on Food Pathogenic Microorganisms. *Packag. Technol. Sci.* **2015**, *28*, 285–292. [CrossRef]
155. Wu, J.; Sun, Q.; Huang, H.; Duan, Y.; Xiao, G.; Le, T. Enhanced physico-mechanical, barrier and antifungal properties of soy protein isolate film by incorporating both plant-sourced cinnamaldehyde and facile synthesized zinc oxide nanosheets. *Colloids Surf. B Biointerfaces* **2019**, *180*, 31–38. [CrossRef]
156. Amjadi, S.; Almasi, H.; Ghorbani, M.; Ramazani, S. Preparation and characterization of TiO2NPs and betanin loaded zein/sodium alginate nanofibers. *Food Packag. Shelf Life* **2020**, *24*, 100504. [CrossRef]
157. Swaroop, C.; Shukla, M. Nano-magnesium oxide reinforced polylactic acid biofilms for food packaging applications. *Int. J. Biol. Macromol.* **2018**, *113*, 729–736. [CrossRef]
158. Kousheh, S.A.; Moradi, M.; Tajik, H.; Molaei, R. Preparation of antimicrobial/ultraviolet protective bacterial nanocellulose film with carbon dots synthesized from lactic acid bacteria. *Int. J. Biol. Macromol.* **2020**, *155*, 216–225. [CrossRef]
159. Bi, F.; Zhang, X.; Liu, J.; Yong, H.; Gao, L.; Liu, J. Development of antioxidant and antimicrobial packaging films based on chitosan, D-α-tocopheryl polyethylene glycol 1000 succinate and silicon dioxide nanoparticles. *Food Packag. Shelf Life* **2020**, *24*, 100503. [CrossRef]
160. Dias, M.V.; de Fátima, F.S.N.; Borges, S.V.; de Sousa, M.M.; Nunes, C.A.; de Oliveira, I.R.N.; Medeiros, E.A.A. Use of allyl isothiocyanate and carbon nanotubes in an antimicrobial film to package shredded, cooked chicken meat. *Food Chem.* **2013**, *141*, 3160–3166. [CrossRef] [PubMed]
161. Liu, Y.; Wang, S.; Lan, W.; Qin, W. Fabrication of polylactic acid/carbon nanotubes/chitosan composite fibers by electrospinning for strawberry preservation. *Int. J. Biol. Macromol.* **2019**, *121*, 1329–1336. [CrossRef] [PubMed]
162. Melendez-Rodriguez, B.; Figueroa-Lopez, K.J.; Bernardos, A.; Martínez-Máñez, R.; Cabedo, L.; Torres-Giner, S.; Lagaron, J.M. Electrospun antimicrobial films of poly (3-hydroxybutyrate-co-3-hydroxyvalerate) containing eugenol essential oil encapsulated in mesoporous silica nanoparticles. *Nanomaterials* **2019**, *9*, 227. [CrossRef]
163. Atef, M.; Rezaei, M.; Behrooz, R. Characterization of physical, mechanical, and antibacterial properties of agar-cellulose bionanocomposite films incorporated with savory essential oil. *Food Hydrocoll.* **2015**, *45*, 150–157. [CrossRef]
164. Meira, S.M.M.; Zehetmeyer, G.; Scheibel, J.M.; Werner, J.O.; Brandelli, A. Starch-halloysite nanocomposites containing nisin: Characterization and inhibition of Listeria monocytogenes in soft cheese. *LWT-Food Sci. Technol.* **2016**, *68*, 226–234. [CrossRef]
165. Hosseini, S.F.; Rezaei, M.; Zandi, M.; Farahmandghavi, F. Development of bioactive fish gelatin/chitosan nanoparticles composite films with antimicrobial properties. *Food Chem.* **2016**, *194*, 1266–1274. [CrossRef]
166. Abdollahi, M.; Rezaei, M.; Farzi, G. A novel active bionanocomposite film incorporating rosemary essential oil and nanoclay into chitosan. *J. Food Eng.* **2012**, *111*, 343–350. [CrossRef]
167. Otoni, C.G.; de Moura, M.R.; Aouada, F.A.; Camilloto, G.P.; Cruz, R.S.; Lorevice, M.V.; de F.F. Soares, N.; Mattoso, L.H.C. Antimicrobial and physical-mechanical properties of pectin/papaya puree/cinnamaldehyde nanoemulsion edible composite films. *Food Hydrocoll.* **2014**, *41*, 188–194. [CrossRef]
168. Salehudin, M.H.; Salleh, E.; Mamat, S.N.H.; Muhamad, I.I. Starch based active packaging film reinforced with empty fruit bunch (EFB) cellulose nanofiber. *Procedia Chem.* **2014**, *9*, 23–33. [CrossRef]
169. Wen, P.; Zhu, D.-H.; Feng, K.; Liu, F.-J.; Lou, W.-Y.; Li, N.; Zong, M.-H.; Wu, H. Fabrication of electrospun polylactic acid nanofilm incorporating cinnamon essential oil/β-cyclodextrin inclusion complex for antimicrobial packaging. *Food Chem.* **2016**, *196*, 996–1004. [CrossRef] [PubMed]

170. Kanmani, P.; Rhim, J.-W. Nano and nanocomposite antimicrobial materials for food packaging applications. In *Future Medicine*; Future Science Ltd.: London, UK; Mokpo National University: Seoul, Korea, 2014; pp. 34–48.
171. Rezaei, M.; Pirsa, S.; Chavoshizadeh, S. Photocatalytic/antimicrobial active film based on wheat gluten/ZnO nanoparticles. *J. Inorg. Organomet. Polym. Mater.* **2020**, *30*, 2654–2665. [CrossRef]
172. Hahn, A.; Fuhlrott, J.; Loos, A.; Barcikowski, S. Cytotoxicity and ion release of alloy nanoparticles. *J. Nanopart. Res.* **2012**, *14*, 1–10. [CrossRef] [PubMed]
173. Vega-Jiménez, A.L.; Vázquez-Olmos, A.R.; Acosta-Gío, E.; Álvarez-Pérez, M.A. In vitro antimicrobial activity evaluation of metal oxide nanoparticles. In *Nanoemulsions Properties, Fabrications and Applications*; IntechOpen: London, UK, 2019; pp. 1–18.

Article

Preparation and Characterization of Carboxymethyl Cellulose-Based Bioactive Composite Films Modified with Fungal Melanin and Carvacrol

Łukasz Łopusiewicz [1,*], Paweł Kwiatkowski [2], Emilia Drozłowska [1], Paulina Trocer [1], Mateusz Kostek [1], Mariusz Śliwiński [3], Magdalena Polak-Śliwińska [4], Edward Kowalczyk [5] and Monika Sienkiewicz [6]

1. Center of Bioimmobilisation and Innovative Packaging Materials, Faculty of Food Sciences and Fisheries, West Pomeranian University of Technology Szczecin, Janickiego 35, 71-270 Szczecin, Poland; emilia_drozlowska@zut.edu.pl (E.D.); p.trocer@gmail.com (P.T.); mkosa9406@gmail.com (M.K.)
2. Chair of Microbiology, Immunology and Laboratory Medicine, Department of Diagnostic Immunology, Pomeranian Medical University in Szczecin, Powstańców Wielkopolskich 72, 70-111 Szczecin, Poland; pawel.kwiatkowski@pum.edu.pl
3. Dairy Industry Innovation Institute Ltd., Kormoranów 1, 11-700 Mrągowo, Poland; mariusz.sliwinski@iipm.pl
4. Chair of Commodity Science and Food Analysis, Faculty of Food Science, University of Warmia and Mazury in Olsztyn, Pl. Cieszyński 1, 10-957 Olsztyn, Poland; m.polak@uwm.edu.pl
5. Department of Pharmacology and Toxicology, Medical University of Łódź, 90-752 Łódź, Poland; edward.kowalczyk@umed.lodz.pl
6. Department of Allergology and Respiratory Rehabilitation, Medical University of Łódź, Żeligowskiego 7/9, 90-752 Łódź, Poland; monika.sienkiewicz@umed.lodz.pl
* Correspondence: lukasz.lopusiewicz@zut.edu.pl; Tel.: +48-91-449-6135

Abstract: Preparation of biodegradable packaging materials and valorisation of food industry residues to achieve "zero waste" goals is still a major challenge. Herein, biopolymer-based (carboxymethyl cellulose—CMC) bioactive films were prepared by the addition, alone or in combination, of carvacrol and fungal melanin isolated from champignon mushroom (*Agaricus bisporus*) agroindustrial residues. The mechanical, optical, thermal, water vapour, and UV-Vis barrier properties were studied. Fourier-transform infrared (FT-IR) spectroscopy studies were carried out to analyse the chemical composition of the resulting films. Antibacterial, antifungal, and antioxidant activities were also determined. Both CMC/melanin and CMC/melanin/carvacrol films showed some antimicrobial activity against *Escherichia coli*, *Staphylococcus aureus*, and *Candida albicans*. The addition of melanin increased the UV-blocking, mechanical, water vapour barrier, and antioxidant properties without substantially reducing the transparency of the films. The addition of carvacrol caused loss of transparency, however, composite CMC/melanin/carvacrol films showed excellent antioxidant activity and enhanced mechanical strength. The developed bioactive biopolymer films have a good potential to be green bioactive alternatives to plastic films in food packaging applications.

Keywords: melanin; carvacrol; agricultural residues; carboxymethyl cellulose; bioactive films; functional films; antioxidant activity; antimicrobial activity

1. Introduction

The packaging industry is currently dominated by synthetic polymers (plastics) due to their low price and excellent functionality (mechanical strength and high barrier properties). Based on available data, the annual plastics production exceeds 400 million tons, and around 40% is used for packaging purposes [1]. Petroleum-based packaging materials are increasingly falling-out of favour due to their unsustainable production and the environmental burden of plastic [1,2]. The development of biodegradable packaging materials is an effective alternative to synthetic packaging materials based on petrochemical products [3,4]. Natural biopolymers and synthetic polymers based on annually renewable resources are the basis of a 21st century portfolio of sustainable, eco-efficient plastics. Biodegradable polymers have a great potential to be used as a green alternative to plastic packaging films as they become more and more affordable [3,5]. Biodegradable polymers from renewable resources have attracted a lot of academic and industrial attention worldwide. They are defined as polymers that undergo microbially-induced chain scission, leading to mineralization. Carbohydrates are used to manufacture biopolymer packaging films due to their excellent film-forming ability, good gas barrier properties, and mechanical properties [1]. Among naturally occurring biopolymers, cellulose is the most abundant one. Its derivatives have several advantages, such as recyclability, high viscosity, non-toxicity, biodegradability, and cost effectiveness [6–11]. Sodium carboxymethyl cellulose is a water-soluble cellulose derivative with carboxy methyl groups attached to some of hydroxyl groups of glucopyranose monomers of cellulose backbone. CMC has received scientific attention due to its polyelectrolyte character. In view of its high transparency, good film-forming property, large mechanical strength, non-toxicity, and biodegradability, it is found to be suitable for applications such as packaging material (for films and coatings), medicine, flocculating agent, chelating agent, emulsifier, thickening agent, water-retaining agent, and sizing agent [6–11].

In recent years, due to the current planet issues, the need for transition to a circular economy model based on the development of new strategies for making the best use of resources and for the elimination of the concept of wastes along the supply chain. In this model materials are recycled (a process in which wastes are transformed into value-added products by making them input elements for other products) and re-circulated during processing created a concept "waste = valuable resource" [12]. The food processing industries produce millions of tons of losses and waste during processing which is becoming a grave economic, environmental, and nutritional problem. These wastes can be a meaningful source of bioactive compounds [2,12–14]. Thus, these by-products can be exploited again in the food industry to develop functional ingredients and new foods or natural additives, or, in other industries, such as the pharmaceutical, agricultural, or chemical industries to obtain bioactive compounds [14].

Melanins is a common name of heterogenous group of dark-coloured biopigments, with high molecular weight. They are derived from the oxidation of monophenols and the subsequent polymerization of intermediate *o*-diphenols and their resulting quinones [15]. Melanins are known from their multifunctionality, including antioxidant, radioprotective, thermo-regulative, chemoprotective, antitumor, antiviral, antimicrobial, immunostimulating, and anti-inflammatory activities [13,16–19]. Recently, melanins have emerged as potential nanofillers and polymer matrix modifiers for food packaging polymers, such as: whey protein concentrate/isolate (WPI/WPC) [2], gelatine [4], poly(lactic acid) (PLA) [3], alginate [20], agar [21], carrageenan [22], cellulose [23], chitosan [24], poly(vinyl alcohol) [25], polypropylene/poly(butylene adipate-co-terephthalate) [26], polyhydroxybutyrate [27], and ethylene-vinyl acetate copolymer [28]. As with many other natural biopolymers, melanins can be obtained from renewable and natural resources. Moreover, they are non-toxic. Due to these characteristics, melanins have potential to be a "green" alternative to many existing commercial food additives. Furthermore, the possibility of melanins production by sustainable extraction from natural agricultural residues (e.g., waste from the harvesting of champignon mushroom, *Agaricus bisporus*, or residual watermelon seeds) has been demonstrated [13,17]. Another potential bioactive functional compound is carvacrol (CV, 5-isopropyl-2-methylphenol), a phenolic compound found primarily in oils of oregano, thyme, and marjoram, and recognized as a safe food additive (Generally Recognized as Safe—GRAS) [7,29–36]. This bioactive compound possesses antimicrobial properties, antioxidant, and a particular aroma which makes an attractive ingredient for certain types of foods [37,38]. Moreover, CV has been also reported to be used for modification of biopolymer [5,30,31,36,39–44], and synthethic films [45,46].

The functional properties of composite films are also essential in active and intelligent food packaging applications. A functional CMC-based film can be developed by adding functional materials such as bioactive compounds. The addition of bioactive compounds is expected to improve the film's physical and functional properties. Considering the potential of melanins and carvacrol as functional materials, some synergies can be expected when these materials are used together. Moreover, the application of bioactive and biodegradable films seem to be an excellent alternative to reduce food loss and waste and to improve food security [5,41].

The main aim of this study was to investigate the effect of adding melanin obtained from *A. bisporus* waste, alone or in combination with carvacrol on the properties of carboxymethyl cellulose films (CMC). To the best of our knowledge, there is a lack of reporting about the modification of CMC with natural melanin and carvacrol to modify the functionality of composite biopolymer films. Fourier-transform infrared (FT-IR) spectroscopy was used to examine the chemical composition of films after active compounds addition. Moreover, we also examined the influence active agents on the colour and optical properties of the films. In order to evaluate the potential of resulted films as bioactive materials, their mechanical, barrier, antioxidant, and antimicrobial properties were determined.

2. Materials and Methods

2.1. Materials and Reagents

Sodium carboxymethyl cellulose (CMC) (degree of substitution = 0.7, M.W. = 90,000), carvacrol (natural, originated from thyme essential oil, 99%, food grade), calcium chloride, sodium chloride, hydrogen peroxide, disodium phosphate, monosodium phosphate, 2,2-diphenyl-1-picrylhydrazyl (DPPH), 2,2′-azino-bis(3-ethylbenzothiazoline-6-sulfonic acid) (ABTS), potassium persulphate, potassium ferricyanide, trichloroacetic acid, ferric chloride, iron sulphate, tris(hydroxymethyl)aminomethane, pyrogallol, resazurin, were purchased from Merck Chemical (Saint Louis, MO, USA). Tween 80, glycerol, ammonia water, hydrochloric acid, sodium hydroxide, chloroform, ethyl acetate, ethanol, and methanol were procured from Chempur (Piekary Śląskie, Poland). Mueller–Hinton broth, Mueller–Hinton agar, and agar-agar were purchased from Merck Chemical (Saint Louis, MO, USA). All chemicals were of analytical grade. *Escherichia coli* ATCC25922, *Staphylococcus aureus* ATCC43300, and *Candida albicans* ATCC10231 were purchased from ATCC (American Type Culture Collection, Manassas, VA, USA).

2.2. Isolation, Purification and Preparation of Melanin from Agaricus Bisporus Waste

Waste from the production of *A. bisporus* (Agaricus Bisporus Waste—ABW) in the form of stipes was obtained from a local producer in Wolsztyn (Wielkopolskie voivodeship, Poland). The isolation and purification of melanin was carried out as described elsewhere [13]. In brief, 500 g of ABW was first homogenised (Heidolph Brinkmann Homogenizer Silent Crusher, Schwabach, Germany) in 500 mL of distilled water and incubated (24 h, 37 °C) to allow acting of tyrosinase. After incubation the homogenate mixture was adjusted to pH = 10 by 1 M NaOH and incubated (24 h, 65 °C) to allow spontaneous polymerization of resulting *o*-diphenols and quinones to form melanin. Then, the mixture was filtered, centrifuged (6000 rpm, 10 min) and alkaline ABW raw melanin (ABW-RM) mixture was used to purify melanin. Alkaline ABW-RM mixture was first adjusted to pH 2.0 with 1 M HCl to precipitate melanin, followed by centrifugation at 6000 rpm for 10 min and a pellet was collected. The acid hydrolysis was then carried out (6 M HCl, 90 °C, 2 h). The resulted melanin was subsequently centrifuged (6000 rpm, 10 min) and washed by distilled water five times to neutral pH, then rinsed with organic solvents (chloroform, ethyl acetate and ethanol) three times to wash away lipids and other residues. Finally, the purified melanin was dried and ground to a fine powder in a mortar.

2.3. Determination of Carvacrol Minimum Inhibitory Concentration (MIC)

The minimal inhibitory concentration (MIC) (the lowest concentration of an antimicrobial that will inhibit the visible growth of a microorganism after overnight incubation) values of the carvacrol against bacteria (*Escherichia coli* ATCC25922 and *Staphylococcus au-*

reus ATCC43300), as well as yeast (*Candida albicans* ATCC10231), were determined by broth microdilution method according to Clinical and Laboratory Standards Institute (CLSI), with the following modification [29,47,48]: a final concentration of 1.0% (v/v) Tween 80 (filter sterilized) was incorporated into the medium to enhance oil solubility. We performed two-fold dilutions (1000–3.91 µg/mL); each well containing 50 µL of the tested carvacrol and 50 µL of bacterial or yeast suspension at a final concentration of 10^6 CFU/mL (CFU–colony forming unit). Bacterial and yeast suspensions were prepared from 18 h cultures using saline. All tests were performed in duplicate. The MIC was estimated after 18 h of incubation at 37 °C in Mueller–Hinton broth (MHB) using resazurin. MIC was determined on the basis of the blue colour appearance in the first tested well after 3 h of incubation with resazurin. The colour change from blue to pink after 3 h of incubation with resazurin at 37 °C indicated the presence of viable microorganisms. To exclude an inhibitory effect of 1.0% Tween 80 on the microbial growth, the control assays with MHB and MHB supplemented with 1.0% Tween 80 were performed. Using the known concentrations of carvacrol, the final result was expressed in µg/mL.

2.4. Preparation of Films

In order to obtain CMC concentration of 2% (w/w) in the final film-forming solutions, CMC was weighed and completely dissolved in distilled water at 50 °C under continuous agitation (in tightly closed glass bottles). The pH of each solution was adjusted to 8.0 with ammonia water. Then, melanin was added to obtain concentrations of 0.1 and 0.5% (w/w). The mixtures were stirred (250 rpm) for 1 h at 50 °C to complete dissolve the melanin. After cooling, glycerol (at 5% (w/w), on a film-forming solution basis) and Tween 80 (0.5% v/v) were added and homogenized. The neat CMC films (without melanin addition) were prepared the same way, and served as reference materials. The film forming solutions were then divided into two batches. The samples only with melanin were described as CMC + 0.1 M and CMC + 0.5 M. CMC/melanin films with carvacrol were prepared using an emulsion method. To one batch of CMC/melanin film forming solutions 0.025% (w/v–250 µg/mL) of carvacrol (CV) was then added and stirred (250 rpm) until a homogenous appearance of solution was obtained (approximately 30 min) and the samples were described as CMC + 0.1 M + CV and CMC + 0.5 M + CV. CMC films with carvacrol (CMC + CV) were also produced following the same procedure. All CMC-based film variants were prepared in 10 repetitions. The film-forming solutions were cast on square (120 × 120 mm) polystyrene plates and dried at 40 °C for 48 h. The dried films were peeled off from the plates and were conditioned at 25 °C and 50% RH (relative humidity) in a temperature and humidity clean room prior to any tests [2].

2.5. Thickness, Moisture Content (MC) and Mechanical Properties of The Films

To determine MC of the films, the samples were dried at 105 °C for 24 h, and the weight change was analysed [2]. The thickness of all samples was measured with a hand-held micrometer (Dial Thickness Gauge 7301, Mitoyuto Corporation, Kangagawa, Japan) with an accuracy of 0.001 mm. Each film was measured in five random points and the results were averaged. The mechanical properties of the samples were determined with Zwick/Roell 2.5 Z equipment (Ulm, Germany). Static tensile testing was carried out to assess tensile strength and elongation at break (the gap between tensile clamps was 25 mm and crosshead speed was 100 mm/min).

2.6. DSC Measurements

A differential scanning calorimetry (DSC) calorimeter (DSC 3 Star System, Mettler-Toledo LLC, Columbus, OH, USA) was used to determine thermal properties of the samples. The tests were carried out over a temperature range from 10 to 300 at $\varphi = 10°$/min and under nitrogen flow (50 mL/min), performing two heating and one cooling scans according to PN-EN ISO 11357-03:2018-06 norm [2].

2.7. The Water Vapour Transmission Rate (WVTR) of the Films

Water vapour transmission rate (WVTR) was performed according to DIN 53122-1 and ISO 2528:1995 norms as described elsewhere [49]. WVTR was measured by means of a gravimetric method that is based on the sorption of humidity by $CaCl_2$ and a comparison of sample weight gain. Initially, the amount of dry $CaCl_2$ inside the container was 9 g. The area of film samples was 8.86 cm². Measurement was carried out for a period of 4 days and each day the containers were weighed to determine the amount of absorbed water vapour through the films. The results were expressed as average values from each day of measurement and each container. Analyses were carried out at 10 independent containers (10 repetitions) for each type films, calculated as a standard unit $g/(m^2 \times Day)$ and presented as a mean ± standard deviation.

2.8. Spectral Analysis

The UV–Vis blocking properties of the film samples were determined using a UV–Vis Thermo Scientific Evolution 220 spectrophotometer (Waltham, MA, USA). The tests were carried out in a range of 300 to 700 nm, by putting particular samples directly on quartz cuvette (Bionovo, Legnica, Poland). FT-IR spectroscopy was used in order to assess the chemical composition of obtained films, as described previously. Firstly, a 4 cm² squares of each film were cut from the samples, then were analysed directly on the ray-exposing stage of the ATR (Attenuated Total Reflectance) accessory of a Perkin Elmer Spectrum 100 FT-IR spectrometer (Waltham, MA, USA) operating in ATR mode. Spectra (64 scans) were recorded over a wavenumber range of 650 to 4000 cm^{-1} at a resolution of 4 cm^{-1} [2]. For analysis, all spectra were baseline corrected and normalized using SPECTRUM™ software v10.

2.9. Colour Analysis

The effect of melanin on the colour of the films was measured using a colorimeter (CR-5, Konica Minolta, Tokyo, Japan). The values measured were L* (white 100 and black 0), a* values (red positive and green negative), and b* values (yellow positive and blue negative). For each film type five samples were analyzed by taking three measurements on both sides of each sample. The Whiteness Index (WI), Yellowness Index (YI), total colour difference (ΔE), chroma (C), and hue angle (H°) were calculated according the following formulas (1–5) [2,4]:

$$WI = 100 - \left[(100 - L^*) + a^2 + b^2\right]^{0.5} \quad (1)$$

$$YI = 142.86 \times b \times L^{-1} \quad (2)$$

$$\Delta E = \left[\left(L_{standard} - L_{sample}\right)^2 + \left(a_{standard} - a_{sample}\right)^2 + \left(b_{standard} - b_{sample}\right)^2\right]^{0.5} \quad (3)$$

$$C = arctg \frac{b_{sample}}{a_{sample}} \quad (4)$$

$$H° = \left[\left(a_{sample}\right)^2 + \left(b_{sample}\right)^2\right]^{0.5} \quad (5)$$

2.10. Antioxidant Activity and Reducing Power

The reducing power, DPPH, $ABTS^{+\cdot}$, and O_2^- radicals scavenging activities were analysed based on methodologies described in our previous study [2]. The reducing power was determined by placing the film samples (100 mg) in 1.25 mL of phosphate buffer (0.2 M, pH 6.6) followed by the addition of 1.25 mL of 1% potassium ferricyanide solution. Samples were then incubated for 20 min at 50 °C followed by the addition of 1.25 mL of trichloroacetic acid. Subsequently, the test tubes were centrifuged at 3000 rpm for 10 min. Then, 1.25 mL of obtained supernatant was diluted with 1.25 mL of deionized water. Finally,

0.25 mL of 0.1% ferric chloride solution was added and the absorbance was measured at 700 nm [2].

To determine DPPH radical scavenging activity 100 mg of each film was placed in 25 mL of 0.01 mM DPPH methanolic solution, incubated for 30 min at room temperature and absorbance at 517 nm was measured. As a control, the same solution was measured, but without any film samples. In total, 10 mL of ABTS$^+$ solution was mixed with 100 mg of the films and the absorbance was measured at 734 nm. To determine O_2^- radical scavenging activity, 3 mL of 50 mmol/L (pH 8.2) Tris-HCl buffer was mixed with 100 mg of the films. Then, a pyrogallol solution (0.3 mL, 7 mmol/L, preheated to 25 °C) was added, and allowed to react for exactly 4 min. Finally, 1 mL od 10 mmol/L of HCl was added to terminate the reaction and absorbance was measured at 318 nm [2].

2.11. Antimicrobial Activity

The film samples were cut into square shapes (3 cm × 3 cm) and their antimicrobial properties were carried out according to the ASTM E 2180-01 standard with modification described elsewhere [50]. As the first step of the experiments, *E. coli*, *S. aureus*, and *C. albicans* cultures originated from 24 h growth (coming from stock cultures) were prepared. The concentrations of the cultures were standardized to 1.5×10^8 CFU/mL. The concentration of each culture was measured using Cell Density Meter (WPA-CB4, Cambridge, UK). The agar slurry was prepared by dissolving 0.85 g of NaCl and 0.3 g of agar–agar in 100 mL of deionized water and autoclaved for 15 min at 121 °C and equilibrated at 45 °C (one agar slurry was prepared for each strain). Then, 1 mL of the culture (separately) was placed into the 100 mL of agar slurry. The final concentration of each culture was 1.5×10^6 CFU/mL in molten agar slurry. The square samples of each film were introduced (separately) into the sterile Petri dishes with a diameter of 55 mm. Inoculated agar slurry (1.0 mL) was pipetted onto each square sample. The samples were incubated 24 h at 30 °C with relative humidity at 90%. After incubation the samples were aseptically removed from the Petri dishes and introduced into the 100 mL of MHB. The samples were dispersed 1 min in the Bag Mixer® CC (Interscience, St Nom la Brètech, France). The dispersion facilitated the complete release of the agar slurry from the samples. Then serial dilutions of the initial inoculum were performed. Each dilution was spread into the Mueller–Hinton Agar and incubated at 30 °C for 48 h. The results were presented as an average value with standard deviations.

2.12. Statistical Analysis

For statistical analysis of the obtained results Statistica software version 10 was used (StatSoft Polska, Kraków, Poland). Differences between means were determined using analysis of variance (ANOVA), followed by Fisher's LSD (Least Significant Difference) post-hoc. The significance of each mean property value was determined ($p < 0.05$). All measurements were carried out in at least three repetitions.

3. Results and Discussion

3.1. Antimicrobial Activity of The Films

Based of MIC determination results it was observed that the MIC values of CV for *E. coli*, *S. aureus*, and *C. albicans* were 256 µg/mL, 128 µg/mL, and 256 µg/mL, respectively. A comparable finding was reported by other authors [32–34]. Therefore, CV was added to CMC filmogenic solutions in concentration of 0.025% (*w/v*–250 µg/mL). The antimicrobial activity of CMC-based films is presented in Figures 1–3. No antimicrobial activity was noticed for neat CMC film. A complete reduction in microbial counts was observed for CMC + CV samples ($p < 0.05$). Moreover, reduction in bacterial and fungal counts was observed for CMC/melanin samples ($p < 0.05$). For sample CMC + 0.5 M the level of *E. coli* was $1.14 \times 10^4 \pm 0.13$ CFU/mL, for *S. aureus* $1.03 \times 10^3 \pm 0.22$ CFU/mL, whereas for *C. albicans*, $3.66 \times 10^3 \pm 0.27$ CFU/mL was noticed. Interestingly, a significant reduction was also observed in case of CMC + 0.1 M + CV and CMC + 0.5 M + CV samples, however, not complete, as noticed for sample CMC + CV. The observed results suggest that this effect can be presumably attributed to some interactions between melanin and carvacrol in CMC matrix, lowering CV diffusion efficiency. The antibacterial activity of chitosan/CV [51], chitosan/CV/pomegranate peel extract [40], polypropylene/CV [31], flaxseed gum/CV [36,44], and PLA/CV [52] films, as well as PLA/CV [53] nanofibers, has been already reported. Carvacrol is a phenolic compound with a hydroxyl group on an aromatic ring. The hydroxyl group of CV plays a crucial role in the antibacterial activity of this phytochemical [32–34]. Indeed, CV interacts with the lipid bilayer of the bacterial cytoplasmic membrane due to its hydrophobic nature and aligns itself between fatty acid chains causing the expansion and destabilization of the membrane structure by increasing its fluidity and permeability for protons and ions (mainly H^+ and K^+) [7,35,36,44]. The loss of the ion gradient leads to bacterial cell death [36,44]. However, the mechanisms of action of CV against *Candida* have been investigated in several studies and seems to exert its antifungal activity by inducing envelope, disrupting membrane integrity, and blocking ergosterol biosynthesis [33]. Fungicidal activity of starch/CV [43], polyethylene/CV [46], and polypropylene/CV [45] films, as well as polyvinyl alcohol/CV coatings [42] was reported. The literature on melanin applications to develop antimicrobial properties of polymer film is relatively limited. It is suggested that melanin antibacterial activity might result from damage of the cell membrane and affect bacteria membrane function [16]. In previous study PLA/melanin film showed antimicrobial effect against some food-borne pathogenic bacteria [3]. Kiran et al. synthetized nanomelanin-polyhydroxybutyrate nanocomposite film which showed a strong protective effect against multidrug-resistant *S. aureus* [27].

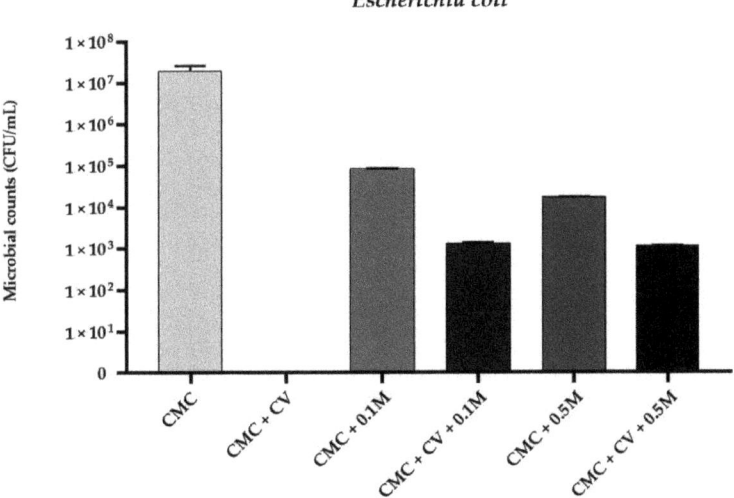

Figure 1. The effect of CMC-based neat and modified films on viability of *Escherichia coli* cells.

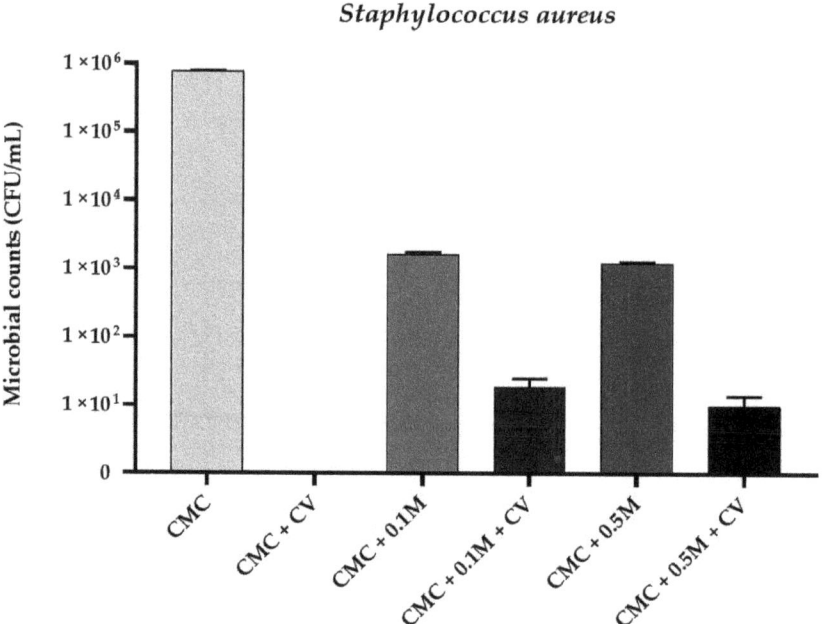

Figure 2. The effect of CMC-based neat and modified films on viability of *Staphylococcus aureus* cells.

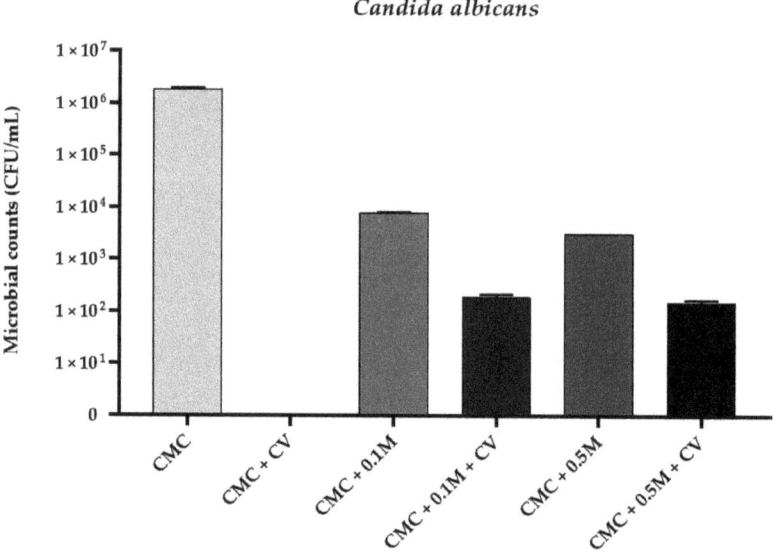

Figure 3. The effect of CMC-based neat and modified films on viability of *Candida albicans* cells.

3.2. Radicals Scavenging Activities and Reducing Power

The antioxidant capacity of the films was determined as the reducing power and radicals (DPPH, ABTS, O_2^-) scavenging activity and the results are summarized in Table 1. The CMC control film showed low reducing power (0.030 ± 0.001) and radical scavenging activities ($15.23 \pm 0.01\%$, $7.98 \pm 0.14\%$, and $9.68 \pm 0.05\%$, for DPPH, ABTS, and O_2^-, respectively) which is consistent with reports of other authors [6,8]. A significant increase in reducing power and radical scavenging activities was noticed when melanin was added ($p < 0.05$). Furthermore, a dose-dependent increment of antioxidant activity was noticed. The antioxidant activity of melanin-modified films has been already observed in numerous studies using melanin in a form of fillers [3,21–23], as well as when dissolved in a film-forming alkaline solutions [2,4]. The antioxidant activity of melanin from *A. bisporus* was previously reported [13]. In general, melanins act as very effective antioxidants due to presence of intramolecular non-covalent electrons, having ability to easily interact with free radicals and other reactive species [23]. It was also observed that CMC + CV sample showed significantly higher reducing power (0.192 ± 0.002) and antioxidant activity ($45.44 \pm 0.02\%$, $69.18 \pm 0.05\%$, and $61.63 \pm 0.09\%$, for DPPH, ABTS, and O_2^-, respectively) when compared with neat CMC ($p < 0.05$). The antioxidant capacity of carvacrol depends on the steric and electronic effect of its ring, besides the presence of the hydroxyl group which is capable of donating hydrogen atoms [7,44,51,52]. This findings are consistent with results of other authors, reporting antioxidant activities of chitosan/CV [51], polypropylene/CV [31], poly(lactic acid)/poly(ε-caprolactone)/CV [52], and flaxseed gum/CV films [36].

Table 1. Reducing power (RP) and radicals scavenging activity of CMC-based films.

Sample	RP (700 nm)	DPPH (%)	ABTS (%)	O_2^- (%)
CMC	0.030 ± 0.001 [f]	15.23 ± 0.01 [f]	7.98 ± 0.14 [f]	9.68 ± 0.05 [f]
CMC + 0.1 M	0.206 ± 0.005 [d]	64.20 ± 0.06 [d]	55.72 ± 0.08 [e]	63.92 ± 0.07 [d]
CMC + 0.5 M	0.244 ± 0.010 [c]	65.73 ± 0.01 [c]	57.65 ± 0.08 [d]	68.08 ± 0.09 [c]
CMC + CV	0.192 ± 0.002 [e]	45.44 ± 0.02 [e]	69.18 ± 0.05 [c]	61.63 ± 0.09 [e]
CMC + 0.1 M + CV	0.259 ± 0.001 [b]	69.70 ± 0.06 [b]	77.61 ± 0.01 [b]	81.83 ± 0.18 [b]
CMC + 0.5 M + CV	0.389 ± 0.006 [a]	72.70 ± 0.05 [a]	93.54 ± 0.03 [a]	89.84 ± 0.07 [a]

Values are means ± standard deviation of triplicate determinations. Means with different letters in the same column are significantly different at $p < 0.05$.

It is worth noting that CMC + 0.1 M + C and CMC + 0.5 M + C samples showed significantly higher antioxidant activity than films only with melanin ($p < 0.05$) with the highest activity of sample CMC + 0.5 M + C ($72.70 \pm 0.05\%$, $93.54 \pm 0.03\%$, and $89.84 \pm 0.07\%$, for DPPH, ABTS, and O_2^-, respectively). Those results are higher than reported for WPI/WPC/melanin [2], as well as flaxseed gum/CV films [36], but comparable with results reported for whey protein isolate nanofibrils–carvacrol films [54]. This effect might be attributed to synergistic effect of both antioxidants present in polymer matrix [40,46]. In fact, it is already established that the antioxidant activity of polymer-based films is directly dependent on the content of antioxidant compounds present in composite materials [2,23,40,46]. A similar synergistic effect of CV with other antioxidants such as pomegranate peel extract [40], whey protein isolate nanofibrils [54], as well as in inclusion complexes with cinnamaldehyde [46] was reported. The hight antioxidant activity of CMC-modified films shown their potential to be used in active antioxidant packaging to prevent oxidation-sensitive food matrices, as well as to increase their shelf life. In fact, the preservative activity against pork lard rancidity of gelatine-based coatings modified with fungal was reported [55]. Similar effect was showed by Wang et al. who applied whey protein isolate nanofibrils-based films with CV on fresh-cut cheese [54]. CV-incorporated flaxseed gum-sodium alginate films reduced formation of total volatile base nitrogen (TVB-N) resulted from activity of aerobic spoilage microflora [44].

3.3. The Thickness, Moisture Content, Mechanical, Thermal and Water Vapour Barrier Properties of The Films

The thickness, TSC (Total Solids Content), mechanical, thermal, and water vapour barrier properties are listed in Table 2. The lowest TSC was noticed for neat CMC film ($89.00 \pm 0.31\%$). The addition of melanin, as well as carvacrol caused the increase in TSC of the films ($p < 0.05$), which is consistent with results reported in other studies, where melanins, as well as carvacrol, were used to modify biopolymers films [2,21,52]. In contrast,

Dhumal et al. found that the impregnation of carvacrol did not influence the moisture content in starch–guar gum films [30]. In terms of thickness, only CMC + CV and CMC + 0.1 M + CV films showed significantly higher values (0.12 ± 0.02 mm and 0.08 ± 0.00 mm, respectively). In the previously reported study, the thickness of WPI/WPC films with the same melanin concentrations was not affected, because of a small amount of melanin used [2]. Hence, as the films only with melanin addition did not show significant increase in thickness, probably the increase in CMC + CV and CMC + 0.1 M+CV samples was resulted by interactions of CV and CMC (caused tighter binding of film matrix), not by higher solids content [8,9].

Table 2. Thickness, total solids content (TSC), water vapour transmission ratio (WVTR), tensile strength (TS), elongation at break (EB), melting temperature (T_m), and melting enthalpy (ΔH_m) of CMC-based films.

Sample	Thickness (mm)	TSC (%)	WVTR (g/(m² × Day))	TS (N)	EB (%)	T_m (°C)	ΔH_m (J/g)
CMC	0.05 ± 0.00 [c]	89.00 ± 0.31 [b]	1098.68 ± 6.74 [a]	18.11 ± 1.25 [d]	9.91 ± 0.25 [a]	130.54	−252.30
CMC + 0.1 M	0.06 ± 0.01 [c]	89.33 ± 0.28 [ab]	962.42 ± 3.51 [c]	19.61 ± 1.03 [d]	8.74 ± 0.76 [b]	105.51	−250.33
CMC + 0.5 M	0.04 ± 0.01 [c]	89.35 ± 0.12 [ab]	933.07 ± 4.32 [d]	25.89 ± 3.06 [c]	7.31 ± 0.66 [cd]	113.04	−283.25
CMC + CV	0.12 ± 0.02 [a]	89.02 ± 0.17 [b]	1092.38 ± 11.73 [ab]	25.30 ± 1.52 [be]	7.88 ± 0.31 [bc]	144.93	−153.91
CMC + 0.1 M + CV	0.08 ± 0.00 [b]	89.43 ± 0.19 [ab]	992.03 ± 3.84 [b]	29.20 ± 2.44 [ab]	6.33 ± 0.49 [d]	102.69	−201.18
CMC + 0.5 M + CV	0.05 ± 0.01 [c]	89.57 ± 0.03 [a]	961.74 ± 9.13 [bc]	34.29 ± 1.83 [a]	4.11 ± 0.72 [e]	107.72	−288.62

Values are means ± standard deviation of triplicate determinations. Means with different letters in the same column are significantly different at $p < 0.05$.

It was noticed that tensile strength of CMC + 0.5 M (25.89 ± 3.06 N) as well as CMC-CV (25.30 ± 1.52 N) films was enhanced in comparison with neat CMC film (18.11 ± 1.25 N) ($p < 0.05$). The improvement of mechanical strength of melanin-modified films has been already reported [2,4]. It should be pointed out, that the mechanical properties of CMC/melanin/CV film was significantly higher than corresponding CMC/melanin films, and can be attributed to synergistic interaction of film constituents. The mechanical properties of films are dependent on the microstructure of the film network, CMC composition, and intermolecular forces. The increase in film mechanical properties might be due to the intermolecular interaction of carboxyl group of CMC and hydroxyl group of melanin and CV molecules [8]. A decrease in the modified films EB in comparison to the control samples was observed ($p < 0.05$). This presumably resulted from strong hydrogen bonding (H-bonding) interactions between melanin, carvacrol, and the polymer matrix, which has been already reported to improve the mechanical properties of the films [2,21,23,24]. However, this results in contrary to results obtained for synthetic films from polypropylene and low-density polyethylene, as well as for chitosan-based films [40], where carvacrol was reported to increase EB due to its plasticizing effect [31,45].

Packaging films must be exposed to environment, hence adsorbtion and penetration of the water vapour is an important factor that controls the final product quality [36]. For better packaging films, low WVTR is required [23]. Although CV is reported to be hydrophobic [35,36,41], no significant WVTR differences of neat CMC (1098.68 ± 6.74 (g/(m² × Day)) and CMC + CV (1092.38 ± 11.73 g/(m² × Day)) were observed ($p > 0.05$), presumably linked with small amount of carvacrol used. Similarly, Medina-Jaramillo et al. reported that the low concentrations of CV added (0.03%, 0.06%, and 0.09%) did not cause changes in the water vapour barrier properties of the alginate coatings [41]. In contrast, Du et al. found that the addition of carvacrol (0.5%, 1.0%, and 1.5%) to tomato puree films increased WVTR compared to the control [39], similarly for chitosan-based films modified with 0.5% and 1.0% of CV as reported by Flores et al. [5]. However, Dhumal et al. who used carvacrol to modify starch–guar gum films, stated that the incorporation of essential oils compounds may reduce the impact of plasticizers within the biopolymer matrix, thereby lowering the moisture transmission rates [30]. On the other hand, a significant decrease in WVTR was observed when melanin was added ($p < 0.05$). Reduction in the WVTR with increasing melanin content results in an improvement of the functional properties of these films, considering the hydrophilic characteristics of the matrix. The lowest WVTR was observed for sample CMC + 0.5 M (933.07 ± 4.32 g/(m² × Day)). Moreover, films only with melanin showed significantly lower WVTR than the corresponding films with melanin and carvacrol ($p < 0.05$). The observed WVTR decrease is consistent with previously reported results for WPC/WPI [2], alginate/poly(vinyl alcohol) [20], cellulose [23], as well as gelatine [4] films modified with various melanins, and could be attributed to highly hydrophobic properties of *A. bisporus* melanin [13]. Furthermore, in the film forming solution the polymeric chains may partially be immobilized at the interface with melanin. Consequently, the polymeric chains become less mobile, reducing the diffusibility of water via the CMC chains interface and leading to a decrease in WVTR. Therefore, the lower WVTR values obtained with melanin addition can be explained by formation of an interconnecting melanin network within the film matrix. The results for CMC modified film are lower than results reported for WPC/WPI films modified with melanin from watermelon seeds [2], but higher than reported for PLA films modified with *A. bisporus* melanin [3]. It was already reported that the effect of melanin on WVTR is concentration dependent. For agar–melanin nanoparticles composite films the incorporation of low melanin-particle content did not affect their WVTR, whereas at higher content, the WVTR increased [21]. When melanin was added to PLA composite films at low content, it was observed that WVTR increased, but decreased at high melanin content [3]. Regarding the thermal properties of the films it was observed that CV significantly improved melting temperature from 130.54 °C (CMC) to 144.93 °C (CMC + CV). A similar observation was reported for flaxseed gum/CV films [36]. Surprisingly, T_m of melanin-modified and melanin/CV-CMC films was lower than of neat CMC films. This result is in contrary to results obtained in previous study, where the addition of melanin increased the melting temperature of WPC/WPI films [2]. On the other hand it was reported that the addition of melanins as a fillers did not change melting temperatures of PLA [3], cellulose [23], and agar [21] films. However, a degradation of all the CMC-based samples was noticed when temperature exceeded 250 °C.

3.4. Appearance, Colour, Opacity, and Transparency Changes

The colour coordinates, the total colour difference (ΔE), the yellowness index (YI), the whiteness index (WI), opacity, and transparency of the films are presented in Table 3, whereas the appearance of the films is illustrated in Figure 4. The neat CMC film was characterized by a high transparency (T_{660} = 82.07%) and was almost colourless. It was noticed that addition of carvacrol did not affect the L*, b*, YI, WI, and C parameters ($p > 0.05$), however a* and H°, increased due to the red–orange colour of pure carvacrol ($p < 0.05$). Moreover CMC-C film was less transparent (T_{660} = 79.10%) and had lower opacity (7.22 ± 0.13) than the neat CMC film (7.62 ± 0.20). The lowering of b* and WI values with the addition of CV was reported for apple puree/pectin/CV films [39], and chitosan/CV [51] films. Generally, films with CV were characterized by low transparency and were not see-through. As can be seen the addition of melanin significantly reduced lightness of the films ($p < 0.05$). Simultaneously, the a* and b* colour coordinates, as well as C and H° values of the modified films increased due to the red-brown colour of melanin ($p < 0.05$) [13]. The increased melanin concentration caused a significant increase in YI of the films ($p < 0.05$), whereas a decrease in WI was noticed ($p < 0.05$). The yellow and brown colour of melanin modified films was also reported in other studies [2–4,21,24]. Moreover, the opacity of the films decreased significantly when melanin concentration increased ($p < 0.05$). A similar effect was reported for PLA and gelatine films modified with *A. bisporus* melanin [3], as well as for chitosan-based coatings modified with CV [5]. Regarding the total colour difference (ΔE) of the films a significant increase was noticed ($p < 0.05$), and ranged from 1.13 ± 0.68 (CMC + C) to 14.24 ± 0.72 (CMC + 0.5 M + C). When ΔE is higher than 1.00, the human eye is able to percept the colour difference, thus modification with carvacrol, as well as melanin caused noticeable colour changes [2,3].

Table 3. Colour, total colour difference (ΔE), chroma (C), hue angle (H°), yellowness index (YI), whiteness index (WI), opacity, and transmittance at 660 nm of CMC-based films.

Sample	L*	a*	b*	ΔE	C	H°	YI	WI	Opacity	T_{660} (%)
CMC	89.77 ± 0.92 [a]	−0.66 ± 0.02 [c]	4.38 ± 0.38 [d]	used as standard	4.43 ± 0.38 [c]	−8.68 ± 0.83 [d]	6.97 ± 0.68 [e]	94.54 ± 0.39 [a]	7.62 ± 0.20 [a]	82.07 [a]
CMC + 0.1 M	86.08 ± 0.61 [ab]	−0.31 ± 0.09 [c]	6.75 ± 0.93 [c]	2.78 ± 0.82 [bc]	6.96 ± 0.94 [b]	−2.71 ± 1.07 [c]	10.98 ± 1.54 [c]	92.84 ± 1.25 [c]	7.50 ± 0.17 [a]	78.72 [c]
CMC + 0.5 M	83.81 ± 1.44 [bc]	0.88 ± 0.03 [b]	16.61 ± 0.38 [b]	13.76 ± 0.64 [a]	17.47 ± 1.40 [a]	3.17 ± 0.64 [a]	28.38 ± 4.47 [a]	82.88 ± 2.37 [d]	7.02 ± 0.13 [b]	67.97 [d]
CMC + CV	89.28 ± 0.50 [a]	−0.08 ± 0.13 [c]	4.94 ± 0.50 [d]	1.13 ± 0.68 [c]	5.04 ± 0.43 [bc]	−0.56 ± 1.13 [b]	7.91 ± 0.84 [de]	94.07 ± 0.46 [a]	7.22 ± 0.13 [b]	79.10 [b]
CMC + 0.1 M + CV	88.62 ± 0.22 [ab]	−0.19 ± 0.09 [c]	6.32 ± 0.93 [cd]	3.62 ± 0.78 [b]	6.06 ± 0.75 [bc]	−1.91 ± 0.83 [bc]	10.20 ± 1.51 [c]	92.82 ± 0.83 [b]	7.14 ± 0.15 [b]	75.29 [e]
CMC + 0.5 M + CV	82.50 ± 0.28 [c]	1.35 ± 0.84 [a]	19.85 ± 0.28 [a]	14.24 ± 0.72 [a]	17.82 ± 3.02 [a]	4.26 ± 1.49 [a]	20.13 ± 0.13 [b]	86.64 ± 0.96 [d]	6.54 ± 0.17 [c]	64.37 [f]

Values are means ± standard deviation of triplicate determinations. Means with different letters in the same column are significantly different at $p < 0.05$.

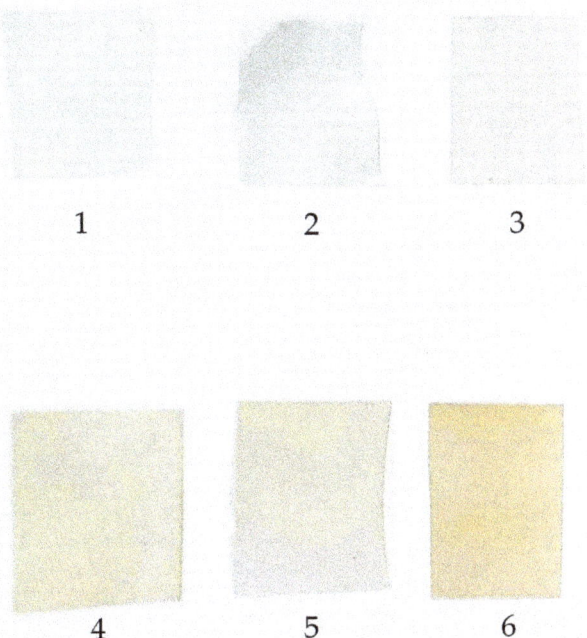

Figure 4. The visual appearance of CMC-based films. **1**–CMC; **2**–CMC + CV; **3**–CMC + 0.1 M; **4**–CMC + 0.5 M; **5**–CMC + 0.1 M +CV; **6**–CMC + 0.5 M + CV.

3.5. The Changes of UV-Vis Blocking Properties

Figure 5 presents the UV-Vis transmittance spectra of the neat and modified CMC films. The neat CMC film showed high transmittance in the range of 200 to 700 nm, indicating that the film is highly transparent to UV and visible light which also is in line with results presented in Table 3, and reports of other authors [6]. Although CMC + C film showed lower opacity, displayed moderate transparency to visible light and no UV-blocking effect. A concentration-dependent decrease in CMC films transmittance was noticed, which indicates that melanin, even at low concentration (0.1%), improved UV–Vis blocking effect of the films. This effect can be attributed to the absorption of UV light by melanin and was reported also for gelatine/melanin [4], as well as WPC/WPI/melanin films [2]. It should be pointed out that one of the main functions of melanins in nature is protection against UV radiation, due to their strong UV-absorbing properties [15]. It was already reported, that melanin from A. bisporus has strong UV–Vis barrier properties [13]. Interestingly, samples CMC + 0.1 M + C and CMC + 0.5 M + C displayed lower transparency than the corresponding CMC + 0.1 M and CMC + 0.5 M films. As mentioned, samples with carvacrol were opaque and the observed higher light-blocking properties seem to be synergistic effect of the absorption of UV light by melanin and light scattering by carvacrol droplets in polymer matrix. A blocking and scattering of the light path by melanins when used as nanofillers, was reported for PLA/melanin [3], and agar/melanin films [21]. High UV-blocking properties of modified films might help to protect packaged food from oxidative deterioration caused by UV radiation which leads to discoloration, nutrient loss and off-flavour production.

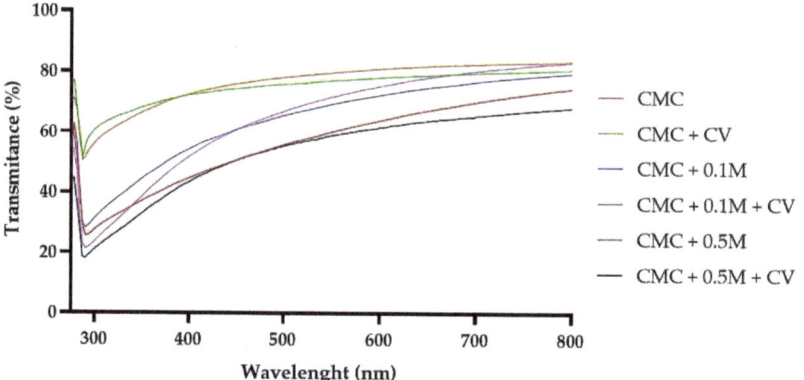

Figure 5. UV-Vis spectra of CMC-based neat and modified films.

3.6. FT-IR Results

FT-IR is a valuable technique used to determine the miscibility and compatibility of polymeric matrices and additives, due to its rapid and non-destructive nature [23]. Figure 6 presents FT-IR spectra of CMC-based films. A broad absorption band at about 3200–3350 cm^{-1} in the CMC films was due to the stretching of hydroxyl groups of cellulose, melanin and carvacrol [2,13,36,52,56,57]. Stretching peaks at approximately 2920 cm^{-1} (CH$_3$) and 2870 cm^{-1} (CH$_2$) were also detected [36]. The two absorption peaks at 1587 and 1412 cm^{-1} were attributed to the asymmetric and symmetric stretching vibration of carboxylic groups, respectively. They correspond also to the vibration of melanin ans carvacrol aromatic C=C bonds [2,13,52]. There were two more bands at 1100 and 1020 cm^{-1} related to the stretching of C–O in polysaccharide. The peak at 1320 cm^{-1} was due to the O–C–H and H–C–H deformation and absorption bands at 1260 and 900 cm^{-1} were related to the C–H vibrations of CMC and plasticizer (glycerol) [2,6,8,57]. Generally, no substantial variations were noted in the functional groups of CMC modified films. The obtained results suggest that there were no structural changes in CMC films due to the addition of melanin and carvacrol. The physical interactions (H–bonding, van der Waals force) between CMC-melanin-carvacrol are presumably the cause of minor shifts and small changes in intensities, which in agreement with other reports [2].

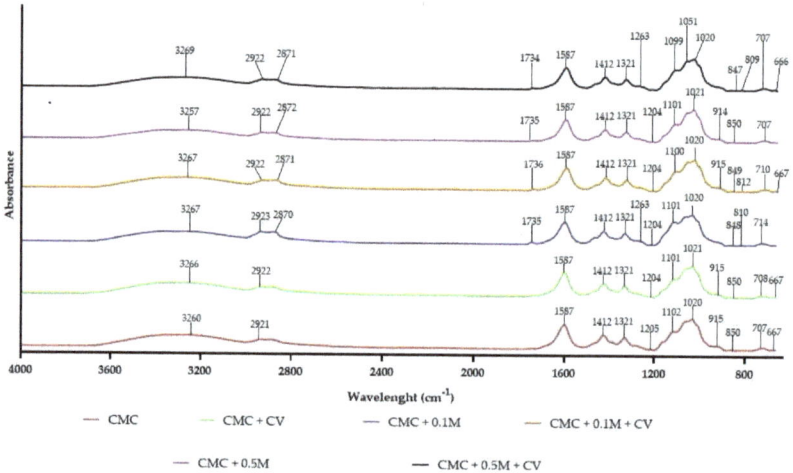

Figure 6. Fourier-transform infrared (FT-IR) spectra of CMC-based neat and modified films.

4. Conclusions

The properties of modified CMC-based films incorporated with fungal melanin and carvacrol (alone or in combination) were explored in this study. The properties of CMC-composites were compared to neat CMC film. Melanin played a vital role in mechanical, antioxidant, antimicrobial, and barrier properties. The increase in tensile strength and water vapour barrier properties as well as decrease in elongation at break were observed for CMC/melanin and CMC/melanin/CV films. CMC/CV films caused the total reduction in viable bacteria and yeast, however, CMC/melanin/CV films also showed good antimicrobial activity. All modified films showed antioxidant activity and the highest level (72.70–93.54% of radicals scavenging activities) was observed for sample CMC + 0.5 M + CV which was attributed to synergistic action of both active compounds. An improvement in the antioxidant and antimicrobial activities of modified CMC films is worth mentioning as an important aspect of the work. Considerable improvement in these properties has been observed in comparison to pure CMC. The results described here are particularly interesting if one considers that the additives used have a natural origin, providing added value in the development of sustainable alternatives to traditional synthetic antioxidants. We conclude that the obtained bioactive CMC-modified films could be potentially used for active food packaging applications. However, further tests of the influence of the developed materials (in the form of films and coatings) on various food products should be carried out to determine their suitability in food technology.

Author Contributions: Ł.Ł.: conceptualization, formal analysis, investigation, methodology, supervision, visualization, writing—original draft, writing-review and editing; P.K.: methodology, visualization; E.D.: methodology, visualization, investigation; P.T.: methodology, investigation; M.K.: methodology, investigation; M.Ś. and M.P.-Ś.: methodology, investigation, formal analysis; E.K. and M.S.: methodology, formal analysis, funding acquisition. All authors have read and agreed to the published version of the manuscript.

Funding: This research received no external funding.

Institutional Review Board Statement: Not applicable.

Informed Consent Statement: Not applicable.

Data Availability Statement: The data presented in this study are available on request from the corresponding authors.

Conflicts of Interest: The authors declare no conflict of interest.

References

1. Roy, S.; Rhim, J.W. Fabrication of copper sulfide nanoparticles and limonene incorporated pullulan/carrageenan based film with improved mechanical and antibacterial properties. *Polymers* **2020**, *12*, 2665. [CrossRef]
2. Łopusiewicz, Ł.; Drozłowska, E.; Trocer, P.; Kostek, M.; Śliwiński, M.; Henriques, M.H.F.; Bartkowiak, A.; Sobolewski, P. Whey protein concentrate/isolate biofunctional films modified with melanin from watermelon (*Citrullus lanatus*) seeds. *Materials* **2020**, *13*, 3876. [CrossRef]
3. Łopusiewicz, Ł.; Jędra, F.; Mizielińska, M. New poly(lactic acid) active packaging composite films incorporated with fungal melanin. *Polymers* **2018**, *10*, 386. [CrossRef] [PubMed]
4. Łopusiewicz, Ł.; Jędra, F.; Bartkowiak, A. New Active Packaging Films Made from Gelatin Modified with Fungal Melanin. *World Sci. News* **2018**, *101*, 1–30.
5. Flores, Z.; San-Martin, D.; Beldarraín-Iznaga, T.; Leiva-Vega, J.; Villalobos-Carvajal, R. Effect of Homogenization Method and Carvacrol Content on Microstructural and Physical Properties of Chitosan-Based Films. *Foods* **2021**, *10*, 141. [CrossRef] [PubMed]
6. Ezati, P.; Rhim, J.W.; Moradi, M.; Tajik, H.; Molaei, R. CMC and CNF-based alizarin incorporated reversible pH-responsive color indicator films. *Carbohydr. Polym.* **2020**, *246*, 116614. [CrossRef]
7. Lei, K.; Wang, X.; Li, X.; Wang, L. The innovative fabrication and applications of carvacrol nanoemulsions, carboxymethyl chitosan microgels and their composite films. *Colloids Surf. B Biointerfaces* **2019**, *175*, 688–696. [CrossRef] [PubMed]
8. Akhtar, H.M.S.; Riaz, A.; Hamed, Y.S.; Abdin, M.; Chen, G.; Wan, P.; Zeng, X. Production and characterization of CMC-based antioxidant and antimicrobial films enriched with chickpea hull polysaccharides. *Int. J. Biol. Macromol.* **2018**, *118*, 469–477. [CrossRef]

9. Jannatyha, N.; Shojaee-Aliabadi, S.; Moslehishad, M.; Moradi, E. Comparing mechanical, barrier and antimicrobial properties of nanocellulose/CMC and nanochitosan/CMC composite films. *Int. J. Biol. Macromol.* **2020**, *164*, 2323–2328. [CrossRef]
10. Tabari, M. Investigation of Carboxymethyl Cellulose (CMC) on Mechanical Properties of Cold Water Fish Gelatin Biodegradable Edible Films. *Foods* **2017**, *6*, 41. [CrossRef]
11. Ghanbarzadeh, B.; Almasi, H.; Entezami, A.A. Physical properties of edible modified starch/carboxymethyl cellulose films. *Innov. Food Sci. Emerg. Technol.* **2010**, *11*, 697–702. [CrossRef]
12. Ancuţa, P.; Sonia, A. Oil press-cakes and meals valorization through circular economy approaches: A review. *Appl. Sci.* **2020**, *10*, 7432. [CrossRef]
13. Łopusiewicz, Ł. Waste from the harvesting of button mushroom (*Agaricus bisporus*) as a source of natural melanin. *Folia Pomeranae Univ. Technol. Stetin. Agric. Aliment. Piscaria Zootech.* **2018**, *343*, 23–42. [CrossRef]
14. Dueñas, M.; García-Estévez, I. Agricultural and food waste: Analysis, characterization and extraction of bioactive compounds and their possible utilization. *Foods* **2020**, *9*, 817. [CrossRef] [PubMed]
15. Solano, F. Melanins: Skin Pigments and Much More—Types, Structural Models, Biological Functions, and Formation Routes. *New J. Sci.* **2014**, *2014*, 498276. [CrossRef]
16. Xu, C.; Chen, T.; Li, J.; Jin, M.; Ye, M. The structural analysis and its hepatoprotective activity of melanin isolated from *Lachnum* sp. *Process. Biochem.* **2020**, *90*, 249–256. [CrossRef]
17. Łopusiewicz, Ł. Antioxidant, antibacterial properties and the light barrier assessment of raw and purified melanins isolated from *Citrullus lanatus* (watermelon) seeds. *Herba Pol.* **2018**, *64*, 25–36. [CrossRef]
18. Al-Tayib, O.A.; Elbadwi, S.M.; Bakhiet, A.O. Cytotoxicity assay for herbal melanin derived from *Nigella sativa* seeds using in vitro cell lines. *IOSR J. Humanit. Soc. Sci.* **2017**, *22*, 43.
19. Ye, M.; Wang, Y.; Guo, G.Y.; He, Y.L.; Lu, Y.; Ye, Y.W.; Yang, Q.H.; Yang, P.Z. Physicochemical characteristics and antioxidant activity of arginine-modified melanin from *Lachnum* YM-346. *Food Chem.* **2012**, *135*, 2490–2497. [CrossRef]
20. Yang, M.; Li, L.; Yu, S.; Liu, J.; Shi, J. High performance of alginate/polyvinyl alcohol composite film based on natural original melanin nanoparticles used as food thermal insulating and UV-vis block. *Carbohydr. Polym.* **2020**, *233*, 115884. [CrossRef]
21. Roy, S.; Rhim, J.W. Agar-based antioxidant composite films incorporated with melanin nanoparticles. *Food Hydrocoll.* **2019**, *94*, 391–398. [CrossRef]
22. Roy, S.; Rhim, J.W. Carrageenan-based antimicrobial bionanocomposite films incorporated with ZnO nanoparticles stabilized by melanin. *Food Hydrocoll.* **2019**, *90*, 500–507. [CrossRef]
23. Roy, S.; Kim, H.C.; Kim, J.W.; Zhai, L.; Zhu, Q.Y.; Kim, J. Incorporation of melanin nanoparticles improves UV-shielding, mechanical and antioxidant properties of cellulose nanofiber based nanocomposite films. *Mater. Today Commun.* **2020**, *24*, 100984. [CrossRef]
24. Roy, S.; Van Hai, L.; Kim, H.C.; Zhai, L.; Kim, J. Preparation and characterization of synthetic melanin-like nanoparticles reinforced chitosan nanocomposite films. *Carbohydr. Polym.* **2020**, *231*, 115729. [CrossRef]
25. Dong, W.; Wang, Y.; Huang, C.; Xiang, S.; Ma, P.; Ni, Z.; Chen, M. Enhanced thermal stability of poly (vinyl alcohol) in presence of melanin. *J. Therm. Anal. Calorim.* **2014**, *115*, 1661–1668. [CrossRef]
26. Bang, Y.J.; Shankar, S.; Rhim, J.W. Preparation of polypropylene/poly (butylene adipate-co-terephthalate) composite films incorporated with melanin for prevention of greening of potatoes. *Packag. Technol. Sci.* **2020**, *33*, 1–9. [CrossRef]
27. Kiran, G.S.; Jackson, S.A.; Priyadharsini, S.; Dobson, A.D.W.; Selvin, J. Synthesis of Nm-PHB (nanomelanin-polyhydroxy butyrate) nanocomposite film and its protective effect against biofilm-forming multi drug resistant *Staphylococcus aureus*. *Sci. Rep.* **2017**, *7*, 1–13. [CrossRef]
28. Di Mauro, E.; Camaggi, M.; Vandooren, N.; Bayard, C.; De Angelis, J.; Pezzella, A.; Baloukas, B.; Silverwood, R.; Ajji, A.; Pellerin, C.; et al. Eumelanin for nature-inspired UV-absorption enhancement of plastics. *Polym. Int.* **2019**, *68*, 984–991. [CrossRef]
29. Kwiatkowski, P.; Pruss, A.; Wojciuk, B.; Dołęgowska, B.; Wajs-Bonikowska, A.; Sienkiewicz, M.; Mężyńska, M.; Łopusiewicz, Ł. The influence of essential oil compounds on antibacterial activity of mupirocin-susceptible and induced low-level mupirocin-resistant MRSA strains. *Molecules* **2019**, *24*, 3105. [CrossRef] [PubMed]
30. Dhumal, C.V.; Ahmed, J.; Bandara, N.; Sarkar, P. Improvement of antimicrobial activity of sago starch/guar gum bi-phasic edible films by incorporating carvacrol and citral. *Food Packag. Shelf Life* **2019**, *21*, 100380. [CrossRef]
31. Ramos, M.; Jiménez, A.; Peltzer, M.; Garrigós, M.C. Characterization and antimicrobial activity studies of polypropylene films with carvacrol and thymol for active packaging. *J. Food Eng.* **2012**, *109*, 513–519. [CrossRef]
32. Magi, G.; Marini, E.; Facinelli, B. Antimicrobial activity of essential oils and carvacrol, and synergy of carvacrol and erythromycin, against clinical, erythromycin-resistant Group A Streptococci. *Front. Microbiol.* **2015**, *6*, 165. [CrossRef]
33. Lima, I.O.; De Oliveira Pereira, F.; De Oliveira, W.A.; De Oliveira Lima, E.; Menezes, E.A.; Cunha, F.A.; De Fátima Formiga Melo Diniz, M. Antifungal activity and mode of action of carvacrol against *Candida albicans* strains. *J. Essent. Oil Res.* **2013**, *25*, 138–142. [CrossRef]
34. Lambert, R.J.W.; Skandamis, P.N.; Coote, P.J.; Nychas, G.J.E. A study of the minimum inhibitory concentration and mode of action of oregano essential oil, thymol and carvacrol. *J. Appl. Microbiol.* **2001**, *91*, 453–462. [CrossRef] [PubMed]
35. Marchese, A.; Arciola, C.R.; Coppo, E.; Barbieri, R.; Barreca, D.; Chebaibi, S.; Sobarzo-Sánchez, E.; Nabavi, S.F.; Nabavi, S.M.; Daglia, M. The natural plant compound carvacrol as an antimicrobial and anti-biofilm agent: Mechanisms, synergies and bio-inspired anti-infective materials. *Biofouling* **2018**, *34*, 630–656. [CrossRef] [PubMed]

36. Fang, S.; Qiu, W.; Mei, J.; Xie, J. Effect of Sonication on the Properties of Flaxseed Gum Films Incorporated with Carvacrol. *Int. J. Mol. Sci.* **2020**, *21*, 1637. [CrossRef] [PubMed]
37. Sharifi-Rad, M.; Varoni, E.M.; Iriti, M.; Martorell, M.; Setzer, W.N.; del Mar Contreras, M.; Salehi, B.; Soltani-Nejad, A.; Rajabi, S.; Tajbakhsh, M.; et al. Carvacrol and human health: A comprehensive review. *Phyther. Res.* **2018**, *32*, 1675–1687. [CrossRef]
38. Suntres, Z.E.; Coccimiglio, J.; Alipour, M. The Bioactivity and Toxicological Actions of Carvacrol. *Crit. Rev. Food Sci. Nutr.* **2015**, *55*, 304–318. [CrossRef] [PubMed]
39. Du, W.X.; Olsen, C.W.; Avena-Bustillos, R.J.; McHugh, T.H.; Levin, C.E.; Friedman, M. Storage stability and antibacterial activity against *Escherichia coli* O157:H7 of carvacrol in edible apple films made by two different casting methods. *J. Agric. Food Chem.* **2008**, *56*, 3082–3088. [CrossRef] [PubMed]
40. Yuan, G.; Lv, H.; Yang, B.; Chen, X.; Sun, H. Physical properties, antioxidant and antimicrobial activity of chitosan films containing carvacrol and pomegranate peel extract. *Molecules* **2015**, *20*, 11034–11045. [CrossRef]
41. Medina-Jaramillo, C.; Quintero-Pimiento, C.; Díaz-Díaz, D.; Goyanes, S.; López-Córdoba, A. Improvement of andean blueberries postharvest preservation using carvacrol/alginate-edible coatings. *Polymers* **2020**, *12*, 2352. [CrossRef]
42. Sapper, M.; Martin-Esparza, M.E.; Chiralt, A.; Martinez, C.G. Antifungal polyvinyl alcohol coatings incorporating carvacrol for the postharvest preservation of golden delicious apple. *Coatings* **2020**, *10*, 1027. [CrossRef]
43. Ochoa-Velasco, C.E.; Pérez-Pérez, J.C.; Varillas-Torres, J.M.; Navarro-Cruz, A.R.; Hernández-Carranza, P.; Munguía-Pérez, R.; Cid-Pérez, T.S.; Avila-Sosa, R. Starch Edible Films/Coatings Added with Carvacrol and Thymol: In Vitro and In Vivo Evaluation against *Colletotrichum gloeosporioides*. *Foods* **2021**, *10*, 175. [CrossRef]
44. Fang, S.; Zhou, Q.; Hu, Y.; Liu, F.; Mei, J.; Xie, J. Antimicrobial Carvacrol Incorporated in Flaxseed Gum-Sodium Alginate Active Films to Improve the Quality Attributes of Chinese Sea bass (*Lateolabrax maculatus*) during Cold Storage. *Molecules* **2019**, *24*, 3292. [CrossRef] [PubMed]
45. Krepker, M.; Prinz-Setter, O.; Shemesh, R.; Vaxman, A.; Alperstein, D.; Segal, E. Antimicrobial carvacrol-containing polypropylene films: Composition, structure and function. *Polymers* **2018**, *10*, 79. [CrossRef] [PubMed]
46. Canales, D.; Montoille, L.; Rivas, L.M.; Ortiz, J.A.; Yañez-S, M.; Rabagliati, F.M.; Ulloa, M.T.; Alvarez, E.; Zapata, P.A. Fungicides Films of Low-Density Polyethylene (LDPE)/Inclusion Complexes (Carvacrol and Cinnamaldehyde) Against *Botrytis cinerea*. *Coatings* **2019**, *9*, 795. [CrossRef]
47. Kwiatkowski, P.; Pruss, A.; Grygorcewicz, B.; Wojciuk, B.; Dołęgowska, B.; Giedrys-Kalemba, S.; Kochan, E.; Sienkiewicz, M. Preliminary study on the antibacterial activity of essential oils alone and in combination with gentamicin against extended-spectrum β-lactamase-producing and New Delhi metallo-β-lactamase-1-producing *Klebsiella pneumoniae* isolates. *Microb. Drug Resist.* **2018**, *24*, 1368–1375. [CrossRef]
48. Kwiatkowski, P.; Łopusiewicz, Ł.; Kostek, M.; Drozłowska, E.; Pruss, A.; Wojciuk, B.; Sienkiewicz, M.; Zielińska-Bliźniewska, H.; Dołęgowska, B. The Antibacterial Activity of Lavender Essential Oil Alone and In Combination with Octenidine Dihydrochloride against MRSA Strains. *Molecules* **2019**, *25*, 95. [CrossRef] [PubMed]
49. Mizielińska, M.; Kowalska, U.; Salachna, P.; Łopusiewicz, Ł.; Jarosz, M. The influence of accelerated UV-A and Q-SUN irradiation on the antibacterial properties of hydrophobic coatings containing *Eucomis comosa* extract. *Polymers* **2018**, *10*, 421. [CrossRef] [PubMed]
50. Mizielińska, M.; Łopusiewicz, Ł.; Mężyńska, M.; Bartkowiak, A. The influence of accelerated UV-A and Q-sun irradiation on the antimicrobial properties of coatings containing ZnO nanoparticles. *Molecules* **2017**, *22*, 1556. [CrossRef]
51. López-Mata, M.A.; Ruiz-Cruz, S.; Silva-Beltrán, N.P.; Ornelas-Paz, J.D.J.; Zamudio-Flores, P.B.; Burruel-Ibarra, S.E. Physicochemical, antimicrobial and antioxidant properties of chitosan films incorporated with carvacrol. *Molecules* **2013**, *18*, 13735–13753. [CrossRef]
52. Lukic, I.; Vulic, J.; Ivanovic, J. Antioxidant activity of PLA/PCL films loaded with thymol and/or carvacrol using scCO$_2$ for active food packaging. *Food Packag. Shelf Life* **2020**, *26*, 100578. [CrossRef]
53. Scaffaro, R.; Lopresti, F.; D'Arrigo, M.; Marino, A.; Nostro, A. Efficacy of poly (lactic acid)/carvacrol electrospun membranes against *Staphylococcus aureus* and *Candida albicans* in single and mixed cultures. *Appl. Microbiol. Biotechnol.* **2018**, *102*, 4171–4181. [CrossRef] [PubMed]
54. Wang, Q.; Yu, H.; Tian, B.; Jiang, B.; Xu, J.; Li, D.; Feng, Z.; Liu, C. Novel Edible Coating with Antioxidant and Antimicrobial Activities Based on Whey Protein Isolate Nanofibrils and Carvacrol and Its Application on Fresh-Cut Cheese. *Coatings* **2019**, *9*, 583. [CrossRef]
55. Łopusiewicz, Ł.; Jędra, F.; Bartkowiak, A. The application of melanin modified gelatin coatings for packaging and the oxidative stability of pork lard. *World Sci. News* **2018**, *101*, 108–119.
56. Tongdeesoontorn, W.; Mauer, L.J.; Wongruong, S.; Sriburi, P.; Rachtanapun, P. Effect of carboxymethyl cellulose concentration on physical properties of biodegradable cassava starch-based films. *Chem. Cent. J.* **2011**, *5*, 1–8. [CrossRef] [PubMed]
57. Oun, A.A.; Rhim, J.W. Preparation and characterization of sodium carboxymethyl cellulose/cotton linter cellulose nanofibril composite films. *Carbohydr. Polym.* **2015**, *127*, 101–109. [CrossRef]

Article

Tannic-Acid-Cross-Linked and TiO$_2$-Nanoparticle-Reinforced Chitosan-Based Nanocomposite Film

Swarup Roy [1], Lindong Zhai [1], Hyun Chan Kim [1], Duc Hoa Pham [1], Hussein Alrobei [2] and Jaehwan Kim [1,*]

[1] Creative Research Center for Nanocellulose Future Composites, Department of Mechanical Engineering, Inha University, Incheon 22212, Korea; swaruproy2013@gmail.com (S.R.); duicaofei@naver.com (L.Z.); Kim_HyunChan@naver.com (H.C.K.); phamduchoa.tdt@gmail.com (D.H.P.)

[2] Department of Mechanical Engineering, Prince Sattam bin Abdul Aziz University, AlKharj 11942, Saudi Arabia; h.alrobei@psau.edu.sa

* Correspondence: jaehwan@inha.ac.kr

Abstract: A chitosan-based nanocomposite film with tannic acid (TA) as a cross-linker and titanium dioxide nanoparticles (TiO$_2$) as a reinforcing agent was developed with a solution casting technique. TA and TiO$_2$ are biocompatible with chitosan, and this paper studied the synergistic effect of the cross-linker and the reinforcing agent. The addition of TA enhanced the ultraviolet blocking and mechanical properties of the chitosan-based nanocomposite film. The reinforcement of TiO$_2$ in chitosan/TA further improved the nanocomposite film's mechanical properties compared to the neat chitosan or chitosan/TA film. The thermal stability of the chitosan-based nanocomposite film was slightly enhanced, whereas the swelling ratio decreased. Interestingly, its water vapor barrier property was also significantly increased. The developed chitosan-based nanocomposite film showed potent antioxidant activity, and it is promising for active food packaging.

Keywords: chitosan; tannic acid; titanium dioxide; nanocomposite film; mechanical properties; antioxidant activity

Citation: Roy, S.; Zhai, L.; Kim, H.C.; Pham, D.H.; Alrobei, H.; Kim, J. Tannic-Acid-Cross-Linked and TiO$_2$-Nanoparticle-Reinforced Chitosan-Based Nanocomposite Film. *Polymers* **2021**, *13*, 228. https://doi.org/10.3390/polym13020228

Received: 19 December 2020
Accepted: 8 January 2021
Published: 11 January 2021

Publisher's Note: MDPI stays neutral with regard to jurisdictional claims in published maps and institutional affiliations.

Copyright: © 2021 by the authors. Licensee MDPI, Basel, Switzerland. This article is an open access article distributed under the terms and conditions of the Creative Commons Attribution (CC BY) license (https://creativecommons.org/licenses/by/4.0/).

1. Introduction

Nowadays, the use of synthetic non-biodegradable plastics in packaging areas has become an ample hazard for our environment. In the year 2015, it was reported that about 6.3 billion tons of plastic waste were formed worldwide [1], and according to the Environmental Protection Agency (EPA), about 40% of municipal waste is generated from plastic packaging [2]. In this context, to overcome the recent plastic-based packaging problem, the effort to develop biodegradable plastics based on biopolymers has gained proper attention in order to replace synthetic polymer-based plastics [3–8]. A recent report showed that bioplastics in the European market have now replaced ~10% of the present plastic market. The European Parliament set a target of using 100% reusable plastic by 2030 [9,10]. Biopolymers have many advantages, including biodegradability, renewability, biocompatibility, and eco-friendliness [6,11–15]. Biopolymers have been extensively used to make films. Chitosan is one of the most promising biopolymers because of its excellent antimicrobial activity [16–19]. Chitosan is a linear polysaccharide formed after chitin's de-acetylation. Chitosan has fair uses in food packaging and product storage, even though weak mechanical and barrier properties restrict its bulk use as an alternative to synthetic plastics [20–24]. Chitosan-based film's limitations can be improved using chemical cross-linkers and physical reinforcement with nanofillers [18,25]. A previous study was carried out based on physical cross-linking of chitosan to improve its physical properties [26–28]. Additionally, varieties of nanofillers and bioactive functional compounds, such as chitin nanowhiskers, cellulose nanofibers, metal nanoparticles (inorganic and organic), essential oils, etc., have already been reinforced to improve the physical and functional characteristics of chitosan-based film [15,29–37].

In this context, many chemical cross-linkers, such as glutaraldehyde, formaldehyde, glyoxal, tannic acid, vanillin, citric acid, genipin, and tripolyphosphate, have been used to improve chitosan films' physical and functional properties [30,38,39]. Tannic acid (TA) has received attention as it is a natural polymer, non-cytotoxic, and widely used in food applications [26]. TA is a water-soluble, polyphenolic, and amphiphilic polymer with excellent cross-linking and antioxidant proficiency [39,40]. On the other hand, different metal nanoparticles and metal oxide nanoparticles have also been used as chemical reinforcing agents so far [41,42]. Among the metal oxide nanoparticles, titanium dioxide (TiO_2) has recently been utilized to prepare different composites. It is inert, non-toxic, eco-friendly, and inexpensive, with excellent biocompatibility [10,43]. The incorporation of TiO_2 is known to improve various physical properties, such as the mechanical, barrier, UV-light barrier, thermal, optical, and antimicrobial properties of composite films [10,44,45]. Previously, TA was used to prepare chitosan-based film [26,39,46]. The effect of TiO_2 on chitosan-based film was also studied [31,47]. To the best of our knowledge, the synergistic effect of TA and TiO_2 in chitosan-based nanocomposite films has not been studied so far. Previous works mostly focused only on cross-linking chitosan with TA or chitosan and various TiO_2 nanoparticles.

Moreover, the cross-linker's concentration and the TiO_2 nanoparticles' size also significantly impact composite films' properties. In the present work, spherical-shape TiO_2 nanoparticles and the optimum TA content were used as reinforcing agents and a cross-linker, respectively. More specifically, this is a more in-depth study of TA-cross-linked and TiO_2-reinforced chitosan nanocomposite film. However, the various properties were not thoroughly studied in previous studies, such as the water vapor barrier properties, hydrodynamic properties, and antioxidant activity. Combining a nano-filler and cross-linker in chitosan-based composite film is expected to improve the film's physical and functional properties. This results obtained in this study are expected to produce new insight into chitosan-based functional composite films for active food packaging applications.

For the fabrication of biobased, eco-friendly, and biodegradable active packaging film, many techniques, such as solution casting, compression molding, extrusion (casting and blowing), etc., are commonly used [48,49]. Solution casting is the most convenient and straightforward method for producing the film on a lab scale. The solution mixing and casting methods have a low cost, are easy to handle, and are efficient for biopolymer-based packaging films [50]. Even though the solution casting method is not useful for industrial-scale production, the ideal composition prepared in this way can be used to further develop environmentally benign packaging films via extrusion methods for mass production.

The present investigation's primary objective is to make a chitosan-based nanocomposite film with improved mechanical and functional properties using a TA cross-linker and a physical reinforcement by TiO_2. The developed nanocomposite film was prepared with a simple solution mixing and casting technique and was characterized using various analytical methods. The film properties, such as the mechanical properties, thermal stability, hydrodynamic properties, water vapor barrier properties, and antioxidant activity, were also assessed.

2. Materials and Methods

2.1. Materials

Chitosan (Chs) (viscosity 200–800 cP at 1% acetic acid, MW: 190,000–310,000 based on viscosity, 75–85% deacetylated), titanium oxide nanopowder, dopamine hydrochloride, sodium hydroxide, 2,2-diphenyl-1-picrylhydrazyl (DPPH), 2,2′-azino-bis(3-ethylbenzothiazoline-6-sulfonic acid) (ABTS), and potassium persulfate were purchased from Sigma-Aldrich, St. Louis, MO, USA. All other chemicals used were of analytical reagent grade.

2.2. Preparation of Chitosan/TA/TiO_2 Nanocomposite Films

The chitosan-based nanocomposite films were prepared by following the solution casting method [15]. Initially, the chitosan solution was designed using 0.5% acetic acid [15,51], and

5 wt% (based on chitosan) of TA was mixed, followed by mixing 0.5 and 1.0 wt% (based on chitosan) of TiO_2 with the prepared chitosan solution at room temperature. The mixture was pulverized using a high-shear homogenizer (T50, IKA Labotechnik, Germany) for 10 min at 5000 rpm, followed by sonication for another 1 h. The completely soluble film-forming solution was cast on a polycarbonate plate using a doctor blade and dried for 48 h in a temperature- and humidity-controlled cleanroom (25 °C and 50% RH). The dried film was peeled off from the plate and left in the cleanroom for 48 h. For comparison, a neat chitosan film without any additives was also prepared by following the same procedure. All film samples were made in triplicate, and the developed films were designated as Chs/TA, Chs/TA/Ti$^{0.5}$, and Chs/TA/Ti$^{1.0}$, respectively, as per the content of additives. The preparation of the nanocomposite film is briefly explained in the following schematics (Scheme 1).

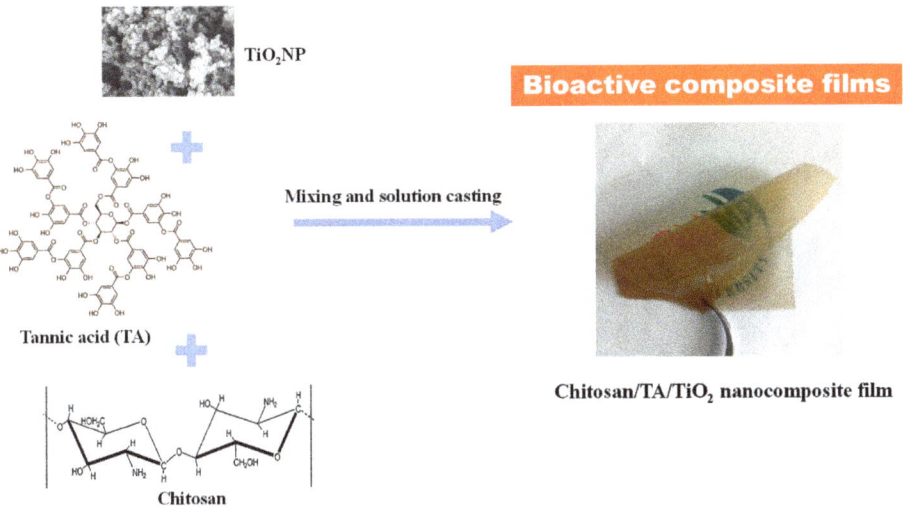

Scheme 1. The preparation process of chitosan/tannic acid (TA)/TiO_2 nanocomposite film.

2.3. Characterization

2.3.1. Morphology

The surface and cross-sectional morphologies of the chitosan-based nanocomposite films and the morphology of TiO_2 were checked with field-emission scanning electron microscopy (FESEM, SU8010, Hitachi, Tokyo, Japan) at accelerating voltages of 5 and 15 kV, respectively. All the film specimens and the TiO_2 powder were sputter-coated with platinum for 90 s before the measurement.

2.3.2. FTIR and Optical Properties

The Fourier-transform infrared (FTIR) spectra of the chitosan-based nanocomposite films were noted in an FTIR spectrometer (Billerica, Bruker Optics) in attenuated total reflection (ATR) mode in the range of 4000–650 cm^{-1} at 16 scan rates with the resolution of 4 cm^{-1}. The nanocomposite films' optical properties were recorded using a UV-vis spectrophotometer (UV-2501PC, Shimadzu, Kyoto, Japan) in 200–800 nm. The UV-barrier property and transparency of the same films were also evaluated by determining the transmittance at 280 nm (T_{280}) and 660 nm (T_{660}), respectively [12].

2.3.3. Moisture Content, Water Solubility, and Swelling Ratio

The moisture content (MC), water solubility (WS), and swelling ratio (SR) of the chitosan-based nanocomposite films were determined by following the standard method [15]. The MC of the specimens (2.5 × 2.5 cm) was measured as the film's weight transformation after dehydrating at 105 °C for 24 h. The MC was calculated using the following equation:

$$\text{MC } (\%) = \frac{W_1 - W_2}{W_1} \times 100 \quad (1)$$

where W_1 and W_2 refer to the initial and dried weight of the film specimens, respectively.

For WS, at first, the tested film specimens (2.5 × 2.5 cm) were dried at 60 °C overnight and weighed (W_1), which was followed by dipping in 30 mL of deionized (DI) water for 24 h with occasional mild shaking at 25 °C; then, the specimens were taken out from the water, dried in an oven at 105 °C for 24 h, and then weighed (W_2). The WS was calculated using the following equation:

$$\text{WS } (\%) = \frac{W_1 - W_2}{W_1} \times 100 \quad (2)$$

To determine the SR, a pre-weighed (W_1) film specimen (2.5 × 2.5 cm) was immersed in 30 mL DI water for 1 h; then, it was taken out of the water and weighed (W_2) after removal of surface water using a blotting paper. The SR was calculated using the following equation:

$$\text{SR } (\%) = \frac{W_2 - W_1}{W_1} \times 100 \quad (3)$$

2.3.4. Water Vapor Permeability

The water vapor permeability (WVP) of the chitosan-based nanocomposite films was determined using a WVP cup by following the ASTM E96-95 standard method. At first, the WVP cup was filled with a prescribed amount of water, covered by the films, sealed, and kept in the controlled environmental chamber at 25 °C and 50% RH. After equilibration, the WVP cup's weight was measured at every 1 h interval, and weight loss was calculated. The water vapor transmission rate can be measured from the ratio of weight loss and the film area. The WVP (g.m/m^2.Pa.s) was determined as follows:

$$\text{WVP} = \frac{\Delta W \times L}{t \times A \times \Delta p} \quad (4)$$

where ΔW is the weight alteration of the WVP cup (g), L is the thickness of the film (m), Δp is the partial water vapor pressure difference across the two sides of the film, A is the permeation area of the film (m^2), and t is the time (s). The same was calculated using the established procedure [52].

2.3.5. Mechanical Properties

The film specimens' thickness was measured using a digital micrometer (Digimatic, Mitutoyo, Kawasaki, Japan) with 1 mm accuracy. For each specimen, two random positions were taken, and the average values were used. Mechanical properties, namely the tensile strength (TS), Young's modulus (YM), and elongation at break (EB), of the prepared nanocomposite films were measured according to the standard (ASTM D-882-97) with a household tensile testing system [53]. The tensile test was conducted in a controlled environment at 25 °C and ~30% RH. The sizes of the film specimens were 5 × 1 cm, and the gauge length and applied pulling rate were 20 mm and 0.005 mm/s, respectively; four specimens were tested for each case, and the values were averaged.

2.3.6. Differential Scanning Calorimetry and Thermogravimetric Analysis

Thermogravimetric analysis (TGA) and differential scanning calorimetry (DSC) were performed to understand the thermal properties. The TGA measurement was carried out using a TGA (STA 409 PC, Netzsch, Selb, Germany), and ~10 mg film specimens were tested at a heating rate of 10 °C/min in a temperature range of 30 to 600 °C under a nitrogen flow of 20 cm^3/min. The maximum disintegration temperature was determined from differential thermogravimetry (DTG) curves [54]. The DSC was observed using a TA instrument (DSC200 F3, Netzsch) at a heating rate of 10 °C/min in a temperature range of 20 to 350 °C under a nitrogen gas atmosphere.

2.3.7. Antioxidant Activity

Antioxidant activities of the chitosan-based nanocomposite films were measured by assessing the free radical scavenging activity. The 2,2-diphenyl-1-picrylhydrazyl radical (DPPH$^\bullet$) and 2,2'-azino-bis(3-ethylbenzothiazoline-6-sulfonic acid) (ABTS$^{\bullet+}$) radical scavenging methods [55,56] were used for the antioxidant test. For the DPPH analysis, a prescribed amount of methanolic solution of DPPH was freshly made, and ~50 mg of the tested film sample was added in a 10 mL DPPH solution and incubated at room temperature for 30 min; then, the absorbance was measured at 517 nm. A control was also tested without adding the film sample in an assay solution. In the ABTS assay, a prescribed amount of potassium sulfate was added to the ABTS solution, followed by overnight incubation in the dark to make the ABTS assay solution. A total of ~50 mg of tested film samples was added to 10 mL of ABTS assay solution and incubated at room temperature for 30 min; then, the absorbance was measured at 734 nm. A control was also tested without adding the film sample in an assay solution. The antioxidative activity was calculated as follows:

$$\text{Free radical scavenging activity (\%)} = \frac{A_c - A_t}{A_c} \times 100 \quad (5)$$

where A_c and A_t were the absorbances of DPPH/ABTS of the control and test films, respectively.

2.3.8. Antioxidant Activity

For statistical analysis of the obtained results, one-way analysis of variance (ANOVA) was performed, and the significance of each mean property value was determined ($p < 0.05$) by Duncan's multiple range test using the SPSS statistical analysis computer program for Windows (SPSS Inc., Chicago, IL, USA).

3. Results and Discussion

3.1. Properties of TiO$_2$ Nanoparticles

Figure 1a shows the UV-vis absorption of aqueous TiO$_2$ solution, and the results obtained from the absorption spectrum showed a distinct absorption profile for TiO$_2$ with a maximum absorption of ~350 nm. The detected UV-vis results are physical characteristics of TiO$_2$, and the obtained results corroborate the findings of a previously published report [57]. The morphology of the TiO$_2$ taken by FESEM is shown in Figure 1b. The TiO$_2$ was roughly spherical and in the size range of 15–45 nm, with an average diameter of 31.3 ± 5.8 nm, as determined with the ImageJ software.

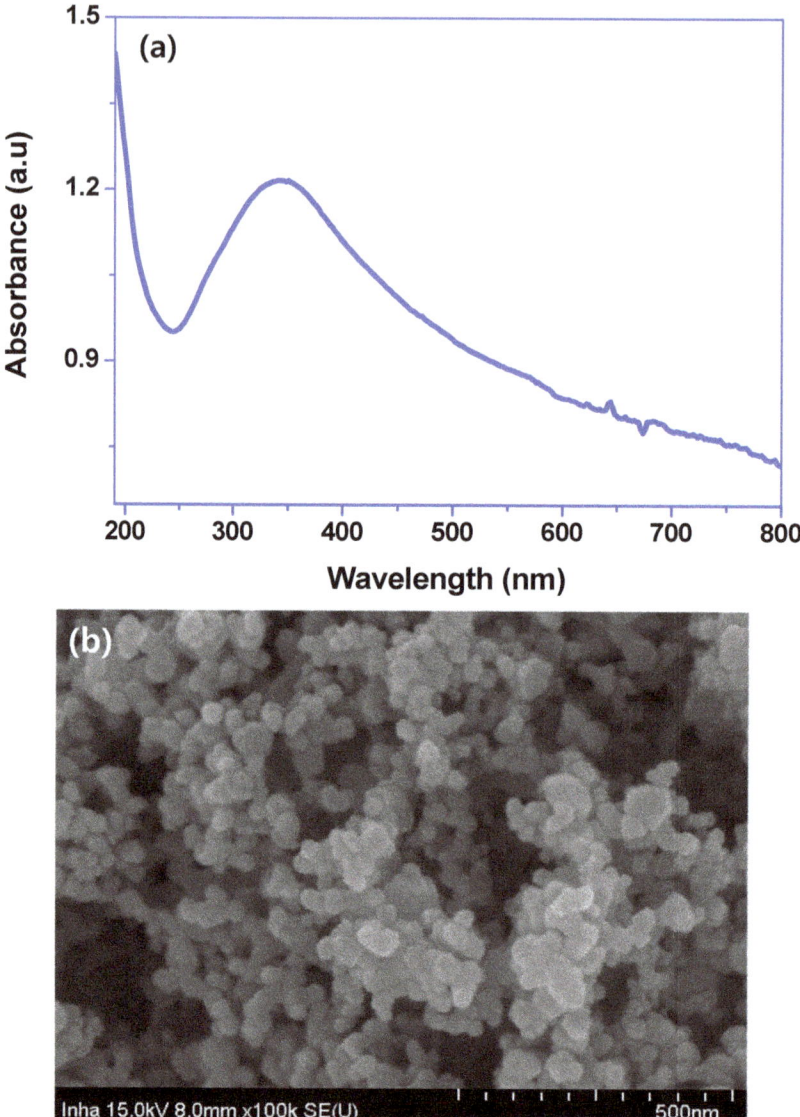

Figure 1. (a) UV-vis spectrum and (b) field-emission scanning electron microscopy (FESEM) image of TiO_2 nanoparticles.

3.2. Properties of Chitosan/TA/TiO$_2$ Nanocomposite Films

3.2.1. Appearance and Optical Properties

The macroscopic appearance of chitosan-based nanocomposite film is displayed in Figure 2a. The neat chitosan film was freestanding, flexible, highly transparent, and colorless. In contrast, the nanocomposite film containing TA exhibited a light brownish color, and the blended TiO_2 films were dark brown depending on the content of TiO_2. The optical properties of the chitosan, Chs/TA, and Chs/TA/TiO_2 nanocomposite films are shown in Figure 2b. The UV-vis spectra exhibit that the neat chitosan film is highly transparent, and in the case of nanocomposite films, the profile nature is different. For better understanding, the UV-light barrier and transparency of the chitosan-based nanocomposite films were scrutinized by measuring the absorbance at 280 and 660 nm. The results are displayed in Table 1. The UV-light transmittance and transparency of the neat chitosan film were 52.5% and 86.3%, respectively, similarly to the previously published report [15]. The UV-light barrier property was almost completely blocked after adding the TA in chitosan, which might be due to TA's strong UV-light absorption [58]. After adding TiO_2, the UV-light barrier remained practically unaltered. The observed results suggest that by adding TA in the chitosan-based nanocomposite films, the UV light can be blocked entirely, benefitting their application in active packaging. The neat chitosan film's transparency was slightly reduced after adding TA, whereas the incorporation of TiO_2 significantly decreased the transparency depending on the nanoparticle content. These results suggest that TA's addition slightly decreases transparency, but addition of TiO_2 reduces the transparency significantly, although well enough for packaging film applications.

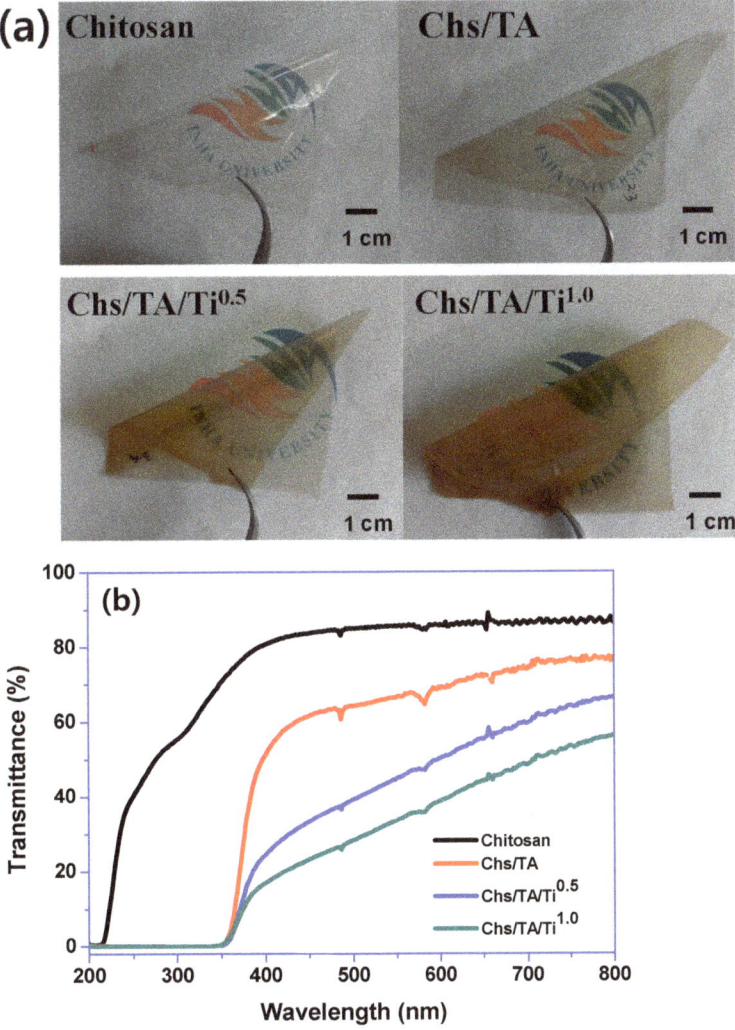

Figure 2. (a) Visual appearance and (b) UV-vis light transmittance spectra of the chitosan/TA/TiO_2 nanocomposite films.

Table 1. Light transmittance, water vapor permeability, and thermogravimetric analysis data of chitosan/TA/TiO$_2$ nanocomposite films. The values are represented as a mean ± standard deviation. Any two means in the same column followed by the same letter are not significantly ($p > 0.05$) different according to Duncan's multiple range tests.

Films	T_{280} (%)	T_{660} (%)	WVP (×10^{-9}) g·m/m^2·Pa·s	T_{onset}/T_{end} (°C)	$T_{0.5}$	Char Content (%)
Chitosan	52.05 ± 2.2 [b]	86.29 ± 0.6 [d]	0.59 ± 0.03 [b]	144/385	270	41.2
Chs/Ta	0.17 ± 0.01 [a]	78.20 ± 2.0 [c]	0.51 ± 0.02 [ab]	150/387	273	43.4
Chs/Ta/Ti$^{0.5}$	0.09 ± 0.01 [a]	56.60 ± 1.0 [b]	0.48 ± 0.07 [ab]	146/386	280	41.2
Chs/Ta^5/Ti$^{1.0}$	0.05 ± 0.01 [a]	42.21 ± 2.2 [a]	0.46 ± 0.03 [a]	152/389	275	42.8

3.2.2. Fourier−Transform Infrared Spectroscopy

The FTIR spectra of the neat chitosan and its nanocomposite films are presented in Figure 3. The peak appeared in the range of 3600–3000 cm^{-1}, mainly due to the O-H and N-H stretching vibrations [15]. The peak observed at 2877 cm^{-1} was attributed to the C-H stretching vibrations of the alkane groups of chitosan. Peaks found at 1645 cm^{-1} and 1548 cm^{-1} referred to the C=O stretching of the acetyl group (amide-I) of chitosan and -NH bending and stretching (amide-II), respectively [59]. Peaks noticed at 1405 cm^{-1} and ~1151 cm^{-1} were ascribed to the O-H bending vibration due to chitosan's saccharide structure [15]. In chitosan-based nanocomposite films, similar peak profiles with a slight alteration in intensity and positions were observed. In conclusion, there was no characteristic alteration in the chitosan-based films' functional groups after adding the cross-linker and reinforcing agent. The minor modifications in intensity and position in the peaks might be because of the physical interaction (H-bonding, van der Waals forces) between chitosan and TA/TiO$_2$ [5].

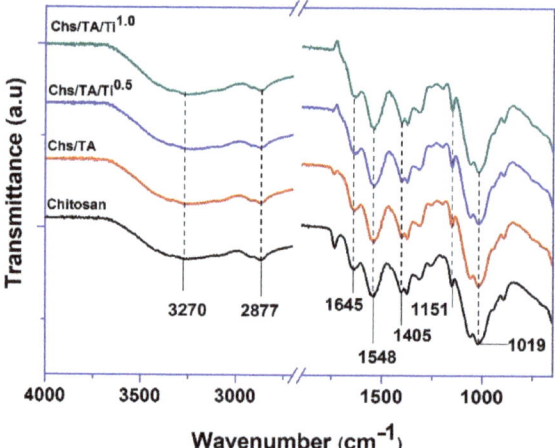

Figure 3. Fourier−transform infrared (FTIR) spectra of the chitosan/TA/TiO$_2$ nanocomposite films.

3.2.3. Morphology

The microstructure (surface and cross-section morphology) of the prepared chitosan-based nanocomposite films found by scanning electron microscopy (SEM) is shown in Figure 4a–h. The surface images of the neat chitosan and its nanocomposite films indicate that all of the films are freestanding, smooth, and without any cracks or voids (Figure 4a–d). The addition of TA/TiO$_2$ did not meaningfully alter the dense structure of the surface morphology. The incorporation of TA/TiO$_2$ did not alter the morphology, and they were uniformly mixed in the matrix polymer, which indicates their good miscibility in the liquid phase. The cross-section morphologies of the chitosan-based films exhibited that the neat chitosan film had a dense layer structure with some cracks.

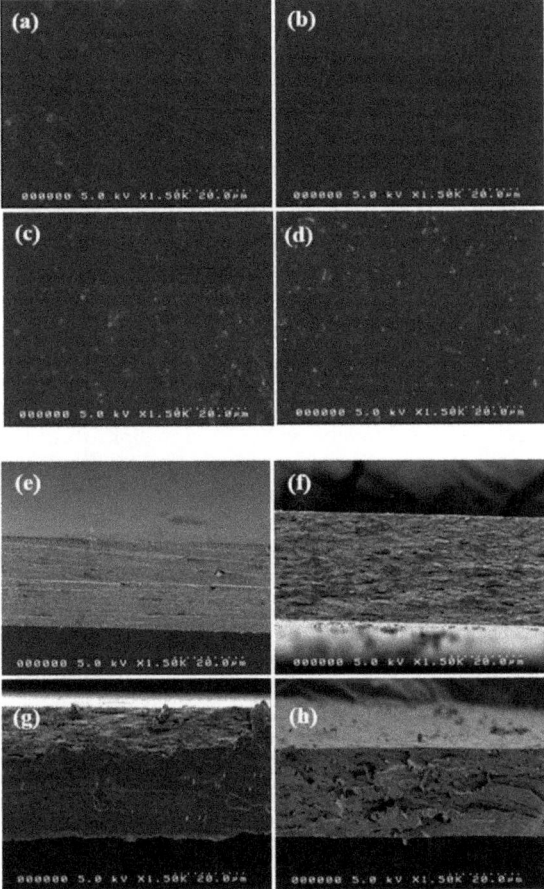

Figure 4. FESEM images of the chitosan/TA/TiO$_2$ nanocomposite films: (**a–d**) surfaces of chitosan, Chs/TA, Chs/TA/Ti$^{0.5}$, and Chs/TA/Ti$^{1.0}$; (**e–h**) cross-sections of Chs/TA, Chs/TA/Ti$^{0.5}$, and Chs/TA/Ti$^{1.0}$.

In contrast, the blended TA/TiO$_2$ films showed similar morphologies, but a more porous structure due to the tannic acid (Figure 4e–h). The addition of TiO$_2$ at 1 wt% showed some ruptures and voids in the chitosan matrix's layered structure. The obtained results indicate that the TA/TiO$_2$ is compatible with the chitosan matrix, and that the fillers are uniformly distributed in the chitosan matrix. These results show the excellent adhesion, intermolecular binding, and affinity between chitosan and TA/TiO$_2$, which might play an essential role in improving the physical properties.

3.2.4. Thermal Properties

The chitosan nanocomposite film's thermal stability was assessed by measuring TGA and DTG thermograms and DSC analysis. The TGA and DTG results are displayed in Figure 5a,b. It was observed that both chitosan and its nanocomposite films show a twofold thermal degradation arrangement. The maximum initial weight loss was demonstrated at ~75–80 °C for all the tested films due to the vaporization of moisture [15]. The onset/endset temperature for the primary degradation was observed at 144/390 °C (Table 1). The maximum thermal degradation was observed around 270–280 °C due to the thermal decomposition of the chitosan polymer matrix [15]. The maximum degradation temperature for the neat chitosan was 270 °C, whereas, after the addition of TA, it slightly increased to 273 °C. The addition of TiO_2 slightly increased the thermal decomposition temperature of the chitosan-based films (Table 1). It was recently reported that the addition of TiO_2 in biopolymer-based composite film improved thermal stability [43]. The char contents observed at 600 °C for neat chitosan, Chs/TA, Chs/TA/Ti$^{0.5}$, and Chs/TA/Ti$^{1.0}$ films were 41.2, 43.4, 41.2, and 42.8%, respectively, which is at the higher side, and most probably comes from the non-ignitable mineral present in the biopolymer used. The DSC spectra of the neat chitosan and its nanocomposite films are shown in Figure 5c. The DSC analysis is a measure of the miscibility and compatibility of the components present in the polymer. The chitosan film showed two major peaks. One was observed in the range of 25 to 150 °C with a maximum at ~88 °C, and this endothermic peak of chitosan was associated with the evaporation of bound and absorbed water [60]. Another peak is shown in the range of 255–320 °C with a maximum of ~286 °C, which might be associated with the thermal degradation of amine units of chitosan. In the case of the chitosan-based nanocomposite films, alteration in both peaks was witnessed. The minor variations of the first peak's position in the nanocomposite films were conceivably due to the different polymer–water interactions. The difference in the TA-cross-linked and TiO_2-added films' peak positions was due to the variable polymer filler interactions. The observed results suggest decent biocompatibility between chitosan and TA/TiO_2.

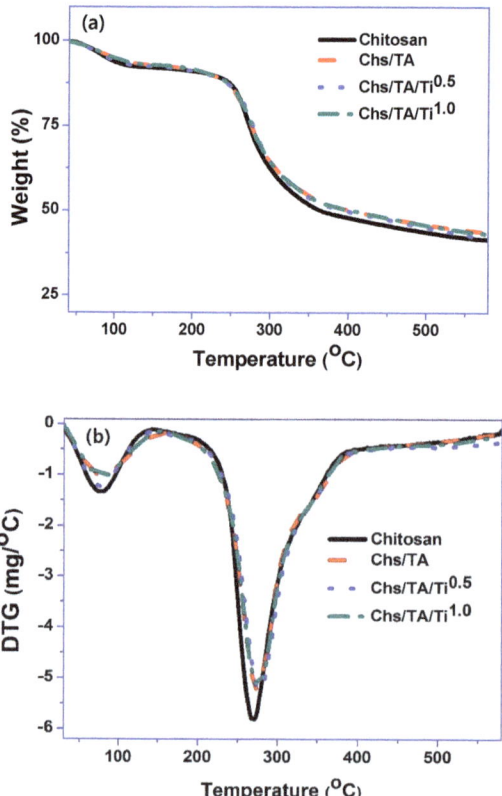

Figure 5. *Cont.*

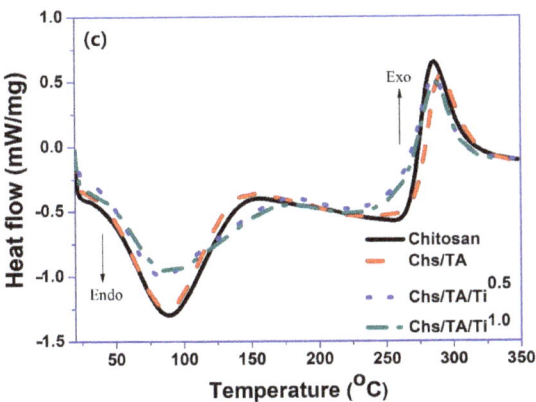

Figure 5. (**a**) Thermogravimetric analysis (TGA), (**b**) DTG, and (**c**) differential scanning calorimetry (DSC) analyses of the chitosan/TA/TiO$_2$ nanocomposite films.

3.2.5. Water Vapor Barrier Properties

The WVP of neat chitosan and chitosan-based nanocomposite films is shown in Table 1. The WVP of the neat chitosan film was $0.59 \pm 0.03 \times 10^{-9}$ g·m/m^2·Pa·s, which decreased to $0.51 \pm 0.03 \times 10^{-9}$ g·m/m^2·Pa·s after adding TA, and the WVP significantly further decreased to $0.46 \pm 0.03 \times 10^{-9}$ g·m/m^2·Pa·s after incorporation of 1 wt% of TiO$_2$. The resulting insight indicates that the cross-linker (TA), as well as reinforcement (TiO$_2$), effectively altered the WVP of the chitosan-based nanocomposite films. The reduction of WVP after adding TA has been reported [26]. Previously, it was also shown that the addition of TiO$_2$ decreased the WVP of biopolymer-based nanocomposite films [43]. In addition, in the case of the TiO$_2$-added chitosan-based film, a similar improvement in water vapor barrier properties was reported previously [43]. The reduction in WVP (increase in the vapor barrier property) was probably due to the creation of a tortuous path for vapor diffusion through the film by decreasing the free -OH groups [61].

3.2.6. Hydrodynamic and Mechanical Properties

The MC, WS, and SR of neat chitosan and its nanocomposite films are presented in Figure 6a–c. The chitosan-based films' MC was significantly increased (~12% to ~18%) in all nanocomposite films compared to the pristine chitosan film. In general, the cross-linking reduces the water interaction with the matrix polymer and affects the water retention inside the films. In contrast to the present observation, it was previously reported that cross-linking can effectively reduce the moisture sensitivity of biopolymer-based films [62]. The probable cause of the noteworthy rise in the MC of the films is due to the reduced polymer network interaction as a result of the availability of increased free -OH groups and, consequently, the greater amount of water absorbed [63]. Like the current outcome, previously, MC was increased after addition of nanofillers in chitosan-based films [15]. The SR of the neat chitosan film was drastically decreased (~300% to ~100%) in the case of all the nanocomposite films. The reduction in SR of the nanocomposite film is perhaps due to the blending of hydrophobic nanofiller in the chitosan [63]. The decrease in the chitosan-based film's swelling degree upon cross-linking was previously reported [25], which validates the present observation. In the biopolymer-based film, the SR mainly depends on the cross-linking material's porosity and nature [15]. Upon increasing the cross-linking, the SR generally decreases [64], and consequently, the incorporation of TA decreases the SR of chitosan-based nanocomposite films. The WS of the chitosan-based nanocomposite film decreased after adding TA, whereas the addition of TiO$_2$ did not much change the WS of the nanocomposite film compared to that of the neat chitosan film. The decrease in WS of the nanocomposite film with TA was possibly due to the cross-linker,

TA [26]. In the TA-cross-linked chitosan nanocomposite film, similar WS variation was previously noticed. The WS of the chitosan-based film was found to be affected by the kind and nature of the fillers. The cross-linking reduced the WS, which is rational, whereas the addition of nanofillers did not much influence or slightly increased the WS, which is probably due to the reduction in polymer network interaction in the presence of nanofillers.

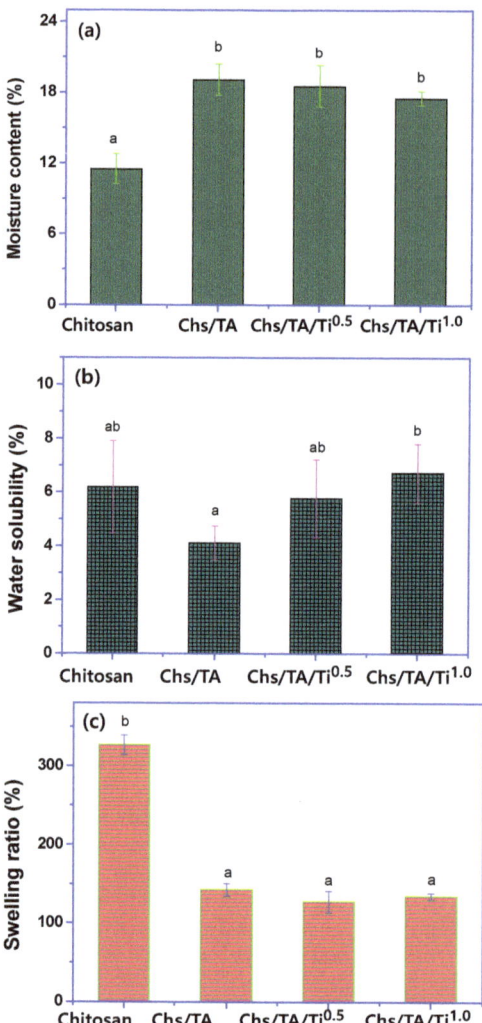

Figure 6. (**a**) Moisture content, (**b**) water solubility, (**c**) and swelling ratio analyses of the chitosan/TA/TiO$_2$ nanocomposite films. The values are represented as a mean ± standard deviation. Any two means on the column bars with the same letter are not significantly different ($p > 0.05$).

The mechanical properties of the chitosan-based nanocomposite films are shown in Figure 7a–d. The average thickness of the chitosan and its nanocomposite films was 20 ± 4 µm. The fabricated films' thickness was comparable with the thickness (26~30 µm) detected from the cross-section FESEM image and analyzed with ImageJ (Figure 4e–h). The mechanical properties were altered by incorporating TA as a cross-linker and TiO$_2$ as a reinforcing nanofiller. The TS, YM, and EB of the neat chitosan film were 95.5 ± 7.7 MPa, 8.8 ± 1.2 GPa, and 2.5 ± 1.2%, respectively, like the previously published data [15]. The TA-cross-linked chitosan-based nanocomposite film's mechanical properties were significantly improved (TS = 108.8 ± 13.6 MPa) due to TA's presence. Previously, it was already shown that the cross-linker TA in chitosan improved the composite film's tensile strength, which agrees with the present findings [26,39]. The mechanical properties were also known to be enhanced by the addition of TiO_2 in biopolymer-based composite films [43]. In addition, in the case of the TiO_2-added chitosan-based film, a similar improvement in water vapor barrier properties was reported before [65]. The addition of 0.5 wt% of TiO_2 improved the TS and YM of the cross-linked chitosan-based film more compared to the addition of 1 wt%. The increase in mechanical properties of the film is predominantly due to the increase in molecular interaction between chitosan and TiO_2 as well as the good interfacial interactions between them. Previously, it was also observed that the addition of a lower content of nanofiller improved the mechanical performance of the nanocomposite film, whereas at a higher content, it starts decreasing depending on the range [61]. In this work, the synergistic effect of TA/TiO_2 was considered for better improvement of the mechanical properties of chitosan-based nanocomposite films. As speculated, the addition of TiO_2 in chitosan-based nanocomposite films significantly improved the mechanical properties, with ~40% improvement in the TS in Chs/TA/Ti$^{0.5}$. In contrast, for Chs/TA/Ti$^{1.0}$, the mechanical properties were reduced, but were still higher than those of the neat chitosan or Chs/TA film. At a 1% TiO_2 concentration, the chitosan-based film's mechanical properties were significantly decreased compared to 0.5%, which might be due to the limited interaction between the chitosan and TiO_2 associated with the phase separation by the aggregated fillers in the polymer matrix. Previously, it was also reported that the addition of TiO_2 in the biopolymer matrix significantly enhanced the film's mechanical properties [43]. Zhang et al. [66] also reported that the addition of TiO_2 in chitosan-based films improved the mechanical properties. The chitosan-based composite film results indicate that incorporating TA/TiO_2 alone or in combination can significantly improve the chitosan-based nanocomposite films' mechanical properties, which might be useful for packaging applications.

3.2.7. Antioxidant Properties

Chitosan's antioxidant actions, as well as those in the nanocomposite films, were determined using the two well-established DPPH and ABTS radical scavenging activity methods, and the obtained results are shown in Figure 8. The neat chitosan film showed considerable antioxidant activity, which is reasonable, as it is an active antioxidant biopolymer. This is due to the hydroxyl and amino groups in chitosan, which can interact with free radicals and show radical scavenging activity [67,68]. The addition of TA in the chitosan-based nanocomposite films showed an enhanced antioxidant action, which might be due to tannic acid's antioxidant potential [69]. The DPPH and ABTS radical scavenging efficiencies of the neat chitosan film were 17.9% and 30.7%, respectively; meanwhile, they were enhanced to 42.4% and 64.5% in the case of the TA-incorporated chitosan-based film, Chs/TA. The nanocomposite films' increased antioxidant activity is mainly due to the potent antioxidant tannic acid, which contains a polyphenol group [69]. In the case of TiO_2-added chitosan-based nanocomposite films, the antioxidant activity was slightly lower than that of the Chs/TA film due to the antioxidant presence of inactive TiO_2. However, the antioxidant results are similar, indicating that the developed Chs/TA/TiO_2 nanocomposite films have good antioxidant potential. It has been known that the antioxidant potential of biopolymer-based films is predominantly reliant on the concentration of antioxidants in the films [70].

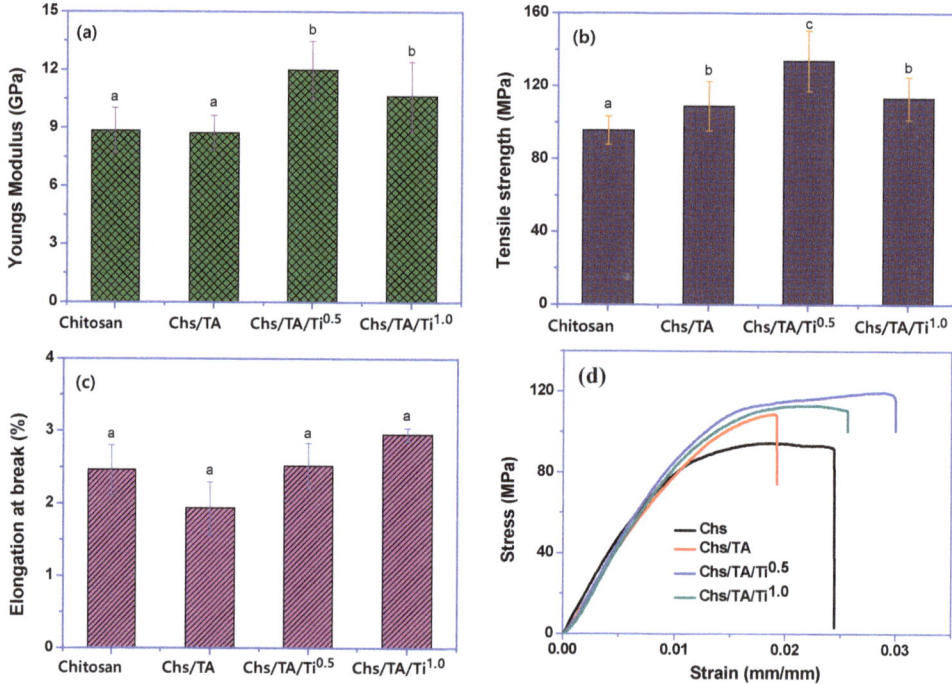

Figure 7. (**a**) Tensile strength, (**b**) Young's modulus, (**c**) elongation at break, and (**d**) stress vs. strain graphs of the chitosan/TA/TiO$_2$ nanocomposite films. The values are represented as a mean ± standard deviation. Any two means on the column bars with the same letter are not significantly different ($p > 0.05$).

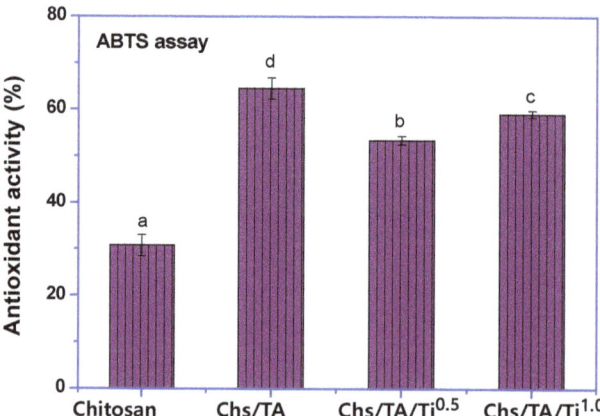

Figure 8. *Cont.*

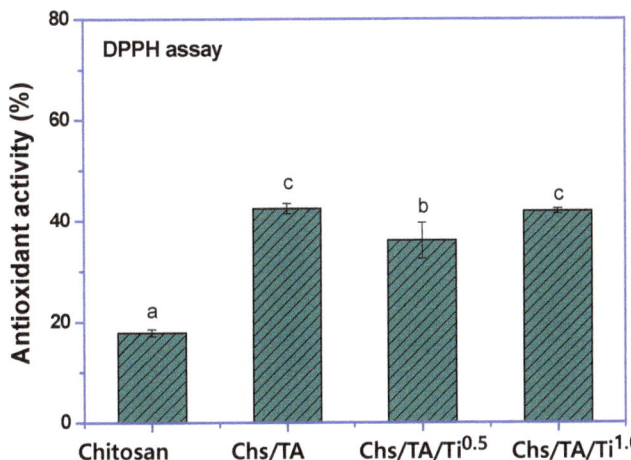

Figure 8. Antioxidant activity (2,2-diphenyl-1-picrylhydrazyl (DPPH) and 2,2'-azino-bis(3-ethylbenzothiazoline-6-sulfonic acid) (ABTS) assay) of the chitosan/TA/TiO$_2$ nanocomposite films. The values are represented as a mean ± standard deviation. Any two means on the column bars with the same letter are not significantly different ($p > 0.05$).

4. Conclusions

Tannic-acid-cross-linked and TiO$_2$-reinforced chitosan-based nanocomposite films prepared with the solution casting process are reported. The cross-linked and nanofiller-added chitosan-based films were compatible and were uniformly spread in the polymer matrix. The developed nanocomposite films were see-through and had excellent UV-light barrier properties, without sacrificing too much transparency. The mechanical properties were meaningfully enhanced due to TA's cross-linking and the reinforcement of TiO$_2$. The films' thermal stability was also slightly improved, whereas the swelling ratio decreased, and the water solubility did not change much. The addition of fillers and a cross-linker slightly improved the water vapor barrier properties of the chitosan-based film. The synergistic effect of TA/TiO$_2$ was considered for better improvement of the physical (mechanical, water vapor barrier) properties of the chitosan-based nanocomposite films. Furthermore, the chitosan/TA/TiO$_2$ nanocomposite films also showed an intense antioxidant activity. The developed chitosan-based nanocomposite films are promising for active food packaging applications based on the improved mechanical properties, UV barrier properties, and antioxidant behavior.

Author Contributions: Conceptualization, S.R. and J.K.; methodology, L.Z. and S.R.; formal analysis, S.R. and H.C.K.; investigation, D.H.P. and J.K.; writing—original draft preparation, S.R. and J.K.; writing—review and editing, S.R., J.K., and H.A.; supervision, J.K.; project administration, H.A.; funding acquisition, H.A. All authors have read and agreed to the published version of the manuscript.

Funding: This project was supported by the Deanship of Scientific Research at Prince Sattam bin Abdul Aziz University under the research project no. 2020/1/17063.

Institutional Review Board Statement: Not applicable.

Informed Consent Statement: Not applicable.

Data Availability Statement: Data available on request due to restrictions eg privacy or ethical.

Conflicts of Interest: The authors declare no conflict of interest.

References

1. Geyer, R.; Jambeck, J.R.; Law, K.L. Production, use, and fate of all plastics ever made. *Sci. Adv.* **2017**, *3*, e1700782. [CrossRef] [PubMed]
2. Robertson, G. *Compatible Food Packaging*; Elsevier: Amsterdam, The Netherlands, 2008; pp. 3–28.
3. Hoffmann, T.; Peters, D.A.; Angioletti, B.; Bertoli, S.; Vieira, L.P.; Reiter, M.G.R.; de Souza, C.K. Potentials Nanocomposites in Food Packaging. *Chem. Eng. Trans.* **2019**, *75*, 253–258.
4. Morin-Crini, N.; Lichtfouse, E.; Torri, G.; Crini, G. Applications of chitosan in food, pharmaceuticals, medicine, cosmetics, agriculture, textiles, pulp and paper, biotechnology, and environmental chemistry. *Environ. Chem. Lett.* **2019**, *17*, 1667–1692. [CrossRef]
5. Shameli, K.; Bin Ahmad, M.; Darroudi, M.; Rahman, R.A.; Jokar, M.; Yunis, W.M.Z.W.; Ibrahim, N.A. Silver/poly (lactic acid) nanocomposites: Preparation, characterization, and antibacterial activity. *Int. J. Nanomed.* **2010**, *5*, 573–579. [CrossRef]
6. Youssef, A.M.; El-Sayed, S.M. Bionanocomposites materials for food packaging applications: Concepts and future outlook. *Carbohydr. Polym.* **2018**, *193*, 19–27. [CrossRef]
7. Mujtaba, M.; Akyuz, L.; Koc, B.; Kaya, M.; Ilk, S.; Cansaran-Duman, D.; Martinez, A.S.; Cakmak, Y.S.; Labidi, J.; Boufi, S. Novel, multifunctional mucilage composite films incorporated with cellulose nanofibers. *Food Hydrocoll.* **2019**, *89*, 20–28. [CrossRef]
8. Mujtaba, M.; Koc, B.; Salaberria, A.M.; Ilk, S.; Cansaran-Duman, D.; Akyuz, L.; Cakmak, Y.S.; Kaya, M.; Khawar, K.M.; Labidi, J.; et al. Production of novel chia-mucilage nanocomposite films with starch nanocrystals; An inclusive biological and physico-chemical perspective. *Int. J. Biolog. Macromol.* **2019**, *133*, 663–673. [CrossRef]
9. Foschi, E.; Bonoli, A. The Commitment of Packaging Industry in the Framework of the European Strategy for Plastics in a Circular Economy. *Adm. Sci.* **2019**, *9*, 18. [CrossRef]
10. Yousefi, A.R.; Savadkoohi, B.; Zahedi, Y.; Hatami, M.; Ako, K. Fabrication and characterization of hybrid sodium montmorillonite/TiO$_2$ reinforced cross-linked wheat starch-based nanocomposites. *Int. J. Biol. Macromol.* **2019**, *131*, 253–263. [CrossRef]
11. Rhim, J.-W.; Wang, L.; Hong, S.I. Preparation and characterization of agar/silver nanoparticles composite films with antimicrobial activity. *Food Hydrocoll.* **2013**, *33*, 327–335. [CrossRef]
12. Roy, S.; Kim, H.C.; Kim, J.W.; Zhai, L.; Zhu, Q.Y.; Kim, J.-H. Incorporation of melanin nanoparticles improves UV-shielding, mechanical and antioxidant properties of cellulose nanofiber based nanocomposite films. *Mater. Today Commun.* **2020**, *24*, 100984. [CrossRef]
13. Adeyeye, O.A.; Sadiku, E.R.; Reddy, A.B.; Ndamase, A.S.; Makgatho, G.; Sellamuthu, P.S.; Perumal, A.B.; Nambiar, R.B.; Fasiku, V.O.; Ibrahim, I.D.; et al. *Materials Research and Applications*; Springer Science and Business Media LLC: Berlin/Heidelberg, Germany, 2019; pp. 137–158.
14. De Azeredo, H.M.C. Nanocomposites for food packaging applications. *Food Res. Int.* **2009**, *42*, 1240–1253. [CrossRef]
15. Roy, S.; Van Hai, L.; Kim, H.C.; Zhai, L.; Kim, J.-H. Preparation and characterization of synthetic melanin-like nanoparticles reinforced chitosan nanocomposite films. *Carbohydr. Polym.* **2020**, *231*, 115729. [CrossRef] [PubMed]
16. Ahsan, S.M.; Thomas, M.; Reddy, K.K.; Sooraparaju, S.G.; Asthana, A.; Bhatnagar, I. Chitosan as biomaterial in drug delivery and tissue engineering. *Int. J. Biol. Macromol.* **2018**, *110*, 97–109. [CrossRef]
17. Liu, Y.; Cai, Y.; Jiang, X.; Wu, J.; Le, X. Molecular interactions, characterization and antimicrobial activity of curcumin–chitosan blend films. *Food Hydrocoll.* **2016**, *52*, 564–572. [CrossRef]
18. Cazón, P.; Vázquez, M. Applications of Chitosan as Food Packaging Materials. *Sustain. Agric. Rev. 48* **2019**, *36*, 81–123. [CrossRef]
19. Guo, Y.; Chen, X.; Yang, F.; Wang, T.; Ni, M.; Chen, Y.; Yang, F.; Huang, D.; Fu, C.; Wang, S. Preparation and Characterization of Chitosan-Based Ternary Blend Edible Films with Efficient Antimicrobial Activities for Food Packaging Applications. *J. Food Sci.* **2019**, *84*, 1411–1419. [CrossRef]
20. Imran, M.; Klouj, A.; Revol-Junelles, A.-M.; Desobry, S. Controlled release of nisin from HPMC, sodium caseinate, poly-lactic acid and chitosan for active packaging applications. *J. Food Eng.* **2014**, *143*, 178–185. [CrossRef]
21. Kumar, S.; Mukherjee, A.; Dutta, J. Chitosan based nanocomposite films and coatings: Emerging antimicrobial food packaging alterna-tives. *Trend. Food Sci. Technol.* **2020**, *97*, 196–209. [CrossRef]
22. Lin, B.; Du, Y.; Liang, X.; Wang, X.; Wang, X.; Yang, J. Effect of chitosan coating on respiratory behavior and quality of stored litchi under ambient temperature. *J. Food Eng.* **2011**, *102*, 94–99. [CrossRef]
23. Souza, V.G.; Pires, J.R.A.; Rodrigues, C.; Coelhoso, I.; Fernando, A.L. Chitosan Composites in Packaging Industry—Current Trends and Future Challenges. *Polymers* **2020**, *12*, 417. [CrossRef] [PubMed]
24. Munteanu, S.B.; Vasile, C. Vegetable Additives in Food Packaging Polymeric Materials. *Polymers* **2020**, *12*, 28. [CrossRef]
25. Liang, J.; Wang, R.; Chen, R. The Impact of Cross-linking Mode on the Physical and Antimicrobial Properties of a Chitosan/Bacterial Cellulose Composite. *Polymers* **2019**, *11*, 491. [CrossRef] [PubMed]
26. Rivero, S.; García, M.A.; Pinotti, A. Cross-linking capacity of tannic acid in plasticized chitosan films. *Carbohydr. Polym.* **2010**, *82*, 270–276. [CrossRef]
27. Wang, H.; Gong, X.; Miao, Y.; Guo, X.; Liu, C.; Fan, Y.-Y.; Zhang, J.; Niu, B.; Li, W. Preparation and characterization of multilayer films composed of chitosan, sodium alginate and carboxymethyl chitosan-ZnO nanoparticles. *Food Chem.* **2019**, *283*, 397–403. [CrossRef] [PubMed]

28. Priyadarshi, R.; Sauraj; Kumar, B.; Negi, Y.S. Chitosan film incorporated with citric acid and glycerol as an active packaging material for extension of green chilli shelf life. *Carbohydr. Polym.* **2018**, *195*, 329–338. [CrossRef] [PubMed]
29. Al-Naamani, L.; Dobretsov, S.; Dutta, J. Chitosan-zinc oxide nanoparticle composite coating for active food packaging applications. *Innov. Food Sci. Emerg. Technol.* **2016**, *38*, 231–237. [CrossRef]
30. Archana, D.; Singh, B.K.; Dutta, J.; Dutta, P. In vivo evaluation of chitosan–PVP–titanium dioxide nanocomposite as wound dressing material. *Carbohydr. Polym.* **2013**, *95*, 530–539. [CrossRef]
31. Bahal, M.; Kaur, N.; Sharotri, N.; Sud, D. Investigations on Amphoteric Chitosan/TiO_2 Bionanocomposites for Application in Visible Light Induced Photocatalytic Degradation. *Adv. Polym. Technol.* **2019**, *2019*, 1–9. [CrossRef]
32. Sani, I.K.; Pirsa, S.; Tağı, Ş. Preparation of chitosan/zinc oxide/Melissa officinalis essential oil nano-composite film and evaluation of physical, mechanical and antimicrobial properties by response surface method. *Polym. Test.* **2019**, *79*, 106004. [CrossRef]
33. Mujtaba, M.; Salaberria, A.M.; Andres, M.A.; Kaya, M.; Gunyakti, A.; Labidi, J. Utilization of flax (Linum usitatissimum) cellulose nanocrystals as reinforcing material for chitosan films. *Int. J. Biol. Macromol.* **2017**, *104*, 944–952. [CrossRef] [PubMed]
34. Kaya, M.; Salaberria, A.M.; Mujtaba, M.; Labidi, J.; Baran, T.; Mulercikas, P.; Duman, F. An inclusive physicochemical comparison of natural and synthetic chitin films. *Int. J. Biol. Macromol.* **2018**, *106*, 1062–1070. [CrossRef] [PubMed]
35. Van Hai, L.; Zhai, L.; Kim, H.C.; Panicker, P.S.; Pham, D.H.; Kim, J.-H. Chitosan Nanofiber and Cellulose Nanofiber Blended Composite Applicable for Active Food Packaging. *Nanomaterials* **2020**, *10*, 1752. [CrossRef]
36. Roy, S.; Rhim, J.-W. Carboxymethyl cellulose-based antioxidant and antimicrobial active packaging film incorporated with curcumin and zinc oxide. *Int. J. Biol. Macromol.* **2020**, *148*, 666–676. [CrossRef]
37. Sharma, S.; Barkauskaite, S.; Jaiswal, A.K.; Jaiswal, S. Essential oils as additives in active food packaging. *Food Chem.* **2021**, *343*, 128403. [CrossRef]
38. Prashanth, K.V.H.; Tharanathan, R.N. Cross-linked chitosan—preparation and characterization. *Carbohydr. Res.* **2006**, *341*, 169–173. [CrossRef]
39. Rubentheren, V.; Ward, T.A.; Chee, C.Y.; Nair, P. Physical and chemical reinforcement of chitosan film using nanocrystalline cellulose and tannic acid. *Cellulose* **2015**, *22*, 2529–2541. [CrossRef]
40. Huang, D.; Zhang, Z.; Quan, Q.; Zheng, Y. Tannic acid: A versatile and effective modifier for gelatin/zein composite films. *Food Packag. Shelf Life* **2020**, *23*, 100440. [CrossRef]
41. Hoseinnejad, M.; Jafari, S.M.; Katouzian, I. Inorganic and metal nanoparticles and their antimicrobial activity in food packaging applications. *Critic. Rev. Microbiol.* **2018**, *44*, 161–181. [CrossRef]
42. Oun, A.A.; Shankar, S.; Rhim, J.-W. Multifunctional nanocellulose/metal and metal oxide nanoparticle hybrid nanomaterials. *Crit. Rev. Food Sci. Nutr.* **2020**, *60*, 435–460. [CrossRef]
43. Balasubramanian, R.; Kim, S.S.; Lee, J.; Lee, J. Effect of TiO_2 on highly elastic, stretchable UV protective nanocomposite films formed by using a combination of k-Carrageenan, xanthan gum and gellan gum. *Int. J. Biolog. Macromol.* **2019**, *123*, 1020–1027. [CrossRef] [PubMed]
44. Chorianopoulos, N.; Tsoukleris, D.; Panagou, E.Z.; Falaras, P.; Nychas, G.-J.E. Use of titanium dioxide (TiO_2) photocatalysts as alternative means for Listeria monocytogenes biofilm disinfection in food processing. *Food Microbiol.* **2011**, *28*, 164–170. [CrossRef] [PubMed]
45. Kubacka, A.; Serrano, C.; Ferrer, M.; Lünsdorf, H.; Bielecki, P.; Cerrada, M.L.; Fernández-García, M.; Fernández-García, M. High-Performance Dual-Action Polymer−TiO_2 Nanocomposite Films via Melting Processing. *Nano Lett.* **2007**, *7*, 2529–2534. [CrossRef] [PubMed]
46. Kaczmarek-Szczepańska, B.; Owczarek, A.; Nadolna, K.; Sionkowska, A. The film-forming properties of chitosan with tannic acid addition. *Mater. Lett.* **2019**, *245*, 22–24. [CrossRef]
47. Díaz-Visurraga, J.; Meléndrez, M.F.; García, A.; Paulraj, M.; Cárdenas, G. Semitransparent chitosan-TiO_2 nanotubes composite film for food package applications. *J. Appl. Polym. Sci.* **2010**, *116*, 3503–3515.
48. Rhim, J.-W.; Ng, P.K.W. Natural Biopolymer-Based Nanocomposite Films for Packaging Applications. *Crit. Rev. Food Sci. Nutr.* **2007**, *47*, 411–433. [CrossRef]
49. Tang, X.Z.; Kumar, P.; Alavi, S.; Sandeep, K.P. Recent Advances in Biopolymers and Biopolymer-Based Nanocomposites for Food Packaging Materials. *Crit. Rev. Food Sci. Nutr.* **2012**, *52*, 426–442. [CrossRef]
50. Roy, S.; Rhim, J.-W. Anthocyanin food colorant and its application in pH-responsive color change indicator films. *Crit. Rev. Food Sci. Nutr.* **2020**, 1–29. [CrossRef]
51. Kamdem, D.P.; Shen, Z.; Nabinejad, O.; Shu, Z. Development of biodegradable composite chitosan-based films incorporated with xylan and carvacrol for food packaging application. *Food Packag. Shelf Life* **2019**, *21*, 100344. [CrossRef]
52. Ramos, Ó.L.; Reinas, I.; Silva, S.I.; Fernandes, J.C.; Cerqueira, M.A.; Pereira, R.N.; Vicente, A.A.; Poças, M.F.; Pintado, M.E.; Malcata, F.X. Effect of whey protein purity and glycerol content upon physical properties of edible films manufactured therefrom. *Food Hydrocoll.* **2013**, *30*, 110–122. [CrossRef]
53. Kafy, A.; Kim, H.C.; Zhai, L.; Kim, J.W.; Van Hai, L.; Kang, T.J.; Kim, J.-H. Cellulose long fibers fabricated from cellulose nanofibers and its strong and tough characteristics. *Sci. Rep.* **2017**, *7*, 17683. [CrossRef] [PubMed]
54. Roy, S.; Shankar, S.; Rhim, J.-W. Melanin-mediated synthesis of silver nanoparticle and its use for the preparation of carrageenan-based antibacterial films. *Food Hydrocoll.* **2019**, *88*, 237–246. [CrossRef]

55. Bishai, M.; De, S.; Adhikari, B.; Banerjee, R. A comprehensive study on enhanced characteristics of modified polylactic acid based versatile biopolymer. *Eur. Polym. J.* **2014**, *54*, 52–61. [CrossRef]
56. Roy, S.; Rhim, J.-W. Fabrication of cellulose nanofiber-based functional color indicator film incorporated with shikonin extracted from Lithospermum erythrorhizon root. *Food Hydrocoll.* **2021**, *114*, 106566. [CrossRef]
57. Panda, J.; Singh, U.P.; Sahu, R. Synthesis, characterization of TiO_2 nano particles for enhancement of electron transport application in DSSC with Cu-BPCA Dye. *IOP Conf. Series: Mater. Sci. Eng.* **2018**, *410*, 012008. [CrossRef]
58. Katwa, L.C.; Ramakrishna, M.; Rao, M.R.R. Spectrophotometric assay of immobilized tannase. *J. Biosci.* **1981**, *3*, 135–142. [CrossRef]
59. Li, Q.; Zhou, J.; Zhang, L. Structure and properties of the nanocomposite films of chitosan reinforced with cellulose whiskers. *J. Polym. Sci. Part B Polym. Phys.* **2009**, *47*, 1069–1077. [CrossRef]
60. Ferrero, F.; Periolatto, M. Antimicrobial Finish of Textiles by Chitosan UV-Curing. *J. Nanosci. Nanotechnol.* **2012**, *12*, 4803–4810. [CrossRef]
61. Roy, S.; Rhim, J.-W.; Jaiswal, L. Bioactive agar-based functional composite film incorporated with copper sulfide nanoparticles. *Food Hydrocoll.* **2019**, *93*, 156–166. [CrossRef]
62. Vartiainen, J.; Harlin, A. Cross-linking as an Efficient Tool for Decreasing Moisture Sensitivity of Biobased Nanocomposite Films. *Mater. Sci. Appl.* **2011**, *2*, 346–354.
63. Roy, S.; Rhim, J.-W. Agar-based antioxidant composite films incorporated with melanin nanoparticles. *Food Hydrocoll.* **2019**, *94*, 391–398. [CrossRef]
64. Li, A.; Ramakrishna, S.N.; Kooij, E.S.; Espinosa-Marzal, R.M.; Spencer, N.D. Poly(acrylamide) films at the solvent-induced glass transition: Adhesion, tribology, and the influence of crosslinking. *Soft Matter* **2012**, *8*, 9092. [CrossRef]
65. Siripatrawan, U.; Kaewklin, P. Fabrication and characterization of chitosan-titanium dioxide nanocomposite film as ethylene scavenging and antimicrobial active food packaging. *Food Hydrocoll.* **2018**, *84*, 125–134. [CrossRef]
66. Zhang, X.; Xiao, G.; Wang, Y.; Zhao, Y.; Su, H.; Tan, T. Preparation of chitosan-TiO_2 composite film with efficient antimicrobial activities under visible light for food packaging applications. *Carbohydre Polym.* **2017**, *169*, 101–107. [CrossRef]
67. Ngo, D.-H.; Kim, S.-K. Antioxidant Effects of Chitin, Chitosan, and Their Derivatives. *Adv. Food Nutr. Res.* **2014**, *73*, 15–31. [CrossRef]
68. Xue, C.; Yu, G.; Hirata, T.; Terao, J.; Lin, H. Antioxidative Activities of Several Marine Polysaccharides Evaluated in a Phosphatidylcholine-liposomal Suspension and Organic Solvents. *Biosci. Biotechnol. Biochem.* **1998**, *62*, 206–209. [CrossRef]
69. Gülçin, I.; Huyut, Z.; Elmastaş, M.; Aboul-Enein, H.Y. Radical scavenging and antioxidant activity of tannic acid. *Arabian J. Chem.* **2010**, *3*, 43–53. [CrossRef]
70. Moradi, M.; Tajik, H.; Rohani, S.M.R.; Oromiehie, A.R.; Malekinejad, H.; Aliakbarlu, J.; Hadian, M. Characterization of antioxidant chitosan film incorporated with Zataria multiflora Boiss essential oil and grape seed extract. *LWT* **2012**, *46*, 477–484. [CrossRef]

Article

Characterization and Application in Packaging Grease of Gelatin–Sodium Alginate Edible Films Cross-Linked by Pullulan

Shuo Li [1], Min Fan [1,2], Shanggui Deng [3] and Ningping Tao [1,2,*]

1 College of Food Science and Technology, Shanghai Ocean University, Shanghai 201306, China
2 Shanghai Engineering Research Center of Aquatic-Product Processing & Preservation, Shanghai 201306, China
3 College of Food and Pharmacy, Zhejiang Ocean University, Zhoushan 316000, China
* Correspondence: nptao@shou.edu.cn; Tel.: +86-021-65-710-706

Abstract: Gelatin–sodium alginate-based edible films cross-linked with pullulan were prepared using the solution casting method. FTIR spectroscopy demonstrated the existence of hydrogen bonding interactions between the components, and scanning electron microscopy observed the component of the films, revealing electrostatic interactions and thus explaining the differences in the properties of the blend films. The best mechanical properties and oxygen barrier occurred at a 1:1 percentage of pullulan to gelatin (GP11) with sodium alginate dosing for modification. Furthermore, GP11 demonstrated the best thermodynamic properties by DSC analysis, the highest UV barrier (94.13%) and the best oxidation resistance in DPPH tests. The results of storage experiments using modified edible films encapsulated in fresh fish liver oil showed that GP11 retarded grease oxidation by inhibiting the rise in peroxide and anisidine values, while inappropriate amounts of pullulan had a pro-oxidative effect on grease. The correlation between oil oxidation and material properties was investigated, and water solubility and apparent color characteristics were also assessed.

Keywords: pullulan; gelatin; sodium alginate; oil oxidation; edible film; grease packaging

1. Introduction

The oxidative stability of oily foods, especially those containing high levels of unsaturated fatty acids, is of significant concern. Traditional stabilization technologies (including the removal of oxygen and catalysts) and addition of antioxidants should be further studied for their safety, synergistic effect due to the combination of different components, as well as the use of packaging material [1]. Furthermore, food packaging systems' primary function is to separate food from the surrounding environment, reducing interaction with spoilage factors (such as microorganisms, water vapor, oxygen and off-flavors) and avoiding losses of desirable compounds [2]. The factors or mechanisms that influence the oxidation of oils and fats are not only oxygen but also temperature [3], light (photosensitive oxidation), ions [4], enzymes (enzymatic oxidation), etc. [5]. Studies on packaging to retard oxidation of fats and oils have focused on foods with high fat content, such as meat [6,7] or systems such as emulsions and microencapsulation [8,9], while no experiments have been reported on packaging materials in direct contact with pure oils in bags.

In China, there is a large market for convenience foods, such as instant noodles or some vermicelli, soups, etc., accompanied by small packages of seasonings or flavored oils. Usually, the material used to seal this type of food is mainly plastic. It is inconvenient to squeeze clean, polluting the environment and causing waste during its usage. This is because these materials are non-biodegradable and non-renewable [10]; less than 3% of waste plastics are recycled globally [11]. If the packaging material of this type of food is edible, it will solve the above problems and make small bags of seasoning with convenience similar to "laundry gel". The material needs to have good hydrophilic and oleophobic properties and meet the strength requirements in transportation to achieve this. Currently, most of the reported raw materials for developing edible packaging materials are edible polymers, mainly composed of polysaccharides, proteins and lipids, which can be easily consumed by animals and humans without causing harm to the body [12]. In addition, improving the mechanical properties, light, water and gas permeability and functionality of the materials are the primary motivations for developing polysaccharide, protein and lipid composites. These materials can be designed to extend the shelf life of food products [13,14]. Edible films are mainly formed using the gelation properties of edible polymers, often held together by electrostatic, hydrogen bonding, hydrophobic, van der Waals forces or a combination thereof, such as gels prepared from botanical hydrocolloids [15] polysaccharides, protein or making starch-based films [16–18].

As a food-grade material, gelatin is widely sourced, it is made from partial hydrolysis of collagen, it has good UV absorption and it is often used as the primary material for edible packagings [19], such as capsule shells and microcapsule wall materials [1]. Films made from pure gelatin are transparent, odorless and have low oxygen permeability, but are fragile such as glass, thus often need to be modified to obtain functionality [20,21], mainly in mechanical properties or antimicrobial applications [22,23], and they are mostly applied as polysaccharide-protein structures such as using cellulose. [24]. Sodium alginate is an anionic marine polysaccharide containing β-(1→4)-linked D-mannuronic acid and α-(1→4)-linked L-gulonate residues [25]. It has good gelation properties. Alginate can form dimers with divalent cations and subsequently form weak inter-dimeric aggregates governed by electrostatic interactions, then alginate will be negatively charged in solution [26]. Thus, people could apply electrostatic interactions to cross-link it with other components.

Pullulan is a microbial polysaccharide consisting of α-(1,6) glycosidic units interlinked with maltotriose units, synthesized from the starch of the sprouting stunt mold [27]. The oleophobic nature of the several hydroxyl groups contained in the structure is the property that we aim to exploit. Pullulan is highly water-soluble and capable of developing edible films. These films are colorless, odorless, heat sealable, water permeable, transparent and have low oxygen permeability. It has been reported that people have used it to keep some foods fresh, for example, developing edible coating for strawberry preservation, edible film for brussels sprouts, etc. [28,29].

Based on the above, this study aims to investigate the interaction of pullulan with gelatin and edible film components, investigate the effect of pullulan on the physicochemical properties of edible gelatin films modified by sodium alginate and provide an option for producing edible seasoning packets. Pullulan itself is not charged, so it would be subjected to charged polymers, and the weakly alkaline environment provided by sodium tripolyphosphate would make gelatin negatively charged; thus, the components would self-assemble and form a film due to electrostatic interaction. FTIR and microstructural observations were used to investigate the interaction of components. Mechanical properties were measured by tensile strength and elongation at break. Oxygen permeability and light barrier were also measured due to the need to prevent the oxidation of oils. Water solubility was measured to meet the film's ready-to-use characteristics. DSC was used to analyze thermal stability and color characteristics, and oxidation resistance was also measured.

2. Materials and Methods

2.1. Materials and Reagents

Pullulan $(C_{37}H_{62}O_{30})_n$, $(M_W \approx 2.5 \times 10^5$ Da) was obtained from Macklin Biochemical Co., Ltd. (Shanghai, China). Gelatin was developed from bovine skin, provided by Solarbio Science & Technology Co., Ltd. (Beijing, China). Sodium alginate $(C_6H_7NaO_6)_n$, F.W. $(198.11)_n$, viscosity (10 g/L, 20 °C) ≥ 0.02, containing 30.0–35.0% ash, was obtained from Sinopharm Chemical Group (Shanghai, China). Sodium tripolyphosphate, Mw = 367.86, and food-grade glycerol were purchased from Macklin Biochemical Co., Ltd. (Shanghai, China). Fish liver was provided by Zhongyang Ecological Fish Co., Ltd. (Jiangsu, China). Soybean oil was purchased from Yihai Jiali Grain & Oil Food Co., Ltd. (Shanghai, China). 2,

2-Diphenyl 1-Picrylhydrazyl (DPPH), P-anisidine, 1, 1, 3, 3-tetra ethoxy propane (purity ≥ 97%), sodium thiosulfate standard solution, trichloroacetic acid, trichloromethane and other reagents were of analytical grade and purchased from Macklin Biochemical Co., Ltd. (Shanghai, China).

2.2. Preparation of Edible Films

The edible films were prepared by the solution casting method [29–31]. Briefly, gelatin, pullulan, sodium alginate and sodium tripolyphosphate were dissolved in distilled water at 50–60 °C, respectively, where sodium tripolyphosphate can be used as a moisture-retaining agent [32–34] and also as a cross-linking agent to enhance protein cross-linking [35–38]. Glycerol was added as a plasticizer. Sodium alginate, sodium tripolyphosphate and glycerol were used in equal amounts in each group, and the gelatin mixed with pullulan solutions were prepared in the proportions of 0:1, 1:3, 1:1, 3:1 and 1:0 (w/w), named GP01, GP13, GP11, GP31 and GP10. Pure gelatin (GEL) and pure pullulan (PUL) membranes were prepared for comparison. The specific formulation for each group is shown in Table 1 (on a dry weight basis). All components were mixed in proportion to develop a film-forming solution, and the film-forming solution was sonicated and degassed for 15 min to remove air bubbles, then pipetted onto clean and dry Petri dishes (Φ = 150 mm), left at room temperature for three days to stabilize and tested within three days. Each was made from 25 mL of film-forming solution per film.

Table 1. Films Component.

Component (%)	GP01	GP13	GP11	GP31	GP10	GEL	PUL
Gelatin	0	19.44	38.89	58.33	77.77	100	0
Pullulan	77.77	58.33	38.89	19.44	0	0	100
Sodium alginate	5.56	5.56	5.56	5.56	5.56	0	0
Sodium tripolyphosphate	11.11	11.11	11.11	11.11	11.11	0	0
Glycerol	5.56	5.56	5.56	5.56	5.56	0	0

2.3. Characterization

2.3.1. Thickness

Film thickness was measured using a digital spiral micrometer (Aladdin Bio-Chem Technology Co., Ltd, Shanghai, China) with 0.001 mm. The average thickness (taken from five random locations on each sample) was used to determine the film's mechanical properties.

2.3.2. Mechanical Property

The tensile strength (TS) and elongation at break (EAB) of the films were tested with an intelligent electronic tensile tester (XLW (EC), Jinan, China) at 25 °C and 90% relative humidity (RH). The samples were cut into 110 × 15 mm strips with an initial clamping distance of 50 mm and stretched at a speed of 50 mm/min. The TS and EAB were calculated by the following equations:

$$TS\ (MPa) = \frac{F}{h \times d}, \quad (1)$$

$$EAB\ (\%) = \frac{L - L_0}{L_0} \times 100\%, \quad (2)$$

where F is the maximum force (N), h is the thickness (mm), d is the width of the membrane (mm), L is the length of the membrane when it breaks (mm) and L_0 is the initial length of the membrane (mm) [39]. Each film sample was measured five times.

2.3.3. Oxygen Transmission Rate (OTR)

The OTR of the films was measured by a gas permeability tester (Labthink, G2/132, Jinan, China). Film samples were prepared with an area of Φ = 97 mm, and the test chamber was set at 23 ± 0.1 °C with 50% ± 1% RH. Each film was determined by the differential oxygen pressure method for 12h. Each film sample was measured five times. The results were presented as OTR (cm^3/(m^2·24 h·0.1 MPa)).

2.3.4. Water Solubility (WS)

The WS was measured using the method of Zhou et al. [39] with some modifications. Briefly, the samples were cut into film sheets (2.0 × 2.0 cm) and placed in 10 mL of deionized water, stirred with a magnetic stirrer until completely dissolved and the dissolution time was recorded. The solubility of the water was calculated from the following equation:

$$WS = \frac{m}{t}, \qquad (3)$$

where m is the mass of the membrane sheet (g), and t is the time taken for it to dissolve (s). Each film sample was measured five times.

2.3.5. SEM

The microscopic morphology of the films was observed with a scanning electron microscope (Hitachi S3400, Tokyo, Japan). Samples were embrittled with liquid nitrogen and then placed in ion sputtering and gold sprayed at 15 mA for 60 s to improve their conductivity. The cross-sectional structure of the different modified films and the compatibility of the components were observed at an accelerating voltage of 5 kV.

2.3.6. Fourier Transform Infrared (FTIR) Spectroscopy Analysis

The chemical composition of the films was analyzed using FTIR spectroscopy (Thermo Nicolet Corporation, Waltham, MA, USA). Each film was scanned 64 times from 4000 to 500 cm^{-1} with a scanning interval of 4 cm^{-1} and the air spectrum was used as a background correction.

2.3.7. Thermal Stability

Differential scanning calorimetry (DSC) tests were carried out using a DSC furnace (Q2000, TA instruments, New Castle, DE, USA). Film samples (5–10 mg) were placed in aluminum pans and heated from 30 °C to 220 °C. The ramp rate was 10 °C/min, and the nitrogen flow rate was 50 mL/min.

2.3.8. Transparency and Color

The transparency of the films was studied in the wavelength range of 365–940 nm using a solar film transmission meter (LS101, Lin Shang Technology Company, Guangdong, China). The visible light transmission rate (VLT, 380–760 nm), ultraviolet light rejection rate (UVR, 365–380 nm) and infrared light rejection rate (IRR, 760–940 nm) were analyzed using air as a blank. Each film sample was measured five times.

The color parameters (L*, a*, b*) of the films were measured with a colorimeter. (Konica Minolta, CR-400, Tokyo, Japan). L* values indicate black/white (0/100), a* values indicate green/red (−80/100) and b* values indicate blue/yellow (−80/70). According to Chen et al. [40], The film sample was placed on top of a white standard plate (L* = 94.69, a* = −1.98, b* = −0.93) for testing. Each film sample was measured five times. The total color difference (ΔE) was calculated according to the following equation:

$$\Delta E = \sqrt{(\Delta L^*)^2 + (\Delta a^*)^2 + (\Delta b^*)^2}. \qquad (4)$$

2.3.9. DPPH Antioxidant Analysis

The DPPH antioxidant analysis was determined according to the method modified by Jiang et al. [41]. Five pieces (25 mg each) of sample film were cut and dissolved in a test tube containing 5 mL of 75% ethanol and then 5 mL of DPPH solution was added. The reaction solution was sealed and protected from light at 60 °C for 2 h. The DPPH radical scavenging ability of the membrane solution was measured at 517 nm. The formula can be expressed as follows:

$$DPPH = \left(1 - \frac{A_i - A_j}{A_0}\right) \times 100\%, \quad (5)$$

where A_i is the absorbance of the DPPH solution mixed with the membrane solution, A_j is the absorbance of the membrane solution and A_0 is the absorbance of the DPPH solution.

2.4. Application of Films in Fish Liver Oil Storage

2.4.1. Preparation and Bagging of Fish Liver Oil

To avoid the antioxidant effect of food additives in the oil products available on the market [9], we used fish liver with oil content (35.23 ± 1.76%) to develop self-made liver oil, which was prepared as follows: 10% soybean oil and 90% fish liver were weighed and fried at 140 °C for 2 min, then the liver oil was collected. The initial peroxide value (PV), anisidine value (An.V), malondialdehyde (MDA) content and total oxidation value (TOV) of the produced liver oil were measured.

The determination of the antioxidant properties of liver oil was carried out in our laboratory in advance, using a Racimat lipid oxidation stability analyzer (Aptar, Switzerland) to determine its oxidation endpoints. The liver oil oxidation curves were obtained by the Q10 extrapolation method [42] as follows:

$$t = 4313 \times e^{-0.06968 \times T} \quad R^2 = 0.99883, \quad (6)$$

where T is the temperature (°C), and t is the storable time (h).

The resulting film materials were cut and heat-sealed using two pieces to develop bags of 8 × 8 cm, filled with 10 g of liver oil, sealed and accelerated in an oven at 40 °C. The experimental temperature was substituted into Equation (6), and we obtained t = 265.65 h. All bags were stored for 12 days, with oil with no encapsulation as control group 1 (blank) and PVC plastic sealed bags as control group 2 (control), and samples were taken every 3 days for testing. The GP01 film was too self-absorbent to meet the bag-making conditions and was not subjected to encapsulation.

2.4.2. Determination of Malondialdehyde (MDA) in Liver Oil

The determination of MDA in oil was conducted according to the China National Food Safety Standard GB5009.181-2016. The absorbance at 532 nm was measured and quantified by comparison with a standard series curve using the following formula [43]:

$$X = \frac{c \times V}{m}, \quad (7)$$

where X is the amount of MDA in the sample (mg/kg); c is the concentration of MDA in the sample solution obtained from the standard series curve (µg/mL); V is the volume of the sample solution (mL); m is the mass of the sample represented by the final sample solution (g). The initial MDA content of the liver oil was 1.25 ± 0.09 mg/kg.

2.4.3. Determination of Peroxide Value (PV), Anisidine Value (An.V) and Total Oxidation Value (TOV) in Cod Liver Oil

The PV was determined by titration with sodium thiosulphate standard solution, reference to the Chinese National Food Safety Standard GB 5009.227-2016, and the results were presented in mmol/kg. The initial PV of the liver oil produced was 6.6 ± 0.35 mmol/kg.

The An.V of liver oil was measured in dimensionless units according to the International Food Safety Standard ISO6885: 2006 and the Chinese National Food Safety Standard GB/T 24304-2009: Animal and vegetable fats and oils—Determination of anisidine value.

The total oxidation value was calculated by the following formula [44], which helps to evaluate the oxidative deterioration of the oil.

$$TOV = PV + 4An.V. \tag{8}$$

The initial $An.V$ of the resulting liver oil was 2.31 ± 0.62. The initial TOV was 28.72 ± 2.00.

2.5. Statistical Analysis

Experimental data during storage were obtained from independent triplicates. The data were compared using one-way analysis of variance (ANOVA) using SPSS 26 software (SPSS Institute Inc., Chicago, IL, USA), and $p < 0.05$ was considered to have significant differences between the data. The data were expressed as mean ± standard deviation (SD). OriginPro 2021 (Origin Lab, Northampton, MA, USA) was used to process and generate images.

3. Results and Discussion

3.1. FTIR Analysis

Figure 1 show the FTIR spectra of the pure gelatin film (GEL), the pure pullulan film (PUL) and the composite films with different gelatin–pullulan ratios. The specific functional group shared by GEL and PUL was a broad peak at 3283 cm^{-1} (O-H stretching vibration). The typical functional group of the GEL was 1636 cm^{-1} (C=O stretching of amide-I) and 1538 cm^{-1} (N-H stretching of amide-II). While the typical functional group of the GEL was at 1108 cm^{-1} (C-O stretching vibration of the primary alcohol), the strong peak at 1010 cm^{-1} of the PUL was caused by the C-O-C stretching of the (1–4) glycosidic bond linkage [22,45].

For PUL and GEL membranes, the peaks around 3300 cm^{-1} and 2937 cm^{-1} are related to the stretching vibration of –OH and the C-H contraction vibration in the methyl group, respectively [27,31]. The intensity of the peaks can reflect the strength of intermolecular hydrogen bonding. It can be seen that the intensity of the broad peak at 3283 cm^{-1} of GP10 co-blended membrane was not high, indicating that the intermolecular hydrogen bonding was not strong, while the intensity of this peak increased after the addition of pullulan, indicating that the intermolecular hydrogen bonding was enhanced [39], Han et al. [27] found that the conjugation of pullulan with egg white protein could increase the absorption of the hydroxyl stretching band (3300 cm^{-1}).

Secondly, changes in the peaks of the blend films at 1636 cm^{-1} and 1538 cm^{-1} were also evident, indicating C=O stretching in the amide-I band, bending and stretching of N-H bonds or C-N stretching in the amide-II band, respectively [6]. The curves of the GP01 showed that C=O bonds could be formed between pullulan and sodium alginate even without the addition of gelatin. In the multiple gelatin structures, the pullulan was able to ensure the structural stability of the amide I and II bands, such as GP11 and GP31. Meanwhile, the effect of pullulan on the amide II band was more evident in the multiple PUL structures, indicating that the microstructure of gelatin was changed. This was demonstrated by subsequent experiments. In turn, the gelatin in the blended films had a similar effect on the characteristic peaks of the pullulan. For example, the change in peak intensity around 1010 cm^{-1} indicating the C-O-C bond, suggested that the structure of the pullulan had also been changed. In addition to this, GP10, GP11 and GP13 also showed new peaks around 887 cm^{-1}, which may be due to the deformation vibration of O-H, while the formation of hydrogen bonds in O-H may be due to the Maillard reaction or the glycosylation of gelatin proteins during the film formation process [27], and the polysaccharide reacting with gelatin in GP10 may be sodium alginate.

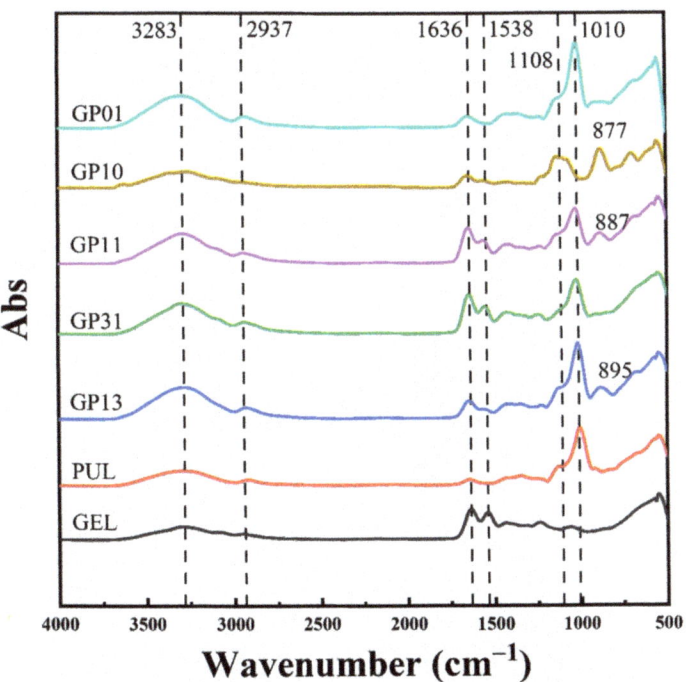

Figure 1. FTIR spectrum of different films.

3.2. Microstructure

Figure 2 show the cross-sectional microstructure of different films. As can be seen from the SEM images, the smooth cross-section of the GP01 indicated that the components were compatible, which was due to a large number of hydrogen bonds between the pullulan and sodium alginate. The GP10 showed a fibrous structure of gelatin protein; meanwhile, the incompatible particles were sodium alginate molecules, mainly on the same side, indicating that the sodium alginate molecules were repelled by electrostatic forces. With the addition of pullulan, the sodium alginate molecules appeared on both sides, indicating that the sodium alginate molecules were still subject to electrostatic repulsion. Due to the co-mixing of the solution, the pullulan and gelatin were bonded to each other through hydrogen bonding and van der Waals forces [16]. However, by observing the GP13 and GP31, we could find that they showed more incompatible granular structures. Our explanation for this is that both the multiple gelatin and pullulan structures could disrupt the electrostatic equilibrium between the charged gelatin molecules and the sodium alginate molecules, resulting in hydrogen bonding being weaker or stronger than the charge interactions. Specifically, the GP13 was a multiple pullulan structure whose hydrogen bonding interaction with gelatin was greater than the electrostatic repulsion of gelatin by sodium alginate leading to agglomeration of gelatin molecules into globular protein rolling, thus reducing compatibility [24]. Comparatively, the higher content of gelatin in the GP31 had a greater effect on the electrostatic attraction and hydrogen bonding of pullulan than that of sodium alginate, resulting in the agglomeration of pullulan molecules and reduced compatibility. In this regard, Figure 3 explain the reasons for this phenomenon.

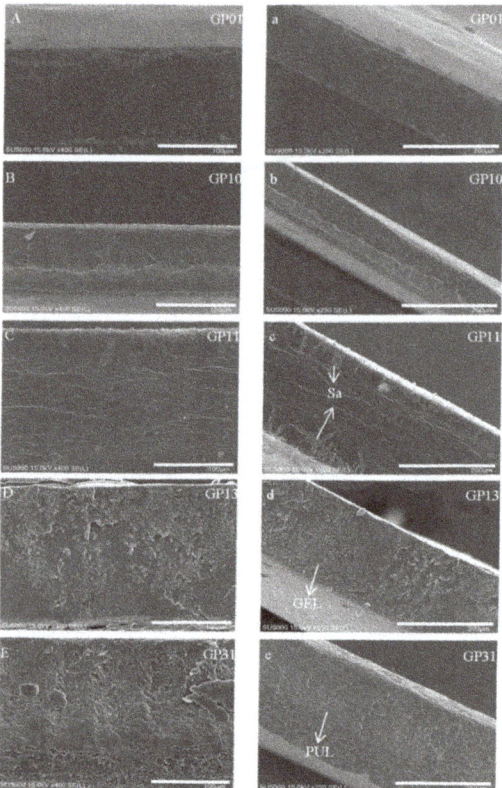

Figure 2. SEM images of the cross-sectional microstructure of different films. (**a**–**e**) are 250× magnification, and (**A**–**E**) are 400× magnification (PUL: Pullulan, GEL: Gelatin, Sa: Sodium alginate).

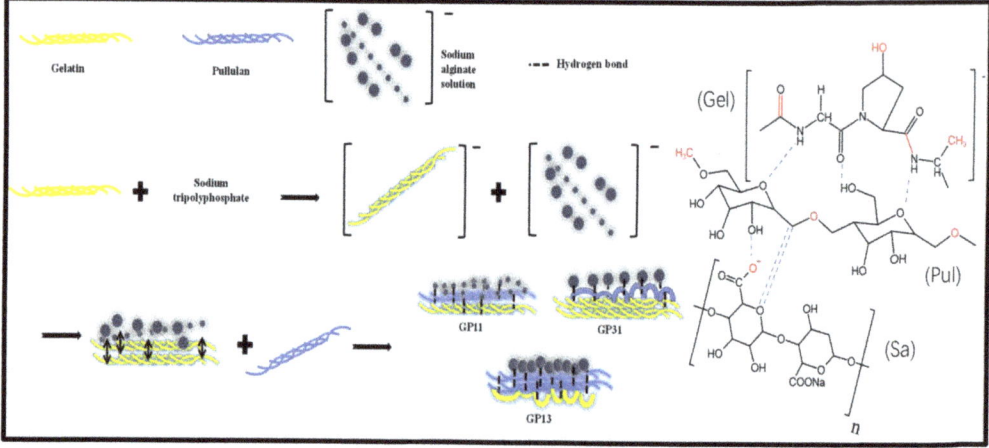

Figure 3. Interaction of pullulan with sodium alginate–gelatin-based films (Schematic diagram of film formation mechanism, Pul: Pullulan, Gel: Gelatin, Sa: Sodium alginate).

3.3. Thickness

The thicknesses of the films developed in each ratio are shown in Table 2. The thicknesses varied when the solutions were poured in the same volume and with the same bottom area, probably due to the agglomeration of pullulan forming a non-uniform structure of the gelatin–sodium alginate film matrix in a non-continuous phase [28] and due to the different amounts of gelatin and the different degrees of electrostatic interaction with the anionic sodium alginate molecules, resulting in their different thicknesses. The GP11 film, for example, was the thickest at 0.201 mm, indicating that the parallel bonding of the polysaccharide molecular chain segments to the fibrin of the gelatin was good and that the uniform interaction gave it a thicker thickness, which had a corresponding effect on its permeability.

Table 2. Thickness, OTR and WS of various films.

Film Groups	Thickness (mm)	OTR (cm^3/(m^2·24 h·0.1 MPa))	WS (g/s)
GP11	0.201 ± 0.018 [a]	47.42 ± 13.73 [ab]	0.10 ± 0.03 [ab]
GP31	0.179 ± 0.008 [b]	54.47 ± 8.80 [a]	0.11 ± 0.01 [a]
GP10	0.149 ± 0.006 [c]	57.69 ± 10.31 [a]	0.07 ± 0.01 [ab]
GP13	0.145 ± 0.004 [c]	49.06 ± 5.26 [ab]	0.10 ± 0.02 [ab]
GP01	0.138 ± 0.016 [c]	-	0.07 ± 0.02 [b]
GEL	0.177 ± 0.009 [b]	35.86 ± 3.37 [b]	0.07 ± 0.01 [b]
PUL	0.143 ± 0.009 [c]	19.09 ± 3.45 [c]	0.09 ± 0.01 [ab]

Abbreviations: Different letters in the same column indicate significant differences ($p < 0.05$). "-" indicates undetectable.

3.4. Oxygen Permeability

Table 2 also show the oxygen permeability of the different films. The determination of the oxygen permeability of the films is of great importance as it is an investigation into the role of preventing the oxidation of grease [1,3]. Of these, the GP01 was not tested this time because it was almost impossible to unfold due to its strong self-adhesive properties. We speculated that the reason for this phenomenon was that the GP01 had a large number of hydrogen bonds in the ratios of pullulan and sodium alginate. Furthermore, glycerol as a plasticizer also had hydrogen bonds, and there was no gelatin to bind the hydrogen bonds; hence the phenomenon of self-adhesion appeared after film formation. From the OTR, we can see that the PUL was a more substantial oxygen barrier than the GEL, which might prove that the pullulan film had a more dense structure than the gelatin film. The oxygen permeability of the GP10 modified with sodium alginate increased by 60.87% compared to the gelatin film, which did not favor the protection of oils and fats. The addition of pullulan increased the oxygen barrier of GP11, GP31 and GP13 by 17.80%, 5.58% and 14.96%, respectively. We found that although pullulan has a dense structure, the higher content of pullulan did not have a better oxygen barrier. The difference between the GP11 and GP13 was not significant ($p > 0.05$). This may be because the unsuitable ratio of gelatin to pullulan led to uneven interaction forces, causing agglomeration of polysaccharide molecules, which affected the compatibility, thus creating gaps in the membrane material and enhancing oxygen permeability [18].

3.5. Solubility

As can be seen from the solubility of the films with different ratios in Table 2, there was no significant difference between most of the films ($p > 0.05$), indicating that the use of sodium alginate, pullulan cross-linked with gelatin did not affect the sensitivity of the material to water. On the one hand, this was because the various raw materials are themselves hydrophilic edible polymers, precisely because the polar part of the membrane material (e.g., hydroxyl functional groups) hydrates with water molecules and forms hydrogen bonds. On the other hand, the membrane material itself also has free hydrogen bonds, which could offer the possibility of subsequent use of such membrane materials in combination with active substances to enrich their functionality. The water solubility of gelatin or pullulan is also a significant requirement for producing packaged edible pouches.

3.6. Mechanical Properties

Figure 4 assess the mechanical properties of the films with different ratios which correlates with the structural integrity, strength and flexibility of the film, including tensile strength (TS, MPa) and elongation at break (EAB, %). Both PUL and GEL exhibited high TS and low EAB, consistent with their inherent brittle nature [19,29]. The GP10 modified film without the addition of pullulan showed a significant decrease in TS ($p < 0.05$) but an increase in EAB to 45%, which was attributed to the modification of gelatin by sodium alginate and glycerol, presumably due to the homo-charge repulsion of the anion-aggregated alginate and negatively charged gelatin molecules resulting in a decrease in strength, while the cross-linked structure of gelatin was the main reason for maintaining its ductility. With the addition of pullulan, the TS of GP11, GP13 and GP31 films all improved, and the groups with more pullulan had higher EAB (GP11, GP13), indicating that pullulan could improve the mechanical properties of gelatin-based films, the multiple pullulan structures had higher EAB, relatively, the multiple gelatin structures had stronger TS, the improvement in mechanical properties was related to the reduction of particles formed by agglomeration of polysaccharide, the formation of hydrogen bonds and the formation of new bonds between polysaccharides and proteins through the Maillard reaction [27]. Interestingly, there was no significant difference in mechanical properties between the GP01 and the GP10, suggesting that the hydrogen bonding enriched by pullulan and sodium alginate in the formulation of the GP01 was the main reason for maintaining the mechanical properties of the film.

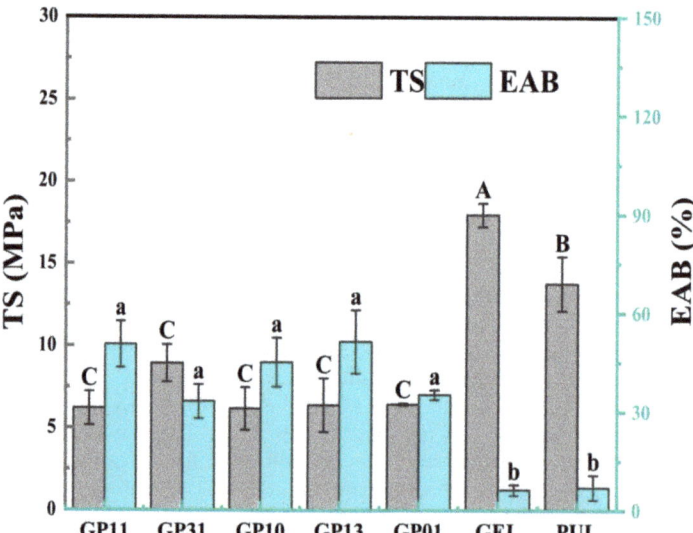

Figure 4. Tensile strength (TS) and elongation at break (EAB) of different films. (Different letters in the same data indicate significant differences ($p < 0.05$).).

3.7. Thermodynamic Properties

The thermal denaturation temperature (T_d), the melting temperature (T_m), the latent heat of thermal denaturation (ΔH_d) and the latent heat of melting (ΔH_m) of different film materials were determined by DSC analysis. The DSC heat flow diagrams of GEL, PUL and composite films with different gelatin–pullulan ratios are shown in Figure 5.

As can be seen from Figure 5, for the GEL and PUL, the first small heat absorption peak was in the evaporated water binding in the membrane, indicating the start of melting, with a T_d of 138.01 °C and 148.14 °C, respectively, while the GP01 had a very high T_d value, reaching 175.66 °C. The increase in ΔH_d indicated a large amount of hydrogen bonding in the membrane molecules. The thermodynamic properties of the GP11 were very similar to those of the GP01, indicating that the GP11 film's molecules also had strong hydrogen bonding. The GP10, GP31 and GP13 showed relatively similar thermodynamic properties. They did not exhibit significant heat absorption peaks, which was considered due to the high content of bound water in the films [46]. The ΔH_d of the GP01, GP10, GP31, GP13 and GP11 modified films were 2.33, 1.48, 2.25, 2.10 and 1.47 times higher than that of the pullulan film and 3.62, 2.30, 3.50, 3.26 and 2.28 times higher than that of the GEL, indicating that the modified films had intermolecular interactions, including but not limited to hydrogen bonding and van der Waals forces [47]. The GP11 had the highest melting and thermal denaturation temperatures, indicating that this ratio of pullulan was the most compatible with gelatin and also had more appropriate hydrogen bonding and intermolecular interactions.

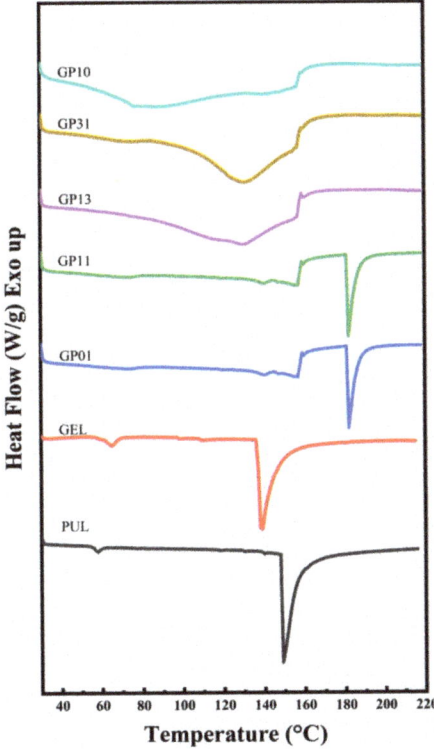

Figure 5. DSC heat flow diagrams for different films.

3.8. Color and Light Transmission

Color plays a critical role in food packaging, influencing consumer choice [27]. Material light transmission, on the other hand, is closely related to the photosensitivity of the packaging contents. Figure 6 and Table 3 demonstrate the optical properties of different films, including color and optical transmittance. As it can be seen from the table, GEL had the smallest color difference values (ΔE), indicating the best transparency. Thus, the transparency gradually increased with the addition of pullulan, the GP01 being the best transparent among the films containing pullulan. The brightness of the films with multiple gelatin structures was lower; for example, the L* value of the GP31 was only 89.07. The b* value of the GEL was higher than that of the PUL, indicating a more yellowish color, which was determined by the nature of gelatin [30]; therefore, yellow was the base color of all films, and the multiple gelatin structures had a more yellow color, which changed from yellow to blue with the addition of pullulan [27]. In terms of a* value, the film color was redder as the pullulan increased, from −2.16 in GP11 to −1.21 in GP13.

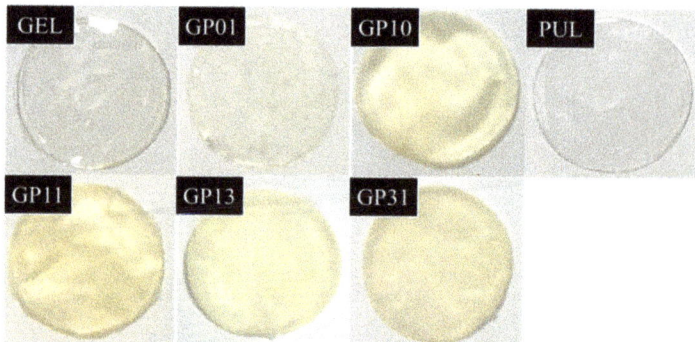

Figure 6. Images of film samples.

Table 3. Color parameters and optical properties of different films.

	L*	a*	b*	ΔE	VLT (%)	UVR (%)	IRR (%)
GP11	90.92 ± 0.52 [ab]	−2.16 ± 0.06 [c]	11.53 ± 0.56 [b]	13.03 ± 0.50 [b]	27.07 ± 4.21 [c]	94.13 ± 3.50 [a]	57.67 ± 4.41 [ab]
GP31	89.07 ± 2.79 [b]	−2.56 ± 0.41 [d]	15.33 ± 3.36 [a]	17.26 ± 4.07 [a]	42.63 ± 14.68 [bc]	75.17 ± 27.89 [a]	69.23 ± 18.08 [a]
GP10	92.78 ± 0.07 [a]	−2.09 ± 0.09 [c]	10.33 ± 0.60 [b]	11.42 ± 0.60 [bc]	56.77 ± 4.01 [b]	72.17 ± 10.96 [a]	28.93 ± 3.72 [cd]
GP13	93.26 ± 0.15 [a]	−1.21 ± 0.10 [b]	6.45 ± 0.50 [c]	7.56 ± 0.51 [d]	66.57 ± 9.94 [ab]	71.9 ± 11.23 [a]	20.93 ± 6.79 [de]
GP01	91.74 ± 1.71 [ab]	−1.14 ± 0.20 [b]	6.96 ± 1.16 [c]	8.53 ± 1.64 [cd]	45.5 ± 32.09 [bc]	69.17 ± 23.69 [a]	44.53 ± 20.74 [bc]
GEL	93.08 ± 0.19 [a]	−1.44 ± 0.07 [b]	7.87 ± 0.47 [c]	8.96 ± 0.43 [cd]	90.7 ± 0.95 [a]	30.6 ± 2.44 [b]	8.63 ± 0.81 [e]
PUL	92.8 ± 2.56 [a]	−0.57 ± 0.10 [a]	3.87 ± 0.17 [d]	5.69 ± 1.00 [d]	91.93 ± 0.06 [a]	10.57 ± 0.21 [b]	8.23 ± 0.55 [e]

Abbreviations: Different letters in the same column indicate significant differences ($p < 0.05$).

The light transmission properties of the films include visible light transmission rate (VLT), ultraviolet light rejection rate (UVR) and infrared light rejection rate (IRR), with the GEL films having the highest visible light transmission, which was consistent with the color analysis results. The GP11 surprised us by having the lowest VLT and the highest UVR among the blended films, which was good news for oil bagging [3]. The UV barrier of the blended films was more significant than that of the GEL and PUL films, but the color difference results did not indicate that the darker the color, the better the UV and light barrier (GP31), which might be related to the structural morphology of the gelatin and pullulan molecules, as the GP11 had a parallel bonded fibrin structure with less agglomeration and fewer gaps created, thus would have greater light-blocking properties.

3.9. Films Antioxidant Activity

The DPPH radical scavenging test is a common method to determine the antioxidant capacity of materials [10]. Figure 7 show the antioxidant activity of GEL, PUL and blended films. The higher the DPPH residual value, the antioxidant activity is weaker. From the figure, the antioxidant activity of GEL was the weakest [22], with a DPPH residual of 97.60% and a scavenging rate of only 2.40% for DPPH radicals, while the scavenging rate of 9.24% for DPPH radicals for PUL, probably due to the presence of phenolic hydroxyl groups in pullulan, resulted in better antioxidant activity ($p < 0.05$), from which it could be concluded that the addition of pullulan would improve the antioxidant properties of gelatin-based edible films. This was demonstrated by the fact that the GP10 without the addition of pullulan had a DPPH residual of 94.33%. In comparison, the GP11 and

GP13 had a scavenging rate of 9.92% and 7.21% for DPPH radicals, which were 4.13 and 3.00 times higher than the antioxidant properties of gelatin, respectively, which confirmed the hydroxyl group that appeared to be hydrogen-bonded in the FTIR analysis, precisely because the hydroxyl group formed by pullulan and gelatin molecules had the phenolic hydroxyl-like effect that enhanced the antioxidant properties of the film [48]. This was evident in the GP31, GP13, GP01 and GP11 films with the addition of pullulan.

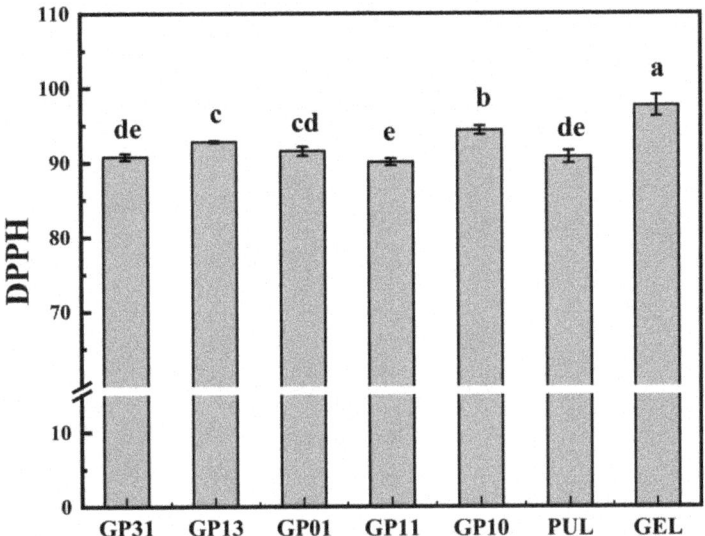

Figure 7. 2,2-Diphenyl-1-picrylhydrazyl (DPPH) radical scavenging activity of different films.

3.10. Protection against Oxidation of Oils and Fats

3.10.1. Changes in Primary Oxidation Products of Liver Oil

Figure 8 show the relevant oxidation indicators during the storage of fish liver oil. Hydroperoxides are the primary oxidation products of oils or fats, and their content is expressed in terms of peroxide values. It can be seen that in the early stage of oxidation (3d), the primary oxidation products were lower in each composite film packaging group than in the blank group (7.82 mmol/kg), probably due to the oxygen barrier. By day 6, the peroxidation values of the GP10 and GP11 groups were significantly lower than those of the other groups. By the middle and late stages (9d and 12d), the GP13 and GP31 groups presented higher peroxidation values, suggesting that an inappropriate pullulan to gelatin ratio could have a pro-oxidation effect [30]. The GP11 group, on the other hand, had a lower peroxide value (7.83 mmol/kg), initially showed a delayed oxidation effect of the oils and fats and had a better effect than the control group, conventional plastic PVC (8.68 mmol/kg) ($p < 0.05$).

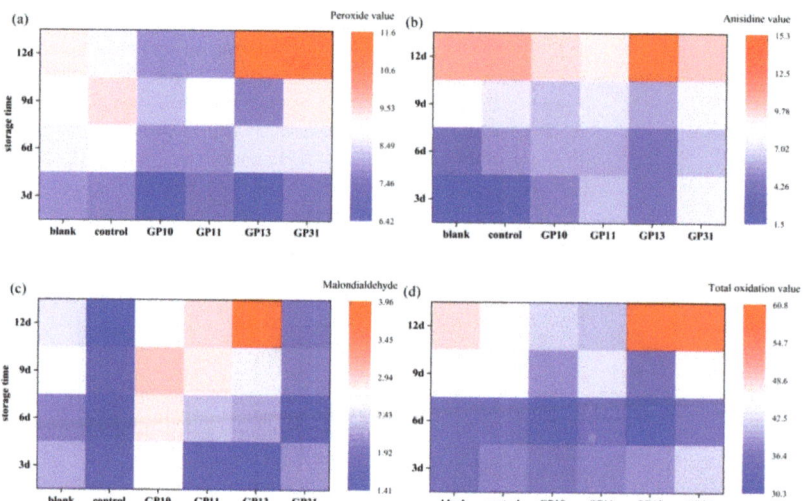

Figure 8. Degree of liver oil oxidation during storage, (**a**) liver oil peroxide value (PV, mmol/kg), (**b**) anisidine value (An.V, dimensionless), (**c**) malondialdehyde (MDA, mg/kg), (**d**) total oxidation value (TOV, dimensionless), at 3d, 6d, 9d, 12d.

3.10.2. Changes in Secondary Oxidation Products of Liver Oil

Hydroperoxides undergo further oxidative decomposition to produce secondary oxidation products. Anisidine is used to detect non-volatile carbonyl compounds such as α and β-unsaturated aldehydes resulting from the oxidation of oils and fats. Malondialdehyde (MDA) is a product of the peroxidation reaction of lipids by the action of oxygen radicals. Both are indicators to characterize the content of secondary oxidation products in oils and fats. As can be seen from Figure 8, the anisidine values (An.V) of 3d for each blended film packaging group were higher than the blank and control, probably because the residual oxygen in the packaging microenvironment of the blended film, after the generation of hydroperoxides, was unable to escape due to the oxygen barrier properties of the films. This residual oxygen continued to remain in the microenvironment for the generation of secondary oxidation products. In contrast, systems where oxygen can diffuse freely were where primary oxidation processes mainly took place, such as the blank group. In the case of MDA, the phenomenon continued until the middle and end of the oxidation. The unsuitable pullulan content of GP13 and GP31 showed a pro-oxidation effect for both primary and secondary oxidation products, while the GP11 group was the better performing group. The GP11 group had the lowest total oxidation value of 41.12, showing good protection against lipids oxidation.

3.11. Correlation Analysis between Resistance to Grease Oxidation and Key Properties of Films

We already know that oxygen and UV light are the two main external factors contributing to grease oxidation [1,3], so it is necessary to explore the effect of the film's shading and oxygen permeability on the four indicators related to grease oxidation. Using Pearson correlation analysis, the results are shown in Figure 9. We found a positive correlation between MDA content and oxygen transmission rate in the early stages of oil bale storage and a negative correlation with UVR, but the effect of OTR was more significant. The better the oxygen permeability, the higher the MDA content in the pre-storage period, which was consistent with the previous experimental results, i.e., the order of oxygen permeability and MDA content on the third day was GP10 > GP31 > GP13 ≈ GP11; meanwhile, the OTR gradually changed to a negative correlation with MDA content as the storage time increased, which explained the higher MDA content in the compounded film packaging group compared to the blank and control groups. The effect of UVR on anisidine was the same as the positive correlation in the early stages, gradually changing to a negative correlation. UVR played a lesser role than OTR in the early stages of oxidation and only started to show its effect in the middle and late stages of oxidation, e.g., by day 9, the higher the UVR, the lower the peroxide value would be. By the late stages of oxidation, UV blocking continued to retard the oxidation of oils and fats.

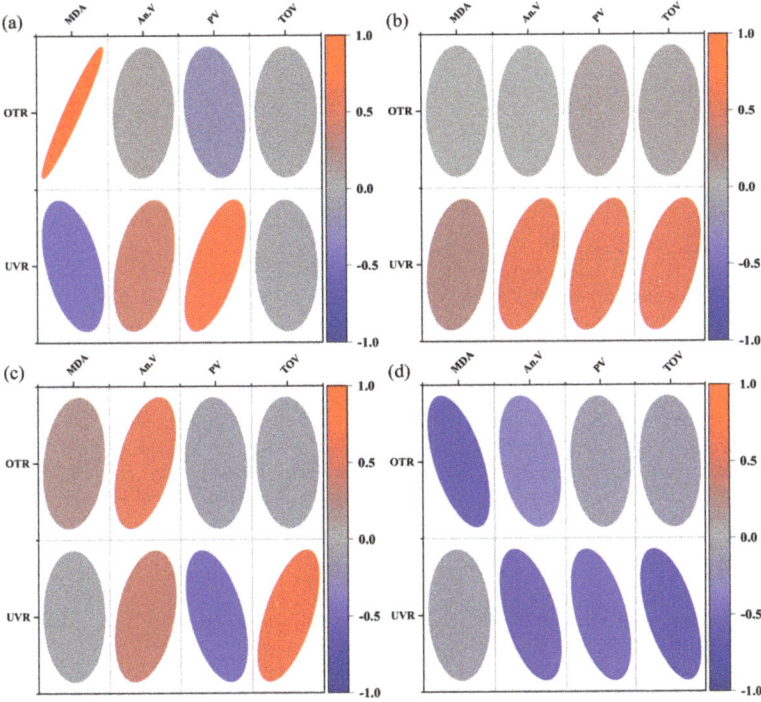

Figure 9. Correlation of grease oxidation indicators with key film properties. (**a**) 3d, (**b**) 6d, (**c**) 9d, (**d**) 12d.

4. Conclusions

In this study, we investigated the film formation mechanism and the effect of resistance to grease oxidation from a practical application. We found that under weak alkaline conditions (provided by sodium tripolyphosphate), the negatively charged gelatin molecules and the polyanionic sodium alginate created space through electrostatic repulsion, providing a place for the cross-linking of pullulan, which would bond with sodium alginate and gelatin molecules through hydrogen bonding. By controlling the ratio of gelatin to pullulan to control the relative strength of hydrogen bonding and electrostatic interactions, we obtained gelatin–sodium alginate–pullulan edible films. A 1:1 ratio of pullulan to gelatin showed the best mechanical strength, oxygen barrier, thermodynamic properties, UV barrier, microstructure observation of parallel bonded fibrous structure and presented the best antioxidant properties. The water solubility also did not differ significantly ($p > 0.05$) from the hydrophilic material. Additionally, it was found that a suitable ratio of pullulan to gelatin (1:1) was able to retard the oxidation of the oil, while the addition of inappropriate pullulan had some pro-oxidation effect. Lastly, the abundant hydrogen bonds in the system can be additionally grafted with antioxidant agents. As a food contact material, the safety and antimicrobial properties of this type of film need to be optimized further, providing alternative options for designing edible inner packaging for convenience foods.

Author Contributions: Conceptualization, S.L. and M.F.; methodology, S.L. and M.F.; formal analysis, S.L.; investigation, S.L.; resources, M.F.; data curation, S.L. and N.T.; writing—original draft preparation, S.L.; writing—review and editing, M.F. and N.T.; visualization, S.L.; supervision, S.D.; project administration, S.D.; funding acquisition, N.T. All authors have read and agreed to the published version of the manuscript.

Funding: This study was supported by the "National Key R&D Program of China" (grant NO. 2020YFD0900905).

Institutional Review Board Statement: Not applicable.

Informed Consent Statement: Not applicable.

Data Availability Statement: Data presented in this study are available on request from the first author.

Conflicts of Interest: The authors declare that there is no conflict of interest in publishing this study.

References

1. Wang, J.; Han, L.; Wang, D.; Sun, Y.; Huang, J.; Shahidi, F. Stability and stabilization of omega-3 oils: A review. *Trends Food Sci. Technol.* **2021**, *118*, 17–35. [CrossRef]
2. Otoni, C.G.; Avena-Bustillos, R.J.; Azeredo, H.M.C.; Lorevice, M.V.; Moura, M.R.; Mattoso, L.H.C.; McHugh, T.H. Recent Advances on Edible Films Based on Fruits and Vegetables—A Review. *Compr. Rev. Food Sci. Food Saf.* **2017**, *16*, 1151–1169. [CrossRef] [PubMed]
3. Choe, E.; Min, D.B. Mechanisms and factors for edible oil oxidation. *Compr. Rev. Food Sci. Food Saf.* **2006**, *5*, 169–186. [CrossRef]
4. Johnson, D.R.; Tian, F.; Roman, M.J.; Decker, E.A.; Goddard, J.M. Development of Iron-Chelating Poly(ethylene terephthalate) Packaging for Inhibiting Lipid Oxidation in Oil-in-Water Emulsions. *J. Agric. Food Chem.* **2015**, *63*, 5055–5060. [CrossRef]
5. Oenel, A.; Fekete, A.; Krischke, M.; Faul, S.C.; Gresser, G.; Havaux, M.; Mueller, M.J.; Berger, S. Enzymatic and Non-enzymatic Mechanisms Contribute to Lipid Oxidation during Seed Aging. *Plant Cell Physiol.* **2017**, *58*, 925–933. [CrossRef]
6. Zhang, M.; Tao, N.; Li, L.; Xu, C.; Deng, S.; Wang, Y. Non-migrating active antibacterial packaging and its application in grass carp fillets. *Food Packag. Shelf Life.* **2022**, *31*. [CrossRef]
7. Sutthasupa, S.; Padungkit, C.; Suriyong, S. Colorimetric ammonia (NH$_3$) sensor based on an alginate-methylcellulose blend hydrogel and the potential opportunity for the development of a minced pork spoilage indicator. *Food Chem.* **2021**, *362*, 130151. [CrossRef]
8. Dierings de Souza, E.J.; Kringel, D.H.; Guerra Dias, A.R.; da Rosa Zavareze, E. Polysaccharides as wall material for the encapsulation of essential oils by electrospun technique. *Carbohydr. Polym.* **2021**, *265*, 118068. [CrossRef]
9. Hamed, S.F.; Hashim, A.F.; Abdel Hamid, H.A.; Abd-Elsalam, K.A.; Golonka, I.; Musial, W.; El-Sherbiny, I.M. Edible alginate/chitosan-based nanocomposite microspheres as delivery vehicles of omega-3 rich oils. *Carbohydr. Polym.* **2020**, *239*, 116201. [CrossRef]
10. Jiang, J.; Dong, Q.; Gao, H.; Han, Y.; Li, L. Enhanced mechanical and antioxidant properties of biodegradable poly (lactic) acid-poly(3-hydroxybutyrate-co-4-hydroxybutyrate) film utilizing α-tocopherol for peach storage. *Packag. Technol. Sci.* **2020**, *34*, 187–199. [CrossRef]
11. Mohamed, S.A.A.; El-Sakhawy, M.; El-Sakhawy, M.A. Polysaccharides, Protein and Lipid -Based Natural Edible Films in Food Packaging: A Review. *Carbohydr. Polym.* **2020**, *238*, 116178. [CrossRef]
12. Kouhi, M.; Prabhakaran, M.P.; Ramakrishna, S. Edible polymers: An insight into its application in food, biomedicine and cosmetics. *Trends Food Sci. Technol.* **2020**, *103*, 248–263. [CrossRef]
13. Umaraw, P.; Munekata, P.E.S.; Verma, A.K.; Barba, F.J.; Singh, V.P.; Kumar, P.; Lorenzo, J.M. Edible films/coating with tailored properties for active packaging of meat, fish and derived products. *Trends Food Sci. Technol.* **2020**, *98*, 10–24. [CrossRef]
14. Gounga, M.E.; Xu, S.Y.; Wang, Z.; Yang, W.G. Effect of whey protein isolate-pullulan edible coatings on the quality and shelf life of freshly roasted and freeze-dried Chinese chestnut. *J. Food Sci.* **2008**, *73*, E155–E161. [CrossRef]
15. Zibaei, R.; Hasanvand, S.; Hashami, Z.; Roshandel, Z.; Rouhi, M.; Guimaraes, J.T.; Mortazavian, A.M.; Sarlak, Z.; Mohammadi, R. Applications of emerging botanical hydrocolloids for edible films: A review. *Carbohydr. Polym.* **2021**, *256*, 117554. [CrossRef]
16. Cao, Y.; Mezzenga, R. Design principles of food gels. *Nat. Food.* **2020**, *1*, 106–118. [CrossRef]
17. Cui, C.; Ji, N.; Wang, Y.; Xiong, L.; Sun, Q. Bioactive and intelligent starch-based films: A review. *Trends Food Sci. Technol.* **2021**, *116*, 854–869. [CrossRef]
18. Chen, Z.; Zong, L.; Chen, C.; Xie, J. Development and characterization of PVA-Starch active films incorporated with β-cyclodextrin inclusion complex embedding lemongrass (Cymbopogon citratus) oil. *Food Packag. Shelf Life* **2020**, *26*. [CrossRef]
19. Etxabide, A.; Kilmartin, P.A.; Maté, J.I.; Gómez-Estaca, J. Characterization of glucose-crosslinked gelatin films reinforced with chitin nanowhiskers for active packaging development. *LWT* **2022**, *154*. [CrossRef]
20. Maryam Adilah, Z.A.; Jamilah, B.; Nur Hanani, Z.A. Functional and antioxidant properties of protein-based films incorporated with mango kernel extract for active packaging. *Food Hydrocoll.* **2018**, *74*, 207–218. [CrossRef]

21. Yin, Y.; Li, Z.; Sun, Y.; Yao, K. A preliminary study on chitosan/gelatin polyelectrolyte complex formation. *J. Mater. Sci.* **2005**, *40*, 4649–4652. [CrossRef]
22. Soltanzadeh, M.; Peighambardoust, S.H.; Ghanbarzadeh, B.; Amjadi, S.; Mohammadi, M.; Lorenzo, J.M.; Hamishehkar, H. Active gelatin/cress seed gum-based films reinforced with chitosan nanoparticles encapsulating pomegranate peel extract: Preparation and characterization. *Food Hydrocoll.* **2022**, *129*. [CrossRef]
23. Thongsrikhem, N.; Taokaew, S.; Sriariyanun, M.; Kirdponpattara, S. Antibacterial activity in gelatin-bacterial cellulose composite film by thermally crosslinking with cinnamaldehyde towards food packaging application. *Food Packag. Shelf Life* **2022**, *31*. [CrossRef]
24. He, B.; Wang, S.; Lan, P.; Wang, W.; Zhu, J. Topography and physical properties of carboxymethyl cellulose films assembled with calcium and gelatin at different temperature and humidity. *Food Chem.* **2022**, *382*, 132391. [CrossRef]
25. Pan, J.; Li, Y.; Chen, K.; Zhang, Y.; Zhang, H. Enhanced physical and antimicrobial properties of alginate/chitosan composite aerogels based on electrostatic interactions and noncovalent crosslinking. *Carbohydr. Polym.* **2021**, *266*, 118102. [CrossRef]
26. Xiao, Q.; Tong, Q.; Lim, L.-T. Pullulan-sodium alginate based edible films: Rheological properties of film forming solutions. *Carbohydr. Polym.* **2012**, *87*, 1689–1695. [CrossRef]
27. Han, K.; Liu, Y.; Liu, Y.; Huang, X.; Sheng, L. Characterization and film-forming mechanism of egg white/pullulan blend film. *Food Chem.* **2020**, *315*, 126201. [CrossRef]
28. Krasniewska, K.; Gniewosz, M.; Kosakowska, O.; Cis, A. Preservation of Brussels Sprouts by Pullulan Coating Containing Oregano Essential Oil. *J. Food Prot.* **2016**, *79*, 493–500. [CrossRef]
29. Chu, Y.; Gao, C.; Liu, X.; Zhang, N.; Xu, T.; Feng, X.; Yang, Y.; Shen, X.; Tang, X. Improvement of storage quality of strawberries by pullulan coatings incorporated with cinnamon essential oil nanoemulsion. *LWT* **2020**, *122*. [CrossRef]
30. Da Nóbrega Santos, E.; Cesar de Albuquerque Sousa, T.; Cassiano de Santana Neto, D.; Brandão Grisi, C.V.; Cardoso da Silva Ferreira, V.; Pereira da Silva, F.A. Edible active film based on gelatin and Malpighia emarginata waste extract to inhibit lipid and protein oxidation in beef patties. *LWT* **2022**, *154*. [CrossRef]
31. Sun, J.; Jiang, H.; Li, M.; Lu, Y.; Du, Y.; Tong, C.; Pang, J.; Wu, C. Preparation and characterization of multifunctional konjac glucomannan/carboxymethyl chitosan biocomposite films incorporated with epigallocatechin gallate. *Food Hydrocoll.* **2020**, *105*. [CrossRef]
32. Jin, H.; Li, P.; Jin, Y.; Sheng, L. Effect of sodium tripolyphosphate on the interaction and aggregation behavior of ovalbumin-lysozyme complex. *Food Chem.* **2021**, *352*, 129457. [CrossRef]
33. Zhang, A.Q.; He, J.L.; Wang, Y.; Zhang, X.; Piao, Z.-H.; Xue, Y.-T.; Zhang, T.-H. Whey protein isolate modified with sodium tripolyphosphate gel: A novel pH-sensitive system for controlled release of Lactobacillus plantarum. *Food Hydrocoll.* **2021**, *120*, 106924. [CrossRef]
34. Szerman, N.; Ferrari, R.; Sancho, A.M.; Vaudagna, S. Response surface methodology study on the effects of sodium chloride and sodium tripolyphosphate concentrations, pressure level and holding time on beef patties properties. *LWT — Food Sci. Technol.* **2019**, *109*, 93–100. [CrossRef]
35. Wang, Y.R.; Zhang, B.; Fan, J.L.; Yang, Q.; Chen, H.Q. Effects of sodium tripolyphosphate modification on the structural, functional, and rheological properties of rice glutelin. *Food Chem.* **2019**, *281*, 18–27. [CrossRef]
36. Li, J.; Wang, J.; Zhai, J.; Gu, L.; Su, Y.; Chang, C.; Yang, Y. Improving gelling properties of diluted whole hen eggs with sodium chloride and sodium tripolyphosphate: Study on intermolecular forces, water state and microstructure. *Food Chem.* **2021**, *358*, 129823. [CrossRef]
37. Chen, J.; Ren, Y.; Zhang, K.; Xiong, Y.L.; Wang, Q.; Shang, K.; Zhang, D. Site-specific incorporation of sodium tripolyphosphate into myofibrillar protein from mantis shrimp (Oratosquilla oratoria) promotes protein crosslinking and gel network formation. *Food Chem.* **2020**, *312*, 126113. [CrossRef]
38. Hu, Y.; Zhang, L.; Yi, Y.; Solangi, I.; Zan, L.; Zhu, J. Effects of sodium hexametaphosphate, sodium tripolyphosphate and sodium pyrophosphate on the ultrastructure of beef myofibrillar proteins investigated with atomic force microscopy. *Food Chem.* **2021**, *338*, 128146. [CrossRef]
39. Zhou, W.; He, Y.; Liu, F.; Liao, L.; Huang, X.; Li, R.; Zou, Y.; Zhou, L.; Zou, L.; Liu, Y.; et al. Carboxymethyl chitosan-pullulan edible films enriched with galangal essential oil: Characterization and application in mango preservation. *Carbohydr. Polym.* **2021**, *256*, 117579. [CrossRef]
40. Chen, C.; Zong, L.; Wang, J.; Xie, J. Microfibrillated cellulose reinforced starch/polyvinyl alcohol antimicrobial active films with controlled release behavior of cinnamaldehyde. *Carbohydr. Polym.* **2021**, *272*, 118448. [CrossRef]
41. Jiang, J.; Gong, L.; Dong, Q.; Kang, Y.; Osako, K.; Li, L. Characterization of PLA-P3,4HB active film incorporated with essential oil: Application in peach preservation. *Food Chem.* **2020**, *313*, 126134. [CrossRef]
42. Ghosh, M.; Upadhyay, R.; Mahato, D.K.; Mishra, H.N. Kinetics of lipid oxidation in omega fatty acids rich blends of sunflower and sesame oils using Rancimat. *Food Chem.* **2019**, *272*, 471–477. [CrossRef]
43. *GB5009.181-2016*; National Standard for Food Safety: Determination of Malondialdehyde in Foods. National Standards of People's Republic of China: China, 2016.
44. *GB/T 24304-2009*; Animal and vegetable fats and oils—Determination of anisidine value. National Standards of People's Republic of China: China, 2009.

45. Li, Y.; Dong, Q.; Chen, J.; Li, L. Effects of coaxial electrospun eugenol loaded core-sheath PVP/shellac fibrous films on postharvest quality and shelf life of strawberries. *Postharvest Biol. Technol.* **2020**, *159*. [CrossRef]
46. Ni, Y.; Sun, J.; Wang, J. Enhanced antimicrobial activity of konjac glucomannan nanocomposite films for food packaging. *Carbohydr. Polym.* **2021**, *267*, 118215. [CrossRef]
47. Zhang, C.; Gao, D.; Ma, Y.; Zhao, X. Effect of gelatin addition on properties of pullulan films. *J. Food Sci.* **2013**, *78*, C805–C810. [CrossRef]
48. Chang, S.C.; Hsu, B.Y.; Chen, B.H. Structural characterization of polysaccharides from Zizyphus jujuba and evaluation of antioxidant activity. *Int. J. Biol. Macromol.* **2010**, *47*, 445–453. [CrossRef]

Article

Development and Characterization of Pectin Films with *Salicornia ramosissima*: Biodegradation in Soil and Seawater

Daniela G. M. Pereira [1], Jorge M. Vieira [2], António A. Vicente [2] and Rui M. S. Cruz [1,3,*]

[1] Department of Food Engineering, Institute of Engineering, Campus da Penha, Universidade do Algarve, 8005-139 Faro, Portugal; a48426@ualg.pt
[2] CEB—Centre of Biological Engineering, Campus de Gualtar, University of Minho, 4710-057 Braga, Portugal; jorgevieirabcd@gmail.com (J.M.V.); avicente@deb.uminho.pt (A.A.V.)
[3] MED—Mediterranean Institute for Agriculture, Environment and Development, Faculty of Sciences and Technology, Campus de Gambelas, Universidade do Algarve, 8005-139 Faro, Portugal
* Correspondence: rcruz@ualg.pt

Abstract: Pectin films were developed by incorporating a halophyte plant *Salicornia ramosissima* (dry powder from stem parts) to modify the film's properties. The films' physicomechanical properties, Fourier-transform infrared spectroscopy (FTIR), and microstructure, as well as their biodegradation capacity in soil and seawater, were evaluated. The inclusion of *S. ramosissima* significantly increased the thickness (0.25 ± 0.01 mm; control 0.18 ± 0.01 mm), color parameters a* (4.96 ± 0.30; control 3.29 ± 0.16) and b* (28.62 ± 0.51; control 12.74 ± 0.75), water vapor permeability ($1.62 \times 10^{-9} \pm 1.09 \times 10^{-10}$ (g/m·s·Pa); control $1.24 \times 10^{-9} \pm 6.58 \times 10^{-11}$ (g/m·s·Pa)), water solubility ($50.50 \pm 5.00\%$; control $11.56 \pm 5.56\%$), and elongation at break ($5.89 \pm 0.29\%$; control $3.91 \pm 0.62\%$). On the other hand, L* (48.84 ± 1.60), tensile strength (0.13 ± 0.02 MPa), and Young's modulus (0.01 ± 0 MPa) presented lower values compared with the control (L* 81.20 ± 1.60; 4.19 ± 0.82 MPa; 0.93 ± 0.12 MPa), while the moisture content varied between 30% and 45%, for the film with *S. ramosissima* and the control film, respectively. The addition of *S. ramosissima* led to opaque films with relatively heterogeneous microstructures. The films showed also good biodegradation capacity—after 21 days in soil (around 90%), and after 30 days in seawater (fully fragmented). These results show that pectin films with *S. ramosissima* may have great potential to be used in the future as an eco-friendly food packaging material.

Keywords: biobased materials; biodegradable; food packaging; pectin film; physicomechanical; *Salicornia ramosissima*; sustainability

Citation: Pereira, D.G.M.; Vieira, J.M.; Vicente, A.A.; Cruz, R.M.S. Development and Characterization of Pectin Films with *Salicornia ramosissima*: Biodegradation in Soil and Seawater. *Polymers* 2021, 13, 2632. https://doi.org/10.3390/polym13162632

Academic Editors: Swarup Roy and Jong-Whan Rhim

Received: 1 July 2021
Accepted: 4 August 2021
Published: 7 August 2021

Publisher's Note: MDPI stays neutral with regard to jurisdictional claims in published maps and institutional affiliations.

Copyright: © 2021 by the authors. Licensee MDPI, Basel, Switzerland. This article is an open access article distributed under the terms and conditions of the Creative Commons Attribution (CC BY) license (https://creativecommons.org/licenses/by/4.0/).

1. Introduction

Packaging plays a key role in containing and protecting food from external influences, such as microorganisms, oxygen, and odors, among others. However, plastic packaging has a negative environmental impact on land and sea, since it generates huge amounts of solid waste. The reduction of this waste can be achieved with the development of new biodegradable packaging systems [1]. In recent decades, a considerable number of packaging films have been developed using biopolymers, such as proteins, and polysaccharides. More recently, the combination of clays, nanostructures, and other innovative materials has been studied for novel packaging applications [2–9]. In addition to biopolymers, plant-derived bioactive compounds—such as essential oils, minerals, carotenoids, vitamins, and polyphenols, among others—have been used to modify the films' antioxidant, antimicrobial, and physicomechanical properties [10–15].

Polysaccharides (e.g., chitosan, cellulose, starch, alginate, and pectin) have gained attention because of their good film-forming capacity [16,17]. Pectin is a heteropolysaccharide found in fruits and vegetables. It is used in food products as a stabilizing and gelling ingredient. These gelling characteristics, together with those of biocompatibility and biodegradability, make pectin an ideal biomaterial for several applications, as in the case of the pharmaceutical and food industries [18]. Several studies have reported the development of pectin films with the incorporation of different compounds. In the study reported by Makaremi et al. [18], pectin–alginate films incorporated with ascorbic and lactic acids were developed, and showed the potential to be used in food packaging. In another study, Chaiwarit et al. [19] developed a pectin film loaded with clindamycin hydrochloride to modify the film's properties and use it as a topical drug delivery system. Nogueira et al. [20] presented a review of several methods of incorporating plant-derived bioactive compounds, among others, into pectin films for application in food packaging. The incorporation of green coffee oil and γ-aminobutyric acid residues affected the color of the films. The same review presented pectin films with *Acca sellowiana* waste byproducts (feijoa peel flour) with increased thickness and different mechanical properties.

An interesting application in the development of pectin films could be the incorporation of *Salicornia ramosissima*, a halophyte plant very rich in sodium, magnesium, potassium, calcium, and manganese. Moreover, this plant also presents good antioxidant properties [21]. Recently, this plant started to be produced by hydroponic systems, and generates a considerable number of byproducts after the cutting process. Thus, the use of this plant as a natural additive could provide a pectin film with modified properties, with a potential preservation effect mainly due to the presence of salts, and could thus be a relevant outcome to obtain an effective biodegradable active packaging material, while also contributing to the valorization of a byproduct.

To our knowledge, no study is available in the literature reporting the use of *S. ramosissima* as an additive in biodegradable packaging films. Thus, the objectives of this study were to develop and characterize a pectin film incorporating a halophyte plant—*S. ramosissima*—as well as to evaluate its biodegradation capacity in soil and seawater.

2. Materials and Methods

2.1. Materials

Apple pectin (50–70% degree of esterification; MW = 60,000–130,000 g mol^{-1}) was purchased from Sigma (Algés, Portugal). Glycerol and NaBr were purchased from JMGS (Odivelas, Portugal). *Salicornia ramosissima* dried powder (from cut stem parts; particle size < 63 µm; moisture content = 7%; lipids 0.9 g, carbohydrates 4.5 g, fiber 16 g, proteins 18 g, and salt 47 g per 100 g) was kindly supplied by RiaFresh (Faro, Portugal). The soil (Eco Grow) was purchased from AKI (Faro, Portugal).

2.2. Preparation of the Films

Pectin films were developed based on the method reported by Mendes et al. [22], with some modifications. First, 0.5 g of apple pectin was added to 25 mL of water previously heated to 75 °C and stirred for 10 min. Next, the film-forming solution was homogenized (Ultra-Turrax T25, Janke & Kunkel, Staufen, Germany) at 20,500 rpm for 5 min, and then 1.5 mL of glycerol was added and mixed with magnetic stirring for 1 h at 40 °C.

The pectin films incorporating *S. ramosissima* were prepared by adding 0.75 g (2.7% w/w) of dried, fine *S. ramosissima* powder (based on preliminary tests to obtain a film with homogeneous appearance) to 25 mL of water previously heated to 75 °C and stirred for 10 min. After this period, the same methodology was followed as previously used for pectin films. Then, the film-forming solutions were degassed for 5 min and placed to dry in Petri dishes for 48 h at 25 °C. After drying, the films were cut and stored at 25 °C and 57% relative humidity (obtained using a saturated NaBr solution) before analysis.

2.3. Thickness and Water Vapor Permeability (WVP)

Three thickness measurements were randomly taken for each testing sample at different points with a digital micrometer (No. 293-5, Mitutoyo, Kawasaki, Japan). Mean values were used to calculate the WVP. The WVP of pectin films was determined gravimetrically, using the ASTM E96-92 procedure, with some modifications [23]. The permeation cell was filled with 50 mL of distilled water to generate a 100% RH (2337 Pa vapor pressure at 20 °C), and the film was sealed on the top of the cells. Then, the cells were weighed using an analytical balance (AE200, Mettler Toledo, Barcelona, Spain) and placed inside a desiccator containing silica (0% RH; 0 Pa water vapor pressure; the air circulation was kept constant by using a fan inside the desiccator). The tests were conducted in triplicate, and changes in weight of the cells were recorded at intervals of 2 h to record moisture loss over time until a steady state was reached. The WVP (g m^{-1} s^{-1} Pa^{-1}) of the films tested was determined by the following equation:

$$WVP = (WVTR \times X)/\Delta P \quad (1)$$

where WVTR = water vapor transmission rate (g m^{-2} s^{-1}) through the film, calculated from the slope of the curve divided by the film's area; X = the film's thickness (m); and ΔP = the partial vapor pressure difference (Pa) across the two sides of the film.

2.4. Color and Opacity

Color measurements were performed using a colorimeter (Minolta CR 400, Tokyo, Japan) previously calibrated with a standard white tile (EU certified; L* = 84.67, a* = −0.55, and b* = 0.68)), which recorded the spectrum of reflected light to determine the parameters L*, a*, and b*. The opacity of the samples was calculated based on the method reported by Martins et al. [24], as the relationship between the opacity of each sample in a black standard (Yb) and the opacity of each sample in a white standard (Yw), as can be seen in the equation:

$$Opacity\ (\%) = Yb/Yw \times 100 \quad (2)$$

Three measurements were taken from each sample, and three samples from each film were tested.

2.5. Solubility and Moisture Content

The water solubility and moisture content of the films were determined according to the method reported by Casariego et al. [25]. The film solubility in water was determined as the percentage of soluble material after 24 h of immersion in water. A disk of the film (2-cm diameter) was dried in an oven at 105 °C until constant weight to obtain the initial dry matter of the films (this part of the method allowed determining the films' moisture content). Then, the sample was immersed into 50 mL of deionized water and gently shaken (20 °C, 24 h). At the end of the 24 h, the insolubilized films were filtered and dried in a drying oven (105 °C, 24 h) to determine the weight of the dried matter that was not solubilized in water. Three replicates were obtained for each sample. The water solubility (%) of the films was calculated as follows:

$$Solubility\ (\%) = (Mi - Mf)/Mi \times 100 \quad (3)$$

where Mi is the initial mass and Mf is the final mass of the sample.

2.6. Texture Measurements

Mechanical properties were evaluated in a texture analyzer (TA.TX Plus Texture Analyzer) with the software "Exponent version 6.1.16" (Stable Micro Systems, Godalming, Surrey, UK), according to the ASTM D 882 standard method [26].

A scalpel was used to cut the samples into 20-mm-wide, 100-mm-long strips, which were mounted between tensile grips. The initial grip separation and the crosshead speed were set at 80 mm and 1.0 mm/s, respectively. The tensile strength (force/initial cross-sectional area) and the elongation at break were computed directly from the strength curves vs. elongation curves by using the "Exponent" software. Young's modulus (Equation (4)) was calculated as the slope of the initial linear portion of this curve. Six measurements were taken of each sample.

$$E = \sigma/\varepsilon \quad (4)$$

where E represents Young's modulus, σ is the tensile stress (force per unit area), and ε is the axial strain (deformation).

2.7. FTIR

ATR-FTIR was used to obtain information about the interactions between components in films. The FTIR spectra of the films were recorded with a PerkinElmer 16 PC (Boston, MA, USA), using attenuated total reflectance mode (ATR). Each spectrum resulted from 16 scans at 4-cm^{-1} resolution, for a spectral range of 400–4000 cm^{-1}. All of the readings were performed at room temperature (20 °C).

2.8. Scanning Electron Microscopy (SEM)

The samples were characterized using a desktop scanning electron microscope. All results were acquired using the ProSuite software.

The samples were added to aluminum pin stubs with electrically conductive carbon adhesive tape (PELCO Tabs™). Samples were coated with 2 nm of Au (20 Angstrom) for improved conductivity. The aluminum pin stub was then placed inside a Phenom Standard Sample Holder (SH) and acquired with 10 kV.

2.9. Biodegradation Tests

2.9.1. Seawater

The biodegradation test in seawater was based on the methodology used by Accinelli et al. [27], with some modifications. The films were cut (3 cm × 2 cm) and submerged in 300 mL of seawater (pH = 7.20). The samples were shaken at 150 rpm (Edmund Buhler-7400 Tubingen shaker) and 25 °C.

The films' appearance was photographed during the time of the experiment. This test was carried out in triplicate for each sample.

2.9.2. Soil

The biodegradation test in soil was based on the methodology used by Altaee et al. [28], with some adaptations. The films were cut (3 cm × 2 cm) and placed inside a perforated polyethylene net (5 cm × 4 cm; mesh opening 4 mm). The films were placed in soil (pH = 5.5–6.5; Humidity = 50–60%; Conductivity = 0.2–1.2 EC; Nitrogen = 80–150 mg L^{-1}; Phosphorus = 80–150 mg L^{-1}; Potassium = 80–150 mg L^{-1}; Organic Matter = >70%) at a distance of 11 cm from the surface in a rectangular vase (71 cm × 26 cm × 25.5 cm), and with a distance of 5 cm between each film. The soil was watered with 500 mL of water every 7 days and maintained at 25 °C throughout the study. The films' appearance was photographed and the area of biodegradation was measured during the time of the experiment. This test was carried out in triplicate for each sample.

2.10. Statistical Analyses

The results were expressed as the mean and standard deviation of at least three replicates. The experimental data were analyzed with IBM SPSS (Statistical Product and Service Solutions) version 26. The analysis was performed with the *t*-test to detect significant differences between the two types of film for each parameter. The significance level used was 0.05.

3. Results and Discussion

3.1. Physicomechanical Properties

Table 1 shows the physical and mechanical parameters of the control and the supplemented pectin films. The thickness of the film with *S. ramosissima* significantly increased ($p < 0.05$). This increase can be explained due to a greater solid mass in the supplemented film.

Table 1. Physico-mechanical parameters of the developed films.

	Control Film	Film with *S. ramosissima*
Thickness (mm)	0.18 ± 0.01 [a]	0.25 ± 0.01 [b]
Color		
L*	81.20 ± 0.70 [a]	48.84 ± 1.60 [b]
a*	3.29 ± 0.16 [a]	4.96 ± 0.30 [b]
b*	12.74 ± 0.75 [a]	28.62 ± 0.51 [b]
Opacity (%)	10.87 ± 0.05 [a]	30.04 ± 1.49 [b]
Water vapor permeability (g/(m·s·Pa))	$1.24 \times 10^{-9} \pm 6.58 \times 10^{-11}$ [a]	$1.62 \times 10^{-9} \pm 1.09 \times 10^{-10}$ [b]
Water solubility (%)	11.56 ± 5.56 [a]	50.50 ± 5.00 [b]
Moisture content (%)	45.79 ± 0.76 [a]	30.11 ± 4.41 [b]
Elongation at break (%)	3.91 ± 0.62 [a]	5.89 ± 0.29 [b]
Young's modulus (MPa)	0.93 ± 0.12 [a]	0.01 ± 0 [b]
Tensile strength (MPa)	4.19 ± 0.82 [a]	0.13 ± 0.02 [b]

[a,b] Different superscript letters indicate significant differences ($p < 0.05$).

According to Hosseini et al. [29], the increase in film thickness may be related to the increase in dry matter content. In terms of color, the control film was clear, colorless, and lighter, as expressed by the highest L* parameter. On the other hand, the incorporation of *S. ramosissima* led to darker and brownish-green films, presenting lower values of the L* parameter and an increase in the a* and b* parameters (Figure 1). These results, and the increase in the film's opacity, are associated with the existing pigments in the dried powder—mainly pheophytins—obtained after drying *S. ramosissima* [30]. In another study reported by Nisar et al. [31], pectin films enriched with thinned young apple polyphenols were produced, and the authors also reported a decrease in the L* values and an increase in the a* and b* values. In another study by Sganzerla et al. [32], Brazilian pine seed starch and pectin films with the incorporation of *Acca sellowiana* extracts also showed lower L* values, together with higher a* and b* parameters.

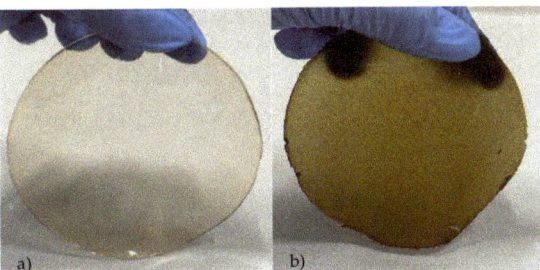

Figure 1. Developed films: (**a**) control film; (**b**) film with *S. ramosissima*.

The film with *S. ramosissima* showed a slight increase in the WVP. This increase may be related to the presence of salts and, thus, to the greater availability of hydroxyl groups to bind to water molecules [33], as well as to the film's capacity to absorb more water molecules due to its higher mass and porous structure [34]. In a study reported by Chen et al. [35], the WVP also presented the highest values in pectin-/tara gum-based edible films with ellagitannins from the unripe fruit of *Rubus chingii* Hu. On the other hand, Nisar et al. [31] showed that the WVP decreased with the addition of components rich in polyphenols. According to the authors, the mobility of pectin chains was negatively affected by the interaction with polyphenols, so that the diffusion or penetration of water molecules through membranes was reduced, with a decrease in hydrophilic groups. The film with *S. ramosissima* also presented a higher water solubility compared with the control film. Aside from any modification to the films' structure, the presence of salts such as sodium, magnesium, or calcium—which are soluble in water [21]—also contributed to a higher solubility of the film in water. According to Nisar et al. [31], the higher values of solubility in water and the degree of swelling of the films can be attributed to the presence of hydrophilic groups; these authors reported a behavior of their system that is identical to the one reported in our study, with a significant increase ($p < 0.05$) in solubility after the addition of apple polyphenols.

The moisture content also showed significant differences ($p < 0.05$) between the control film and the film with S. ramosissima. This difference is related to the higher mass of dry matter present in the film with S. ramosissima, contributing to lower moisture content. According to Pereda et al. [36], the moisture content is related to the total void volume occupied by water molecules in the network microstructure of the film. Similar results were obtained by Yehuala and Emire [37], where the addition of Aloe debrana extract and papaya leaf extract affected the moisture content of gelatin films.

Regarding the films' mechanical properties, the control film presented a more resistant structure, showing a higher value of tensile strength, while the film with S. ramosissima was more brittle/fragile. On the other hand, the elongation at break was higher for the film with S. ramosissima. According to Shaw et al. [38], the decrease in film resistance (tensile strength) and the increase in stretching capacity (elongation at break) can be attributed to the reduction in the number of intermolecular crosslinks between pectin molecules within the films, thus contributing to a weaker material. Similar results to this study were obtained by Gouveia et al. [39], who reported that the addition of CHCl to the pectin films caused a decrease in traction resistance. The results of Meerasri and Sothornvit [40] also showed the same behavior, with a decrease in tensile strength and an increase in elongation at break with the addition of γ-aminobutyric acid and glycerol. Moreover, Kang et al. [41] reported similar results to the ones obtained in the film with S. ramosissima. Pectin–polyvinyl alcohol–glycerol films combined with gamma irradiation and $CaCl_2$ immersion presented tensile strength values between 0.09 and 0.27 MPa, and elongation at break values between 1.02 and 3.45%.

3.2. Fourier-Transform Infrared Spectroscopy (FTIR)

FTIR spectroscopy was performed to determine the intermolecular interactions within the film matrix. The FTIR spectra of the control film and the pectin film with S. ramosissima are shown in Figure 2. A broad peak ranging from 3700 to 3000 cm^{-1} corresponds to the stretching of O–H because of hydrogen bonding interactions in the galacturonic acid [42]. The peak at 2935 cm^{-1} is attributed to stretching of C–H bonds [43], while the bands at 1744 and 1612 cm^{-1} are attributed to the absorptions by esterified and free carboxyl groups of pectin, respectively [44,45]. Moreover, the bands at 1103 and 1026 cm^{-1} were assigned to C–O–C stretching vibrations of the polymer chain structure [46]. In general, the pectin film with S. ramosissima exhibited a similar pattern compared to the control film, but showed a weaker response. S. ramosissima dry powder is a complex mixture of diverse salts and cellulosic compounds, and its incorporation probably contributed to weakening the intermolecular forces between the chains of adjacent macromolecules [47]. This result is corroborated, as previously referred, by the lower tensile strength obtained for the film with S. ramosissima.

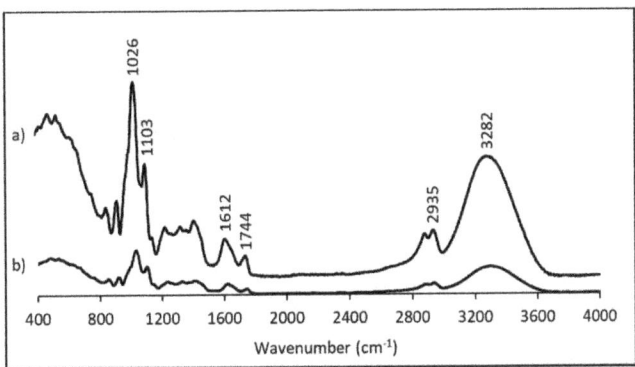

Figure 2. FTIR spectra: (**a**) control film; (**b**) film with *S. ramosissima*.

3.3. Scanning Electron Microscopy (SEM)

SEM micrographs of the films' surface are presented in Figure 3. The pectin film presented a homogeneous, compact, and smooth surface, while the film with *S. ramosissima* presented a heterogeneous and rough surface. This result is related to the presence of salts (mainly sodium, magnesium, and potassium) from *S. ramosissima* powder that changed the structure of the pectin film. The possible presence of faults or microholes also contributed to the lower tensile strength values obtained, and facilitated the migration of water vapor, causing an increase in the WVP [48], as previously shown in Table 1.

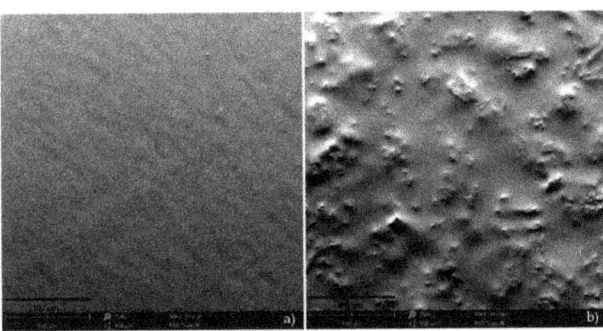

Figure 3. SEM micrographs: (**a**) control film; (**b**) film with *S. ramosissima* (scale bar = 100 µm).

3.4. Biodegradation Properties

3.4.1. Seawater

During the biodegradation test in seawater, both films underwent several changes (Figure 4). After the first day, the samples kept their initial appearance, and after this period the seawater in which they were submerged began to show signs of clouding for the film with *S. ramosissima*, while for the control film this change was not so evident. This result can be explained by the transfer of salts and pigments from the film with *S. ramosissima* to the seawater.

Figure 4. Biodegradation test in seawater.

On the 8th day, it was possible to verify that both films began to fragment into small pieces. In this period, the clouding of the water for the control film was also noticeable, possibly caused by the release of pectin. On the 16th day, greater changes were observed in the films' structure, as they began to dissolve considerably in the seawater. This can be explained by the swelling of the film, as both the swelling and the solubility of the film can directly affect the water resistance properties of the films, particularly if it occurs in a humid environment [31].

On the 22nd day, the seawater began to lose signs of clouding, especially for the film with *S. ramosissima*. After 30 days, it was found that both films were quite fragmented, presenting a "flaky" appearance, and completely losing their initial rectangular shape and, consequently, most of their initial structure. The rate of biodegradation of the film is related to different factors, including the swelling, the movement/agitation of the seawater, the presence of microorganisms in the seawater, the ratio volume of seawater/film, and the existence of oxygenation [27,49]. Data obtained by Alvarez-Zeferino et al. [49] corroborate those of this study by verifying low levels of biodegradation in the first days of testing and presenting a high rate of loss of physical integrity. Furthermore, Nakayama et al. [50] performed experiments to test the biodegradation of aliphatic polyesters in seawater; the authors concluded that, aside from microorganisms, several factors—such as physicochemical effects from the sunlight, waves, and inorganic salts, among others—contribute to faster biodegradation in seawater.

3.4.2. Soil

After 24 h in soil, there were no changes in the films' structure, although the films showed signs of having some water absorption (Figure 5). After two days, the control film remained identical compared with day 0; however, the film with *S. ramosissima* absorbed water from the soil, and began to show some changes in its structure. This result is related to the fact that the film with *S. ramosissima*, as previously noted, presented a higher solubility in water.

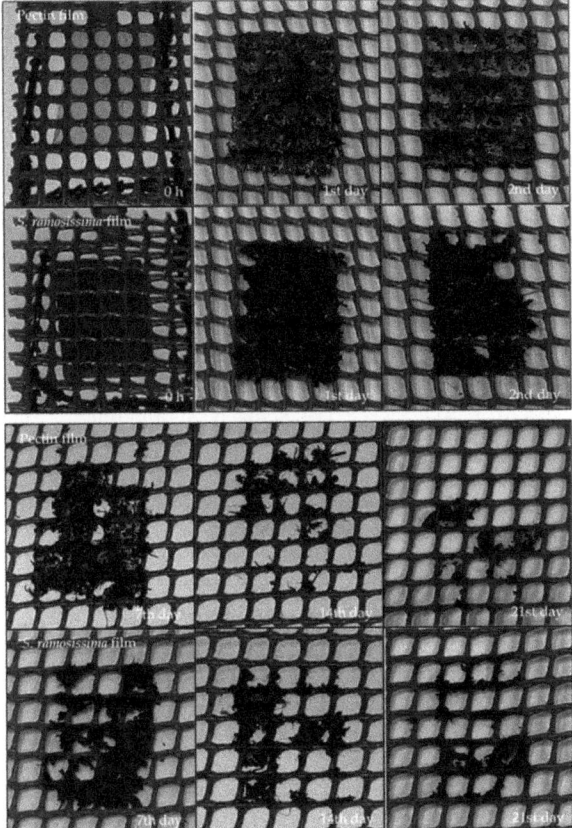

Figure 5. Biodegradation test in soil.

After 14 days, there was a loss of more than 50% of the structure in both films, and after 21 days this loss was ~90%—the value indicated by the European Standard EN 13432 [51] for packaging to be considered biodegradable by biological action in a period of 6 months. This degradation is related, on the one hand, to the action of microorganisms existing in the soil. According to Shah et al. [52], the microorganisms responsible for biodegradation are bacteria and fungi; *Acidovorax facilis*, *Aspergillus fumigatus*, *Comamonas* sp., *Pseudomonas lemoignei*, and *Variovorax paradoxus* are among those normally found in soil. Our results are in agreement with the study reported by Jaramillo et al. [53], in which cassava starch films presented signs of biodegradation after 6 days. After 12 days, there was a greater change in the decomposition of films. On the other hand, the particular characteristics of the soil—such as the availability of phosphorus, which contributes to a higher load of fungi—are also responsible for the biodegradation phenomenon [54].

The degradation mechanism can first result from a variety of physical and biological forces—such as heating/cooling, freezing/thawing, or wetting/drying—causing mechanical damage, such as the cracking of the polymeric materials. In our study, the presence of water contributed to the initial breakdown of the films. Then, during degradation, depolymerization occurs when the extracellular enzymes from the microorganisms break down the polymer, yielding short chains or smaller molecules—e.g., oligomers, dimers, and monomers—that are small enough to pass the semipermeable outer bacterial membranes. These short-chain molecules are then converted to biodegradation end-products, such as CO_2, H_2O, and biomass [52,55].

In general, biodegradation in both soil and seawater was practically obtained in a short period, with an average degradation rate of 4.3% and 3.3% per day in soil and seawater, respectively. Conversely, plastic or even paper packaging degradation is a very slow process, and it can take several years for those materials to be fully degraded, depending on the type of plastic or paper and the used conditions [56]. According to Chamas et al. [57], low-density polyethylene (LDPE) bags are estimated to decompose by 50% after 4.6 and 3.4 years, in inland (buried) and marine environments, respectively. In a study reported by Olaosebikan et al. [58], by the 10th–12th week of exposure in soil, brown newspapers started to degrade, and only pieces of the papers were found remaining, while plastic bags had thinned off and become transparent.

4. Conclusions

The inclusion of *S. ramosissima* led to the development of opaque films with improved elongation at break (more flexible), but with less tensile strength (less rigid)—the latter being one of the films' limitations. Although the films presented a homogeneous appearance, the SEM analysis revealed a heterogeneous and rough surface across the film matrix. These modifications, as verified by FTIR analysis, could be credited to the interaction between functional groups of pectin and *S. ramosissima*. Generally, the properties are promising, and the presence of salts in the films may also contribute to a preservation effect, although further characterization and improvements of the antimicrobial and antioxidant properties are necessary to achieve the desired features. Finally, the biodegradation results demonstrated that pectin films with *S. ramosissima* can also be utilized as an eco-friendly food packaging material.

Author Contributions: Conceptualization, R.M.S.C.; validation, A.A.V. and R.M.S.C.; formal analysis, D.G.M.P., J.M.V., and R.M.S.C.; investigation, D.G.M.P., J.M.V., A.A.V., and R.M.S.C.; resources, A.A.V. and R.M.S.C.; writing—original draft preparation, D.G.M.P. and J.M.V.; writing—review and editing, A.A.V. and R.M.S.C.; supervision, R.M.S.C. All authors have read and agreed to the published version of the manuscript.

Funding: This research received no external funding.

Institutional Review Board Statement: Not applicable.

Informed Consent Statement: Not applicable.

Data Availability Statement: The data presented in this study are available on request from the corresponding author.

Conflicts of Interest: The authors declare no conflict of interest.

References

1. Bonilla, J.; Talón, E.; Atarés, L.; Vargas, M.; Chiralt, A. Effect of the incorporation of antioxidants on physicochemical and antioxidant properties of wheat starch–chitosan films. *J. Food Eng.* **2013**, *118*, 271–278. [CrossRef]
2. Bertolino, V.; Cavallaro, G.; Milioto, S.; Lazzara, G. Polysaccharides/Halloysite nanotubes for smart bionanocomposite materials. *Carbohydr. Polym.* **2020**, *245*, 116502. [CrossRef] [PubMed]
3. Kalateh-Seifari, F.; Yousefi, S.; Ahari, H.; Hosseini, S.H. Corn starch-chitosan nanocomposite film containing nettle essential oil nanoemulsions and starch nanocrystals: Optimization and characterization. *Polymers* **2021**, *13*, 2113. [CrossRef] [PubMed]
4. Lisuzzo, L.; Cavallaro, G.; Milioto, S.; Lazzara, G. Effects of halloysite content on the thermo-mechanical performances of composite bioplastics. *Appl. Clay Sci.* **2020**, *185*, 105416. [CrossRef]
5. Oliver-Ortega, H.; Tresserras, J.; Julian, F.; Alcalà, M.; Bala, A.; Espinach, F.X.; Méndez, J.A. Nanocomposites materials of PLA Reinforced with nanoclays using a masterbatch technology: A study of the mechanical performance and its sustainability. *Polymers* **2021**, *13*, 2133. [CrossRef]
6. Abe, M.M.; Martins, J.R.; Sanvezzo, P.B.; Macedo, J.V.; Branciforti, M.C.; Halley, P.; Botaro, V.R.; Brienzo, M. Advantages and disadvantages of bioplastics production from starch and lignocellulosic components. *Polymers* **2021**, *13*, 2484. [CrossRef]

7. Lionetto, F.; Esposito Corcione, C. Recent applications of biopolymers derived from fish industry waste in food packaging. *Polymers* **2021**, *13*, 2337. [CrossRef]
8. Ediyilyam, S.; George, B.; Shankar, S.S.; Dennis, T.T.; Wacławek, S.; Černík, M.; Padil, V.V.T. Chitosan/gelatin/silver nanoparticles composites films for biodegradable food packaging applications. *Polymers* **2021**, *13*, 1680. [CrossRef]
9. Boey, J.Y.; Mohamad, L.; Khok, Y.S.; Tay, G.S.; Baidurah, S. A review of the applications and biodegradation of polyhydroxyalkanoates and poly(lactic acid) and its composites. *Polymers* **2021**, *13*, 1544. [CrossRef]
10. Brito, T.B.; Carrajola, J.F.; Gonçalves, E.C.B.A.; Martelli-Tosi, M.; Ferreira, M.S.L. Fruit and vegetable residues flours with different granulometry range as raw material for pectin-enriched biodegradable film preparation. *Food Res. Int.* **2019**, *121*, 412–421. [CrossRef]
11. Nogueira, G.F.; Fakhouri, F.M.; de Oliveira, R.A. Incorporation of spray dried and freeze-dried blackberry particles in edible films: Morphology, stability to pH, sterilization and biodegradation. *Food Packag. Shelf Life* **2019**, *20*, 100313. [CrossRef]
12. Zarandona, I.; Barba, C.; Guerrero, P.; de la Caba, K.; Maté, J. Development of chitosan films containing β-cyclodextrin inclusion complex for controlled release of bioactives. *Food Hydrocoll.* **2020**, *104*, 105720. [CrossRef]
13. Riaz, A.; Lagnika, C.; Luo, H.; Dai, Z.; Nie, M.; Hashim, M.M.; Liu, C.; Song, J.; Li, D. Chitosan-based biodegradable active food packaging film containing Chinese chive (*Allium tuberosum*) root extract for food application. *Int. J. Biol. Macromol.* **2020**, *150*, 595–604. [CrossRef]
14. Nogueira, G.F.; Soares, C.T.; Cavasini, R.; Fakhouri, F.M.; de Oliveira, R.A. Bioactive films of arrowroot starch and blackberry pulp: Physical, mechanical and barrier properties and stability to pH and sterilization. *Food Chem.* **2019**, *275*, 417–425. [CrossRef]
15. Jancikova, S.; Dordevic, D.; Jamroz, E.; Behalova, H.; Tremlova, B. Chemical and physical characteristics of edible films, based on κ- and ι-carrageenans with the addition of lapacho tea extract. *Foods* **2020**, *9*, 357. [CrossRef] [PubMed]
16. Liang, T.; Wang, L. A pH-sensing film from tamarind seed polysaccharide with *Litmus lichen* extract as an indicator. *Polymers* **2017**, *10*, 13. [CrossRef]
17. Cazón, P.; Velazquez, G.; Ramírez, J.A.; Vázquez, M. Polysaccharide-based films and coatings for food packaging: A review. *Food Hydrocoll.* **2017**, *68*, 136–148. [CrossRef]
18. Makaremi, M.; Yousefi, H.; Cavallaro, G.; Lazzara, G.; Goh, C.B.S.; Lee, S.M.; Solouk, A.; Pasbakhsh, P. Safely dissolvable and healable active packaging films based on alginate and pectin. *Polymers* **2019**, *11*, 1594. [CrossRef] [PubMed]
19. Chaiwarit, T.; Rachtanapun, P.; Kantrong, N.; Jantrawut, P. Preparation of clindamycin hydrochloride loaded de-esterified low-methoxyl mango peel pectin film used as a topical drug delivery system. *Polymers* **2020**, *12*, 1006. [CrossRef] [PubMed]
20. Nogueira, G.F.; Oliveira, R.A.; Velasco, J.I.; Fakhouri, F.M. Methods of incorporating plant-derived bioactive compounds into films made with agro-based polymers for application as food packaging: A brief review. *Polymers* **2020**, *12*, 2518. [CrossRef]
21. Barreira, L.; Resek, E.; Rodrigues, M.J.; Rocha, M.I.; Pereira, H.; Bandarra, N.; da Silva, M.M.; Varela, J.; Custódio, L. Halophytes: Gourmet food with nutritional health benefits? *J. Food Compos. Anal.* **2017**, *59*, 35–42. [CrossRef]
22. Mendes, J.F.; Norcino, L.B.; Manrich, A.; Pinheiro, A.C.M.; Oliveira, J.E.; Mattoso, L.H.C. Characterization of pectin films integrated with cocoa butter by continuous casting: Physical, thermal and barrier properties. *J. Polym. Environ.* **2020**, *28*, 2905–2917. [CrossRef]
23. Bourtoom, T.; Chinnan, M.S. Preparation and properties of rice starch–chitosan blend biodegradable film. *LWT Food Sci. Technol.* **2008**, *41*, 1633–1641. [CrossRef]
24. Martins, J.T.; Cerqueira, M.A.; Souza, B.W.S.; Carmo Avides, M.d.; Vicente, A.A. Shelf life extension of ricotta cheese using coatings of galactomannans from nonconventional sources incorporating nisin against *Listeria monocytogenes*. *J. Agric. Food Chem.* **2010**, *58*, 1884–1891. [CrossRef] [PubMed]
25. Casariego, A.; Souza, B.W.S.; Cerqueira, M.A.; Teixeira, J.A.; Cruz, L.; Díaz, R.; Vicente, A.A. Chitosan/clay films' properties as affected by biopolymer and clay micro/nanoparticles' concentrations. *Food Hydrocoll.* **2009**, *23*, 1895–1902. [CrossRef]
26. ASTM D 882. *Standard Test Method for Tensile Properties of Thin Plastic Sheeting*; ASTM International: West Conshohocken, PA, USA, 2002.
27. Accinelli, C.; Saccà, M.L.; Mencarelli, M.; Vicari, A. deterioration of bioplastic carrier bags in the environment and assessment of a new recycling alternative. *Chemosphere* **2012**, *89*, 136–143. [CrossRef] [PubMed]
28. Altaee, N.; El-Hiti, G.A.; Fahdil, A.; Sudesh, K.; Yousif, E. Biodegradation of different formulations of polyhydroxybutyrate films in soil. *SpringerPlus* **2016**, *5*. [CrossRef]
29. Hosseini, S.F.; Rezaei, M.; Zandi, M.; Farahmandghavi, F. Fabrication of bio-nanocomposite films based on fish gelatin reinforced with chitosan nanoparticles. *Food Hydrocoll.* **2015**, *44*, 172–182. [CrossRef]
30. Heaton, J.W.; Marangoni, A.G. Chlorophyll degradation in processed foods and senescent plant tissues. *Trends Food Sci. Technol.* **1996**, *7*, 8–15. [CrossRef]
31. Nisar, T.; Wang, Z.-C.; Alim, A.; Iqbal, M.; Yang, X.; Sun, L.; Guo, Y. Citrus pectin films enriched with thinned young apple polyphenols for potential use as bio-based active packaging. *CyTA J. Food* **2019**, *17*, 695–705. [CrossRef]
32. Sganzerla, W.G.; Paes, B.B.; Azevedo, M.; Ferrareze, J.; Da Rosa, C.G.; Nunes, M.R.; De Lima Veeck, A.P. Bioactive and biodegradable films packaging incorporated with *Acca sellowiana* extracts: Physicochemical and antioxidant characterization. *Chem. Eng. Trans.* **2019**, *75*, 445–450. [CrossRef]
33. Dias, A.B. Desenvolvimento e caracterização de filmes biodegradáveis obtidos de amido e de farinha de arroz. *Univ. Fed. St. Catarina* **2008**, 1–116. Available online: http://repositorio.ufsc.br/xmlui/handle/123456789/92138 (accessed on 10 February 2021).

34. Meneguin, A.; Cury, B.; Evangelista, R. Films from resistant starch-pectin dispersions intended for colonic drug delivery. *Carbohydr. Polym.* **2014**, *99*, 140–149. [CrossRef]
35. Chen, Y.; Xu, L.; Wang, Y.; Chen, Z.; Zhang, M.; Chen, H. Characterization and functional properties of a pectin/tara gum based edible film with ellagitannins from the unripe fruits of *Rubus chingii* Hu. *Food Chem.* **2020**, *325*, 126964. [CrossRef]
36. Pereda, M.; Ponce, A.G.; Marcovich, N.E.; Ruseckaite, R.A.; Martucci, J.F. Chitosan-gelatin composites and bi-layer films with potential antimicrobial activity. *Food Hydrocoll.* **2011**, *25*, 1372–1381. [CrossRef]
37. Yehuala, G.A.; Emire, S.A. Antimicrobial activity, physicochemical and mechanical properties of Aloe (*Aloe debrana*) based packaging films. *BJAST* **2014**, *3*, 1257–1275. [CrossRef]
38. Shaw, N.B.; Monahan, F.J.; O'Riordan, E.D.; O'Sullivan, M. Physical properties of WPI films plasticized with glycerol, xylitol, or sorbitol. *J. Food Sci.* **2002**, *67*, 164–167. [CrossRef]
39. Gouveia, T.I.A.; Biernacki, K.; Castro, M.C.R.; Gonçalves, M.P.; Souza, H.K.S. A New approach to develop biodegradable films based on thermoplastic pectin. *Food Hydrocoll.* **2019**, *97*, 105175. [CrossRef]
40. Meerasri, J.; Sothornvit, R. Characterization of bioactive film from pectin incorporated with gamma-aminobutyric acid. *Int. J. Biol. Macromol.* **2020**, *147*, 1285–1293. [CrossRef] [PubMed]
41. Kang, H.; Jo, C.; Lee, N.; Kwon, J.; Byun, M. A Combination of gamma irradiation and CaCl immersion for a pectin-based biodegradable film. *Carbohydr. Polym.* **2005**, *60*, 547–551. [CrossRef]
42. Cerna, M.; Barros, A.S.; Nunes, A.; Rocha, S.M.; Delgadillo, I.; Copikova, J.; Coimbra, M.A. Use of FT-IR spectroscopy as a tool for the analysis of polysaccharide food additives. *Carbohydr. Polym.* **2003**, *51*, 383–389. [CrossRef]
43. Lorevice, M.V.; Otoni, C.G.; Moura, M.R.d.; Mattoso, L.H.C. Chitosan nanoparticles on the improvement of thermal, barrier, and mechanical properties of high and low-methyl pectin Films. *Food Hydrocoll.* **2016**, *52*, 732–740. [CrossRef]
44. Ye, S.; Zhu, Z.; Wen, Y.; Su, C.; Jiang, L.; He, S.; Shao, W. Facile and green preparation of pectin/cellulose composite films with enhanced antibacterial and antioxidant behaviors. *Polymers* **2019**, *11*, 57. [CrossRef] [PubMed]
45. Syntsya, A.; Čpíková, J.; Marounek, M.; Mlčochová, P.; Sihelková, L.; Blafková, P.; Tkadlecová, M.; Havlíček, J. Preparation of N-alkylamides of highly methylated (HM) citrus pectin. *Czech J. Food Sci.* **2011**, *21*, 162–166. [CrossRef]
46. Singthong, J.; Cui, S.W.; Ningsanond, S.; Goff, H.D. Structural characterization, degree of esterification and some gelling properties of krueo ma noy (*Cissampelos pareira*) pectin. *Carbohydr. Polym.* **2004**, *58*, 391–400. [CrossRef]
47. Yu, H.; Peng, C.; Li, F.-C.; Yu, P. Effect of chloride salt type on the physicochemical, mechanical and morphological properties of fish gelatin film. *Mater. Res. Express* **2019**, *6*, 126414. [CrossRef]
48. Ju, A.; Song, K.B. Active biodegradable films based on water soluble polysaccharides from white jelly mushroom (*Tremella fuciformis*) containing roasted peanut skin extract. *LWT Food Sci. Technol.* **2020**, *126*, 109293. [CrossRef]
49. Alvarez-Zeferino, J.C.; Beltrán-Villavicencio, M.; Vázquez-Morillas, A. Degradation of plastics in seawater in laboratory. *Open J. Polym. Chem.* **2015**, *05*, 55–62. [CrossRef]
50. Nakayama, A.; Yamano, N.; Kawasaki, N. Biodegradation in seawater of aliphatic polyesters. *Polym. Degrad. Stab.* **2019**, *166*, 290–299. [CrossRef]
51. EN 13432:2000. *Packaging—Requirements for Packaging Recoverable Through Composting and Biodegradation—Test. Scheme and Evaluation Criteria for the Final Acceptance of Packaging*; European Committee for Standardization: Brussels, Belgium, 2000.
52. Shah, A.A.; Hasan, F.; Hameed, A.; Ahmed, S. Biological degradation of plastics: A comprehensive review. *Biotechnol. Adv.* **2008**, *26*, 246–265. [CrossRef]
53. Jaramillo, M.C.; Gutiérrez, T.J.; Goyanes, S.; Bernal, C.; Famá, L. Biodegradability and plasticizing effect of yerba mate extract on cassava starch edible films. *Carbohydr. Polym.* **2016**, *151*, 150–159. [CrossRef]
54. Rech, C.R.; da Silva Brabes, K.C.; Bagnara e Silva, B.E.; Bittencourt, P.R.S.; Koschevic, M.T.; da Silveira, T.F.S.; Martines, M.A.U.; Caon, T.; Martelli, S.M. Biodegradation of eugenol-loaded polyhydroxybutyrate films in different soil types. *Case Stud. Chem. Environ. Eng.* **2020**, *2*, 100014. [CrossRef]
55. Folino, A.; Karageorgiou, A.; Calabrò, P.S.; Komilis, D. Biodegradation of wasted bioplastics in natural and industrial environments: A Review. *Sustainability* **2020**, *12*, 6030. [CrossRef]
56. Webb, H.K.; Arnott, J.; Crawford, R.J.; Ivanova, E.P. Plastic degradation and its environmental implications with special reference to poly(ethylene terephthalate). *Polymers* **2013**, *5*, 1–18. [CrossRef]
57. Chamas, A.; Moon, H.; Zheng, J.; Qiu, Y.; Tabassum, T.; Jang, J.H.; Abu-Omar, M.; Scott, S.L.; Suh, S. Degradation rates of plastics in the environment. *ACS Sustain. Chem. Eng.* **2020**, *8*, 3494–3511. [CrossRef]
58. Olaosebikan, O.O.; Alo, M.N.; Ugah, U.I.; Olayemi, A.M. Environmental effect on biodegradability of plastic and paper bags. *IOSR J. Environ. Sci. Toxicol. Food Technol.* **2014**, *8*, 22–29.

Article

Fabrication of Copper Sulfide Nanoparticles and Limonene Incorporated Pullulan/Carrageenan-Based Film with Improved Mechanical and Antibacterial Properties

Swarup Roy and Jong-Whan Rhim *

Department of Food and Nutrition, BioNanocomposite Research Institute, Kyung Hee University, 26 Kyungheedae-ro, Dongdaemun-gu, Seoul 02447, Korea; swaruproy2013@gmail.com
* Correspondence: jwrhim@khu.ac.kr

Received: 28 October 2020; Accepted: 10 November 2020; Published: 12 November 2020

Abstract: Edible biopolymer (pullulan/carrageenan) based functional composite films were fabricated by the addition of copper sulfide nanoparticles (CuSNP) and D-limonene (DL). The DL and CuSNP were compatible with the pullulan/carrageenan biopolymer matrix. The addition of CuSNP significantly increased the UV-blocking properties without substantially reducing the transparency of the film. The addition of CuSNP improved the film's tensile strength by 10%; however, the DL addition did not significantly influence the strength, while the combined addition of CuSNP and DL increased the strength by 15%. The addition of the fillers did not significantly affect the thermal stability of the film, but the water vapor barrier property was slightly improved. There was no significant change in the moisture content and hydrophobicity of the composite film. Besides, the composite film showed some antimicrobial activity against food-borne pathogenic bacteria. The fabricated pullulan/carrageenan-based film with antimicrobial and UV-barrier properties is likely to be used in active food packaging applications.

Keywords: pullulan/carrageenan; CuSNP; limonene; composite film; mechanical property; antibacterial activity

1. Introduction

Today, the use of plastics is rapidly increasing not only in the food packaging sector, but also in all other industries, and they generate enormous plastic waste and environmental pollution due to their non-degradability [1,2]. Since the industrialization of plastics in the 1950s, its production has increased exponentially every year, and now around 400 million tons of plastic are produced annually worldwide [3]. This increase in production is expected to reach about 180 million tons in 2050 [4,5]. In Korea alone, the annual consumption of disposable plastics is more than 600,000 tons, and accordingly, plastic waste also increases significantly, generating about 3 million tons of plastic waste every year. Only 22.3% of these were managed by the extended producer responsibility [6]. In general, there are two ways to solve this problem: one is to reduce plastic waste and increase the recycling rate, and the other is to use biodegradable materials that can replace plastics. In this context, biopolymer plastics or bioplastics, which uses annually renewable resources, is an eco-friendly material that can replace petroleum-based plastics [7,8]. Carbohydrates are used to manufacture biopolymer packaging films due to their excellent film-forming ability, good gas barrier properties, and mechanical properties. Pullulan, a carbohydrate produced by microorganisms, is a homo-polysaccharide composed of repeating units of maltotriose. It is a highly water-soluble, colorless, and odorless substance. It has excellent gas and oil barrier properties, making it an ideal choice to make edible films and coatings material. The pullulan-based film has high hydrophilicity, low transparency, and poor mechanical properties [9]. The high cost is another problem with pullulan. One way to solve this problem is to make a composite film by blending pullulan with another compatible polymer such as carrageenan. Carrageenan is a linear sulfated polysaccharide extracted from seaweed, a cost-effective biopolymer known to produce transparent films with moderate mechanical and barrier properties, but the film is very brittle [10,11]. Combining the pullulan and carrageenan blend film is expected to be an excellent film that can improve physical properties and supplement each biopolymer's problems. Previously, blending various biopolymers such as alginate, carboxymethyl cellulose, chitosan, and casein has been tested to improve the physical properties of pullulan film [9,12–14].

The functional properties of composite films are also essential in active and intelligent food packaging applications. A functional pullulan/carrageenan film can be developed by adding functional materials such as nanofillers and bioactive compounds. The addition of nanofillers and bioactive compounds is expected to improve the film's physical and functional properties.

Recently, copper sulfide nanoparticles (CuSNP) have emerged as a potential nanofiller that enhances films' physical and functional properties [15,16]. CuSNP has received considerable attention due to its low toxicity and potential applications in biological fields such as drug delivery, photothermal therapy, antimicrobial agents, in vitro bio-sensing, etc. [17–21]. Another potential bioactive functional compound is limonene, a monoterpene found primarily in citrus fruits such as lemons, grapefruits, oranges, etc. [22]. Limonene, belonging to the GRAS material, is also used in many applications such as flavorings, food preservatives, etc. [23,24]. Recently, some functional films reinforced with CuSNP have been reported [15,16,25], and there have been several reports of functional films with added limonene [22,26–28]. Considering the potential of CuSNP and limonene as functional materials, synergies can be expected when these materials are used together. To the best of our knowledge, so far, there are no reports of manufacturing pullulan/carrageenan-based functional films by combining CuSNP with D-limonene.

Therefore, the primary purpose of this study was to prepare a pullulan/carrageenan-based functional film by integrating CuSNP and limonene as fillers for active food packaging applications. The effect of CuSNP and limonene alone or in combination with pullulan/carrageenan-based films were closely investigated.

2. Materials and Methods

2.1. Materials

Pullulan powder was obtained from Korea Bio Polymer Co. Ltd. (Bucheon, Gyeonggi-do, Korea). Food grade carrageenan was purchased from MSC Co., Ltd. (Sungnam City, Gyeonggi-do, Korea). Glycerol was acquired from Daejung Chemicals & Metals Co., Ltd. (Siheung, Gyeonggi-do, Korea). D-limonene (DL) was purchased from Sigma-Aldrich (St. Louis, MO, USA). Brain heart infusion broth (BHI), tryptic soy broth (TSB), and agar powder were obtained from Duksan Pure Chemicals Co., Ltd. (Ansan, Gyeonggi-do, Korea). *Escherichia coli* O157: H7 ATCC 43895 and *Listeria monocytogenes* ATCC 15313 were procured from the Korean Collection for Type Culture (KCTC, Seoul, Korea). The copper sulfide nanoparticles (CuSNP) used in this study were prepared as previously reported [25].

2.2. Preparation of Films

The pullulan/carrageenan-based film was prepared using a solution casting method [29,30], as shown schematically in Scheme 1. For the preparation of film solution, DL (5 wt.% based on biopolymer) was mixed with 150 mL distilled water with vigorous mixing using a magnetic stirrer. The CuSNP (0.5 wt.% based on biopolymer) was dispersed in 150 mL of distilled water using a magnetic stirrer and then ultrasonicated at 60% amplitude for 3 min with pulses 5 s on and 2 s off in a probe ultrasonicator (Model VCX 750, Sonics & Materials, Inc., New Town, CT, USA). Mixed solutions of CuSNP (0.5 wt.%) and DL (5 wt.%) were also prepared. To the CuSNP and DL dispersed solutions, 1.2 g of glycerol (30 wt.% based on polymers) was added with continuous stirring. Then, 4 g of biopolymers (2 g each of pullulan and carrageenan) was dissolved slowly and heated for 20 min at 95 °C with constant stirring. The film-forming solution was cast on a flat Teflon film-coated glass plate and dried at room temperature for 48 h. The dried film was peeled from the container and conditioned at 25 °C and 50% RH for at least 48 h. For comparison, control pullulan/carrageenan was prepared following the same procedure without adding CuSNP and DL. The produced films were designated as Pul/Carr, Pul/Carr/DL, Pul/Carr/CuSNP, and Pul/Carr/DL/CuSNP depending on the type of biopolymer and filler material.

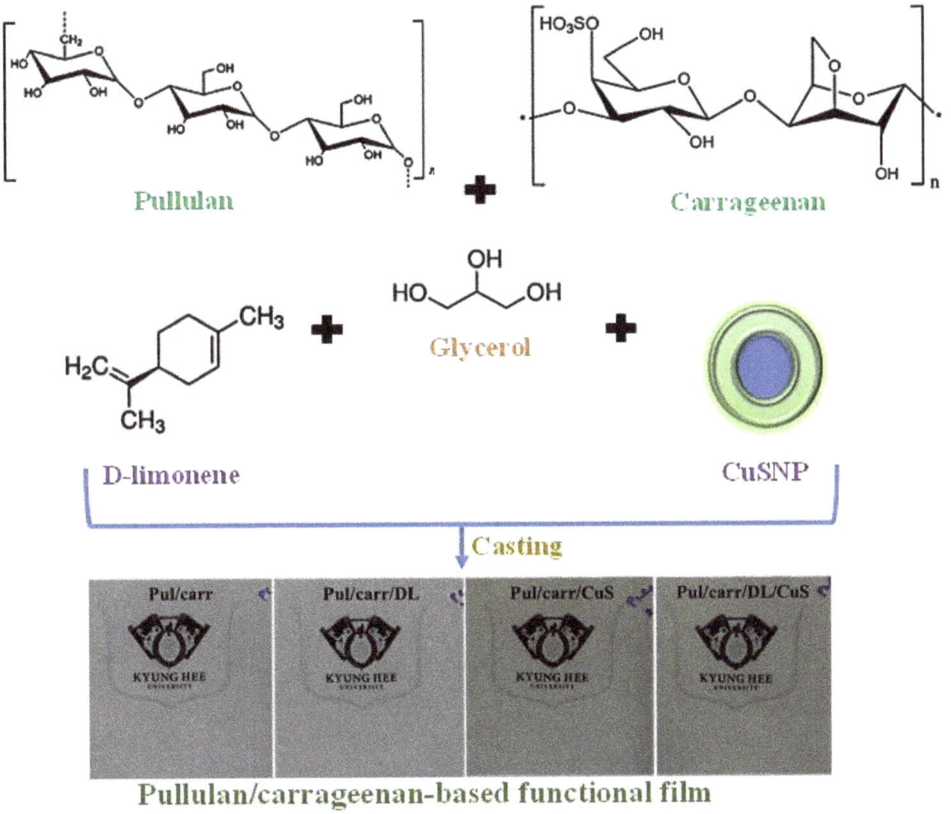

Scheme 1. Schematic presentation for the preparation of Pul/Carr-based films.

2.3. Characterization and Properties of the Film

2.3.1. Optical Properties

The films' surface color (L, a, and b-values) was evaluated using a Chroma meter (Konica Minolta, CR-400, Tokyo, Japan). The total color difference (ΔE), whiteness index (WI), Chroma (C), and yellowness index (YI) of the film were calculated as follows:

$$\Delta E = \sqrt{(\Delta L)^2 + (\Delta a)^2 + (\Delta b)^2} \qquad (1)$$

where ΔL, Δa, and Δb are the difference between each color value of the standard color plate and film sample, respectively.

$$YI = (142.86 \times b)/L \tag{2}$$

$$WI = 100 - \sqrt{(100 - L) + a^2 + b^2} \tag{3}$$

$$C = \sqrt{a^2 + b^2} \tag{4}$$

The UV-vis spectra of the Pul/Carr-based film were recorded using a UV-vis spectrophotometer (Mecasys Optizen POP Series UV/Vis, Seoul, Korea), and the UV-barrier and transparency properties of the film were assessed by measuring the light transmittance at 280 (T_{280}) and 660 nm (T_{660}), respectively [16].

2.3.2. Morphology, FTIR, and XRD

The film samples' microscopic morphological view was inspected using a FESEM (FE-SEM, SU 8010, Hitachi Co., Ltd., Matsuda, Japan). FTIR spectra of the film samples were noted using an attenuated total reflectance-FTIR spectrophotometer (TENSOR 37 Spectrophotometer with OPUS 6.0 software, Billerica, MA, USA) at a wavenumber of 4000–500 cm^{-1} with the resolution of 32 scans at 4 cm^{-1}. The XRD pattern of the film sample was determined at 2θ = 20–80° using an XRD diffractometer (PANalytical X'pert Pro MRD Diffractometer, Amsterdam, Netherlands) at a scan rate of 0.4°/min.

2.3.3. Mechanical Properties

The film's thickness was measured using a hand-held digital micrometer (Digimatic Micrometer, QuantuMike IP 65, Mitutoyo, Japan) with an accuracy of 1 μm. The film thickness was measured at five random locations of each film, and their average value was reported. Each film's mechanical properties were determined according to the standard method of ASTM D 882-88 using an Instron Universal Testing Machine (Model 5565, Instron Engineering Corporation, Canton, MA, USA). For the measurement, the film sample was cut with (2.54 × 15 cm) dimension using a precision double blade cutter (model LB.02/A, Metrotec, S.A., San Sebastian, Spain) [31]. The machine was operated in the tensile mode with an initial grip separation and crosshead speed set at 50 mm and 50 mm/min, respectively.

2.3.4. Water Vapor Permeability (WVP), Water Contact Angle (WCA), and Moisture Content (MC)

The WVP of the films was determined gravimetrically using the ASTM E96-95 standard method [16]. At first, the WVP cup was first filled with a prescribed amount of water and then covered with the film and sealed and kept in the controlled environmental chamber at 25 °C and 50% RH. After equilibration, the WVP cup's weight was measured at every one-hour interval, and weight loss was calculated. The WVTR (g/m^2 s) was determined from the slope (linear) of the steady-state portion of the weight loss vs. time curve. Then, the WVP of the films was calculated in g m/m^2 Pa s as follows:

$$WVP = (WVTR \times L)/\Delta p \tag{5}$$

where L was the film thickness (m), and Δp was water vapor partial pressure difference (Pa) across the film.

The film's surface wettability was determined by computing the film surface's water contact angle using a WCA analyzer (Phoneix 150, Surface Electro Optics Co., Ltd., Kunpo, Gyeonggi-do, Korea). For the measurement, the film sample was fixed on the film holder, and then a drop of water (~10 μL) was added to the surface of the film and immediately determined the WCA.

The MC of the film was determined by following the previously published methodology [32]. The MC of the film samples was determined by checking the weight of the film sample (W_1), and then the same film was dried in a hot air oven at 105 °C for 24 h and then weighed again (W$_2$). The MC of the film was expressed as a percentage of the initial weight of the film:

$$\text{MC (\%)} = \frac{W_2 - W_1}{W_1} \times 100 \tag{6}$$

2.3.5. Thermal Analysis

The films' thermal stability was determined using a thermogravimetric analyzer (Hi-Res TGA 2950, TA Instrument, New Castle, DE, USA). For measurement, ~10 mg of film sample was taken in a standard aluminum pan and scanned at a heating rate of 10 °C/min in a temperature range of 30–600 °C under a nitrogen flow of 50 cm^3/min with an empty pan as a reference [32].

2.4. Antibacterial Activity

The film specimen's antibacterial activity was tested against food-borne pathogenic bacteria, *E. coli*, and *L. monocytogenes* by following a previously published method [33]. The microorganisms were inoculated in the TSB and BHI broth, respectively, and subsequently cultured overnight at 37 °C with agitation at 100 rpm. After diluting the inoculum appropriately, 100 μL of the diluted inoculum was transferred to 20 mL of TSB and BHI broth containing 100 mg of film samples and incubated at 37 °C for 12 h with agitation at 120 rpm. Samples were taken out at a predetermined time interval and plated on agar plates after appropriate dilution to evaluate the viable colonies. For calculation, an antibacterial test was performed using culture medium without film and neat film as negative control and positive control, correspondingly. The antimicrobial tests were done in triplicate.

2.5. Statistical Analysis

For statistical analysis of the obtained results, one-way analysis of variance (ANOVA) was performed, and the significance of each mean property value was determined ($p < 0.05$) by Duncan's multiple range test using the SPSS statistical analysis computer program (SPSS, Inc., Chicago, IL, USA).

3. Results and Discussion

3.1. Apparent Color, Morphology, and Optical Properties

The macroscopic appearance of all the prepared films is displayed in Figure 1a–d. The Pul/Carr and Pul/Carr/DL films were transparent without color, but the CuSNP-added films were greenish. The films' morphology was observed by the FESEM (Figure 1e–h), and it shows all the films were intact and without any apparent defects. The blending of carrageenan in pullulan makes a uniformly distributed homogenous and compatible film. The fillers were also evenly mixed in the matrix polymer, indicating their good miscibility in the liquid phase. In addition, there were no cracks or voids in the surface image and no apparent accumulation of particles, indicating both CuSNP and DL are compatible with the polymer matrix. The excellent compatibility was due to the strong adhesion, intermolecular binding, and affinity between the fillers and the matrix polymer.

UV-visible transmission spectra determined the film's optical properties, and the results are displayed in Figure 2. The UV-vis spectra exhibit that the Pul/Carr and Pul/Carr/DL films were transparent. On the other hand, the CuSNP-added (alone or combined) composite film's absorption profile was completely different due to the near-infrared absorption of CuSNP [15]. For further understanding, the film's UV-light barrier and transparency were scrutinized by measuring the absorbance at 280 nm and 660 nm, respectively, and the results are displayed in Table 1. The UV light transmittance and transparency of the control film were 70.1% and 88.3%, respectively. The addition of DL did not much influence the optical properties but was greatly affected by the incorporation of CuSNP. The UV-light barrier property was increased ~75%, which is very significant, whereas the transparency remains ~70%, which is suitable for packaging application. The substantial increase in

UV-light barrier property was due to the UV-light absorption of CuSNP [25]. CuSNP and DL's present findings on Pull/Carr films' color properties are consistent with the previously published results [16,26]. Therefore, the current finding suggested that the blending of CuSNP/DL improved the UV-light barrier properties of the film without greatly affecting the transparency.

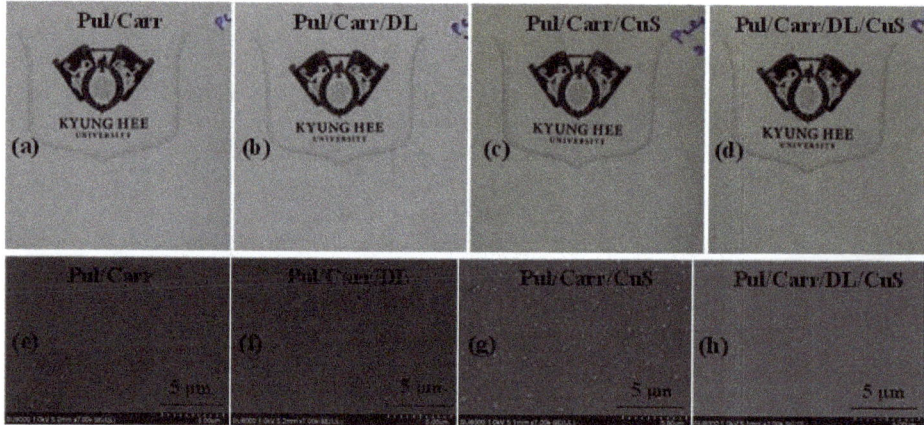

Figure 1. The visual appearance (**a–d**) and microstructure (**e–h**) of the pullulan/carrageenan-based films.

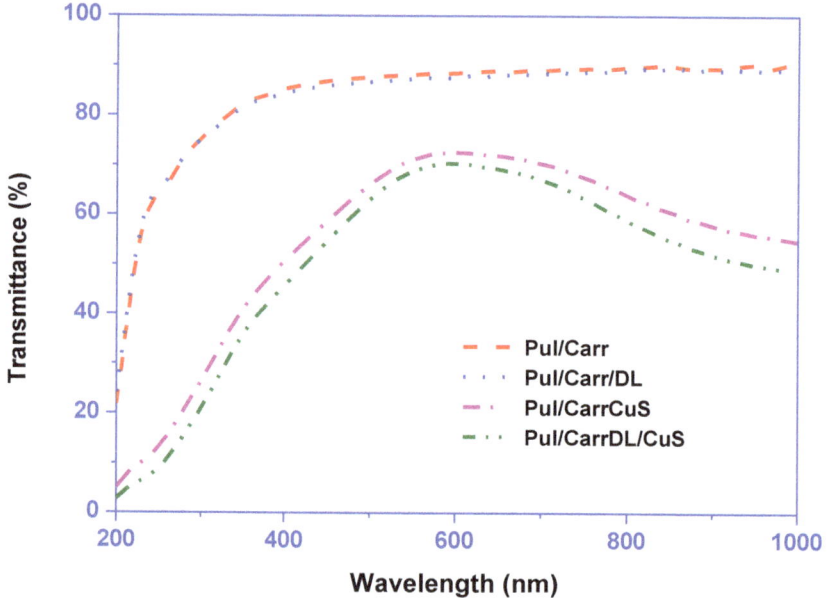

Figure 2. Light transmittance spectra of pullulan/carrageenan-based films.

Table 1. Surface color and transmittance of pullulan/carrageenan-based composite films.

Films	L	a	b	ΔE	WI	C	YI	T_{280} (%)	T_{660} (%)
Pul/Carr	91.3 ± 0.1 [b]	−0.6 ± 0.0 [b]	5.6 ± 0.1 [a]	1.6 ± 0.1 [a]	93.6 ± 0.1 [c]	5.6 ± 0.1 [a]	8.8 ± 0.1 [a]	70.1 ± 1.2 [b]	88.3 ± 0.4 [b]
Pul/Carr/DL	91.3 ± 0.0 [b]	−0.6 ± 0.0 [b]	5.6 ± 0.1 [a]	1.6 ± 0.1 [a]	93.7 ± 0.1 [c]	5.6 ± 0.1 [a]	8.7 ± 0.1 [a]	71.0 ± 1.6 [b]	88.4 ± 0.7 [b]
Pul/Carr/CuS	76 ± 0.6 [a]	−4.9 ± 0.7 [a]	17.2 ± 0.8 [c]	20.9 ± 0.8 [b]	81.5 ± 0.6 [a]	17.9 ± 0.6 [c]	32.1 ± 1.8 [c]	18.5 ± 3.2 [a]	69.5 ± 1.9 [a]
Pul/Carr/DL/CuS	76.3 ± 0.5 [a]	−5.2 ± 0.4 [a]	15.8 ± 0.8 [b]	20.2 ± 0.3 [b]	82.7 ± 0.7 [b]	16.6 ± 0.7 [b]	29.6 ± 1.4 [b]	15.6 ± 2.1 [a]	68.1 ± 1.4 [a]

The values are represented as a mean ± standard deviation. In the same column, any two means, followed by the same letter, are not significantly ($p > 0.05$) different from Duncan's multiple range tests.

Table 1 also shows the surface color and light transmittance properties of the films. The lightness (Hunter L-value) of the film was not significantly affected by the DL, whereas the addition of CuSNP slightly reduced the lightness. The lightness, a-value, b-value, and ΔE were not much changed for DL, whereas it decreased for the CuSNP-added film. The drastic changes in the color parameter were mainly due to the greenish color of CuSNP. The color variation of the CuSNP/DL-added Pul/Carr film was further perceived by analyzing the C, WI, and YI values. The obtained results are consistent with the Hunter color parameter. For DL-added film, there was no significant alteration in C, WI, and YI value. For the CuSNP-added film, the WI was decreased slightly, whereas the C increased similar to the ΔE. The YI of the film also increased significantly in the case of CuSNP-added film. Overall, the luminosity of the film was high, which could be useful for packaging applications. The effect of CuSNP on the color parameter was similar to the previously published data [16].

3.2. FTIR and XRD

The FTIR spectra of CuSNP, DL, and the pullulan/carrageenan-based films are shown in Figure 3. The characteristic peaks of limonene were obtained at 2918, 1645, 1436, and 885 cm^{-1} due to the –C–H stretching, –C=C stretching, –C–H bending, and out-of-plane bending, respectively [34]. In the case of CuSNP, peaks were seen at 3200, 1615, 1100, and 597 cm^{-1} due to the O–H stretching, C–O bending, –C–O stretching, and the pyranoid ring of corn starch capped CuSNP, respectively [35]. The broad peaks in the range of 3600–3000 cm^{-1} observed in all films were due to O–H stretching vibration and intermolecular or intramolecular H-bonding [36]. The peak at 2920 cm^{-1} was ascribed to the C–H stretching vibration of the methylene group of the matrix biopolymer chain [36]. The peak found at 1648 cm^{-1} was attributed to the amide-I of carrageenan [37]. A peak found at 1235 cm^{-1} was due to the presence of sulfate ester groups in carrageenan [38], and the peak noticed at 1156 cm^{-1} was ascribed to the stretching vibration of the (α-1-4) glycosidic bond of pullulan [39]. The peak at 1023 cm^{-1} corresponded to the C–O stretching bond of pullulan [12]. The peak seen at 921 cm^{-1} was due to 3, 6-anhydro-D-galactose of carrageenan, and the peak observed at 842 cm^{-1} was due to galactose-4-sulfate of carrageenan and α-glucopyranose units of pullulan [37,39]. The observed results indicate that the blending of carrageenan in pullulan makes a compatible composite film. In addition, the addition of filler did not change the chemical structure of neat biopolymers. The observed FTIR results conclude that, except for slight alteration in peak intensity, there was no significant alteration in the film's functional groups, suggesting that the film's chemical structure was not transformed after incorporating the filler.

The XRD pattern of the Pul/Carr-based films are presented in Figure 4. All the tested films showed a characteristic broad peak ~20° and some unknown peaks, which is presumably due to the amorphous nature of the biopolymers (pullulan and carrageenan). Except for some slight alteration, the XRD data of the Pul/Carr-based composite films do not show any clear change in peak pattern after the addition of DL and CuSNP. Although CuSNP show characteristic XRD patterns [25], they did not appear in the present findings as only 0.5% of CuSNP was used.

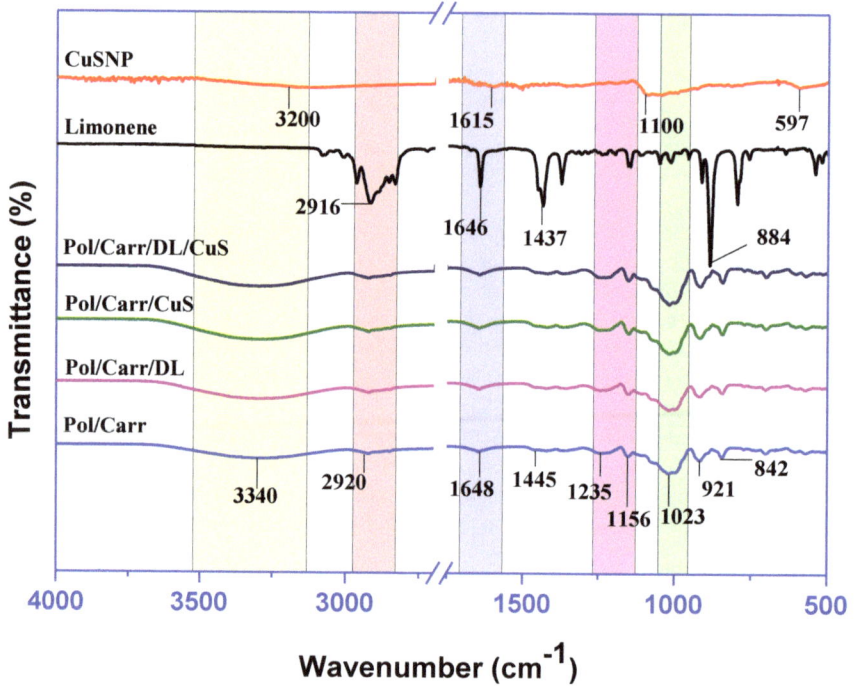

Figure 3. FTIR spectra of CuSNP, DL, and pullulan/carrageenan-based films.

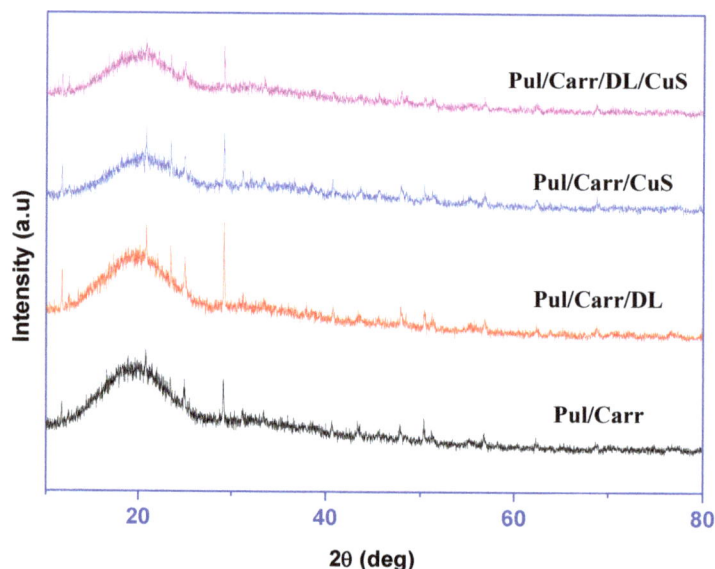

Figure 4. XRD pattern of pullulan/carrageenan-based films.

3.3. Mechanical Properties

The thickness and mechanical properties of the films are presented in Table 2. The film's thickness was ~47–49 µm and was not much altered after blending with the fillers. The addition of fillers meaningfully influenced the mechanical properties of the films (Table 2). The tensile strength (TS) of the film increased significantly by adding CuSNP alone or in combination, although it was not greatly affected by the addition of DL. The flexibility (EB) and stiffness (EM) of the CuSNP/DL incorporated film also increased meaningfully. The addition of only CuSNP improved the TS by 10%. The combined addition of CuSNP/DL improved 15% of the TS, probably due to the increase in molecular interaction between the matrix polymer and fillers and good interfacial surface interactions. Overall, the fillers' addition in the Pul/carr-based film improved the film's strength, flexibility, and stiffness. The CuSNP is already recognized as an efficient nanofiller for enhancing the mechanical properties of the biopolymer-based film. In the case of CuSNP-added agar, alginate, and carrageenan film, a similar improvement in mechanical properties was reported previously [15,16,25]. Similar to the current findings, the tensile strength of the LDPE film was found to not have changed significantly with the addition of lemon aroma [27]. On the other hand, it has been reported that the addition of limonene significantly reduces the mechanical properties of the PLA film [26]. Conversely, it has been reported that PVA/chitosan films have significantly improved mechanical properties by adding DL [22]. Overall, the addition of fillers improves the mechanical properties of the Pul/carr-based film.

Table 2. Mechanical properties, WVP, WCA, and MC of pullulan/carrageenan-based films.

Films	Thickness (μm)	TS (MPa)	EB (%)	EM (GPa)	MC (%)	WVP ($\times 10^{-9}$ g m/m^2 Pa s)	WCA (deg.)
Pul/Carr	48.3 ± 3.0 [a]	54.0 ± 6.5 [a]	2.7 ± 0.3 [a]	3.4 ± 0.3 [a]	10.9 ± 0.6 [a]	1.0 ± 0.1 [ab]	61.6 ± 4.2 [a]
Pul/Carr/DL	47.4 ± 1.8 [a]	54.5 ± 7.3 [a]	3.0 ± 0.4 [a]	3.2 ± 0.2 [a]	11.5 ± 0.3 [a]	0.93 ± 0.1 [ab]	64.0 ± 3.2 [a]
Pul/Carr/CuS	48.8 ± 2.8 [a]	66.5 ± 6.6 [b]	4.0 ± 0.3 [b]	3.4 ± 0.2 [a]	12.0 ± 1.0 [a]	1.06 ± 0.1 [b]	61.7 ± 3.0 [a]
Pul/Carr/DL/CuS	47.5 ± 3.5 [a]	70.5 ± 3.2 [b]	4.1 ± 0.3 [b]	3.7 ± 0.3 [b]	11.6 ± 0.3 [a]	0.87 ± 0.1 [a]	63.7 ± 2.4 [a]

The values are represented as a mean ± standard deviation. In the same column, any two means, followed by the same letter, are not significantly ($p > 0.05$) different from Duncan's multiple range tests.

3.4. WVP, WCA, and MC

The WVP of all the films is also presented in Table 2, and the results indicate the good water vapor barrier properties of the film (~1.0×10^{-9} g m/m^2 Pa s). The WVP of the Pul/Carr-based film was also affected by the addition of filler. The effect was more significant in the combined addition of CuSNP/DL than only CuSNP or DL. The film's increased water vapor barrier properties might be due to creating a tortuous water vapor diffusion path formed by the well-distributed water vapor impermeable filler [22]. Similar to this observation previously, it was also described that DL's addition improved the water vapor barrier property of the PVA/chitosan film [22]. Earlier, it was explained that the addition of CuSNP at a lower content in alginate and agar-based films improved the water vapor barrier properties [16,25].

The WCA of the films is also shown in Table 3. The WCA of the control film and all other films were < 65°, representing their hydrophilic nature [40]. The incorporation of CuSNP and DL alone or in-combination into the film slightly increased the WCA, although the changes were statistically insignificant ($p < 0.05$). The increase in the WCA of the films was mainly due to the presence of CuSNP and DL, which are hydrophobic. Previously in the case of CuSNP-added agar and alginate film, an increase in the WCA was reported [16,25]. In contrast to the current observation in the case of DL-added PVA/chitosan film and PLA-based film, a significant increase in WCA was reported recently [22,26].

Table 3. Thermogravimetric data of pullulan/carrageenan-based films.

Films	T_{onset}/T_{end} (°C)	$T_{0.5}$ (°C)	Char Content (%)	T_{max} (°C)
Pul/Carr	160/445	360	36.2	240
Pul/Carr/DL	160/440	361	37.1	239
Pul/Carr/CuS	159/441	370	38.1	240
Pul/Carr/DL/CuS	159/445	372	38.3	240

The moisture content of all the tested films was ~11–12%, which is relatively low and suitable for packaging film. The moisture content of the filler added film was not altered meaningfully compared to the control film. Similar MC results were reported in previously published pullulan-based packaging film recently [36].

3.5. Thermal Stability

The thermal stability (TGA/DTG) of the films are shown in Figure 5. During thermal decomposition, all the film showed a multi-step weight loss. The first weight loss occurred at 50–115 °C, mostly due to the film's humidity. The second and main weight loss was observed at 160–280 °C with a maximum of around 240 °C due to glycerol and biopolymer degradation [38]. The final stage of weight loss occurred at 290–410 °C due to the remaining biopolymers' decomposition [37,41]. The decomposition temperatures (onset/endset), 50% decomposition temperature ($T_{0.5}$), and the films' char content are shown in Table 3. The $T_{0.5}$ of the CuSNP added film was increased ~10 °C, whereas it remains unchanged for DL. The maximum decomposition temperature of the film was not affected by the presence of fillers. Similar thermal stability results were reported in the case of DL-added PLA, and CuSNP-added alginate film previously [16,26]. TGA test results designated that the incorporation of fillers does not alter the Pul/carr-based film's thermal stability. The char content at 600 °C for the film was varied in the range of 36–38% depending on the type of filler, and the relatively high char content of the films was due to the non-ignitable minerals present in the biopolymer [38].

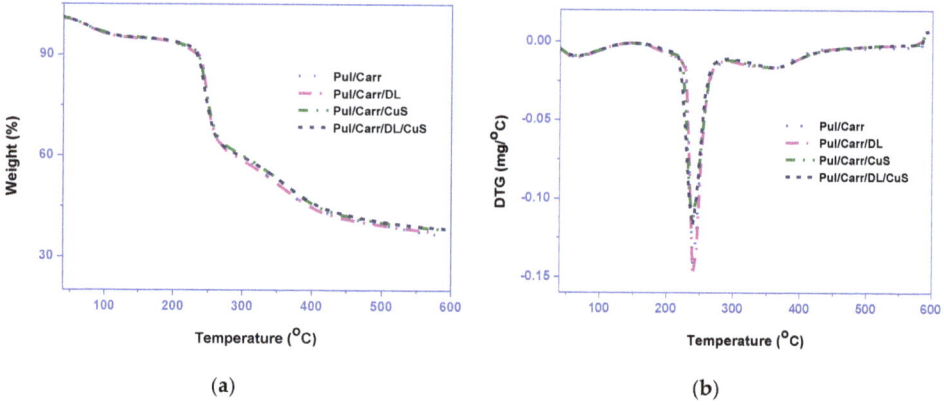

Figure 5. (a) TGA and (b) DTG thermograms of pullulan/carrageenan-based films.

3.6. Antimicrobial Activity

The antimicrobial activity of all the film samples is presented in Figure 6. The control and neat polymer film do not display any antimicrobial activity against tested bacteria, whereas the films comprising CuSNP and DL showed some antibacterial activity against tested strains. The antibacterial activity of CuSNP was more effective compared to DL. However, the composite film containing the combined filler displayed a superficial bactericidal effect. It was observed that the activity of CuSNP is more comprehensive and showed a significant reduction in the growth of *E. coli* (about 7 log cycles) after culturing for 12 h compared to the control, whereas, in the case of *L. monocytogenes*, a slight decrease in growth witnessed (about two log cycles). The variation of antimicrobial activity against *E. coli* and *L. monocytogenes* was most probably due to the cell wall structure's difference. The obtained results suggested that DL showed some antibacterial activity against *E. coli*, but CuSNP exhibited efficient antibacterial activity against *E. coli* and some activity against *L. monocytogenes*. It is previously reported that the CuSNP showed antibacterial activity only at higher content (2%), and in this work, as only 0.5% of CuSNP was used, intense antibacterial action was not observed. Like the current findings, the potent antimicrobial activity of CuSNP-loaded agar, alginate, and carrageenan composite film was reported previously [15,16,25]. The antibacterial activity of CuSNP was not exactly elucidated yet but believed to damage the cell membrane and oxidative damage through the formation of reactive oxygen species [42]. The DL-containing film's antimicrobial activity was comparatively lower than predictable; however, a similar low antimicrobial activity has been reported previously [22]. In the case of DL, the antibacterial activity is dependent on the content and also on the microbial strain [28]. The antibacterial mechanism of DL relies on the ability to penetrate through the microorganism's cell wall, which disturbs the cell viability [22]. Overall, the Pul/Carr-based film with moderate antimicrobial action can delay the growth of the microorganism.

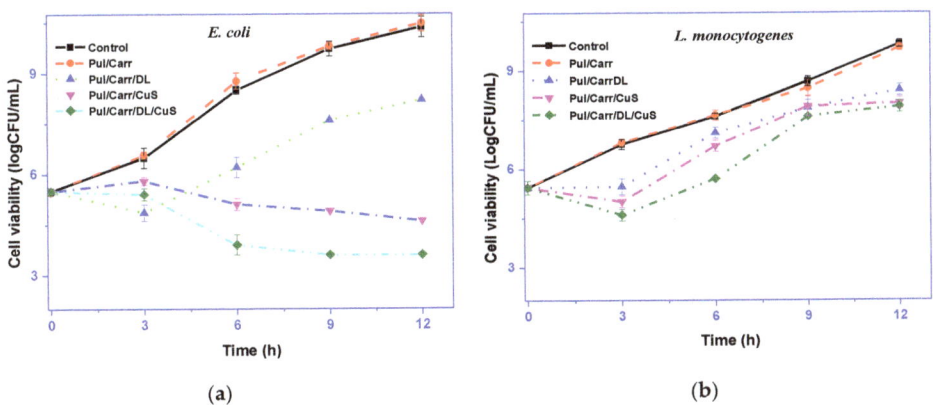

Figure 6. Antimicrobial activity of pullulan/carrageenan-based films against (a) *E. coli* and (b) *L. monocytogenes*.

4. Conclusions

CuSNP and limonene were reinforced to produce a pullulan/carrageenan-based novel composite film. The fillers were compatible and homogeneously spread in the matrix film. The incorporation of fillers meaningfully enhanced (>75%) the UV-light shielding property of the films without greatly losing the transparency (~20%). The mechanical properties (tensile strength, elongation at break, and elastic modulus) of the film are also meaningfully improved (~15%), while the thermal stability remains unaltered. The water vapor barrier property was increased slightly (~10%), and hydrophobicity remains unchanged. The developed film also showed good antimicrobial activity against the pathogenic food microbe, which can be used to control the growth of the microorganism in packaged foods. The fabricated pullulan/carrageenan-based functional film with increased film properties can be used for active food packaging applications. For practical application of the edible active packaging film, additional research is required for mass production of functional fillers, industrial production of functional films using biopolymers, and application of developed films.

Author Contributions: Conceptualization, S.R and J.-W.R.; formal analysis, S.R.; investigation, S.R.; resources, J.-W.R.; writing—original draft preparation, S.R.; review and editing, J.-W.R; supervision, J.-W.R.; funding acquisition, J.-W.R. All authors have read and agreed to the published version of the manuscript.

Funding: This work was supported by the National Research Foundation of Korea (NRF) grant funded by the Korea government (MSIT) (No. 2019R1A2C2084221).

Conflicts of Interest: The authors declare no conflict of interest.

References

1. Groh, K.J.; Backhaus, T.; Carney-Almroth, B.; Geueke, B.; Inostroza, P.A.; Lennquist, A.; Leslie, H.A.; Maffini, M.; Slunge, D.; Trasande, L.; et al. Overview of known plastic packaging-associated chemicals and their hazards. *Sci. Total Environ.* **2019**, *651*, 3253–3268. [CrossRef] [PubMed]
2. Foschi, E.; Bonoli, A. The Commitment of packaging industry in the framework of the European strategy for plastics in a circular economy. *Adm. Sci.* **2019**, *9*, 18. [CrossRef]
3. Geyer, R.; Jambeck, J.R.; Law, K.L. Production, use, and fate of all plastics ever made. *Sci. Adv.* **2017**, *3*, e1700782. [CrossRef] [PubMed]
4. Ryan, P.G. A brief history of marine litter research. In *Marine Anthropogenic Litter*; Springer International Publishing: Berlin/Heidelberg, Germany, 2015; pp. 1–25, ISBN 9783319165103.
5. Lau, W.W.Y.; Shiran, Y.; Bailey, R.M.; Cook, E.; Stuchtey, M.R.; Koskella, J.; Velis, C.A.; Godfrey, L.; Boucher, J.; Murphy, M.B.; et al. Evaluating scenarios toward zero plastic pollution. *Science* **2020**, *369*, 1455–1461. [CrossRef] [PubMed]
6. Jang, Y.C.; Lee, G.; Kwon, Y.; Lim, J.H.; Jeong, J.H. Recycling and management practices of plastic packaging waste towards a circular economy in South Korea. *Resour. Conserv. Recycl.* **2020**, *158*, 104798. [CrossRef]
7. Ramesh, M.; Narendra, G.; Sasikanth, S. A review on biodegradable packaging materials in extending the shelf life and quality of fresh fruits and vegetables. In *Waste Management as Economic Industry Towards Circular Economy*; Springer: Singapore, 2020; pp. 59–65.
8. Hoffmann, T.; Peters, D.A.; Angioletti, B.; Bertoli, S.; Vieira, L.P.; Reiter, M.G.R.; Souza, C.K. De Potentials nanocomposites in food packaging. *Chem. Eng. Trans.* **2019**, *75*, 253–258. [CrossRef]
9. Tong, Q.; Xiao, Q.; Lim, L.T. Preparation and properties of pullulan-alginate-carboxymethylcellulose blend films. *Food Res. Int.* **2008**, *41*, 1007–1014. [CrossRef]
10. Yadav, M.; Chiu, F.C. Cellulose nanocrystals reinforced κ-carrageenan based UV resistant transparent bionanocomposite films for sustainable packaging applications. *Carbohydr. Polym.* **2019**, *211*, 181–194. [CrossRef]
11. Jancikova, S.; Dordevic, D.; Jamroz, E.; Behalova, H.; Tremlova, B. Chemical and physical characteristics of edible films, based on κ- and ι-carrageenans with the addition of lapacho tea extract. *Foods* **2020**, *9*, 357. [CrossRef]
12. Li, Y.; Yokoyama, W.; Wu, J.; Ma, J.; Zhong, F. Properties of edible films based on pullulan-chitosan blended film-forming solutions at different pH. *RSC Adv.* **2015**, *5*, 105844–105850. [CrossRef]
13. Kristo, E.; Biliaderis, C.G.; Zampraka, A. Water vapor barrier and tensile properties of composite caseinate-pullulan films: Biopolymer composition effects and impact of beeswax lamination. *Food Chem.* **2007**, *101*, 753–764. [CrossRef]
14. Yang, Y.; Xie, B.; Liu, Q.; Kong, B.; Wang, H. Fabrication and characterization of a novel polysaccharide based composite nanofiber films with tunable physical properties. *Carbohydr. Polym.* **2020**, *236*, 116054. [CrossRef] [PubMed]
15. Li, F.; Liu, Y.; Cao, Y.; Zhang, Y.; Zhe, T.; Guo, Z.; Sun, X.; Wang, Q.; Wang, L. Copper sulfide nanoparticle-carrageenan films for packaging application. *Food Hydrocoll.* **2020**, *109*, 106094. [CrossRef]
16. Roy, S.; Rhim, J.-W. Effect of CuS reinforcement on the mechanical, water vapor barrier, UV-light barrier, and antibacterial properties of alginate-based composite films. *Int. J. Biol. Macromol.* **2020**, *164*, 37–44. [CrossRef]
17. Xiao, Z. CuS nanoparticles: Clinically favorable materials for photothermal applications? *Nanomedicine* **2014**, *9*, 373–375. [CrossRef]
18. Ku, G.; Zhou, M.; Song, S.; Huang, Q.; Hazle, J.; Li, C. Copper sulfide nanoparticles as a new class of photoacoustic contrast agent for deep tissue imaging at 1064 nm. *ACS Nano* **2012**, *6*, 7489–7496. [CrossRef]
19. Dharsana, U.S.; Varsha, M.K.N.S.; Behlol, A.A.K.; Veerappan, A.; Thiagarajan, R.; Sai Varsha, M.K.N.; Khan Behlol, A.A.; Veerappan, A.; Thiagarajan, R. Sulfidation modulates the toxicity of biogenic copper nanoparticles. *RSC Adv.* **2015**, *5*, 30248–30259. [CrossRef]

20. Goel, S.; Chen, F.; Cai, W. Synthesis and biomedical applications of copper sulfide nanoparticles: From sensors to theranostics. *Small* **2014**, *10*, 631–645. [CrossRef]
21. Saldanha, P.L.; Brescia, R.; Prato, M.; Li, H.; Povia, M.; Manna, L.; Lesnyak, V. Generalized one-pot synthesis of copper sulfide, selenide-sulfide, and telluride-sulfide nanoparticles. *Chem. Mater.* **2014**, *26*, 1442–1449. [CrossRef]
22. Lan, W.; Wang, S.; Chen, M.; Sameen, D.E.; Lee, K.; Liu, Y. Developing poly(vinyl alcohol)/chitosan films incorporate with d-limonene: Study of structural, antibacterial, and fruit preservation properties. *Int. J. Biol. Macromol.* **2020**. [CrossRef]
23. Umagiliyage, A.L.; Becerra-Mora, N.; Kohli, P.; Fisher, D.J.; Choudhary, R. Antimicrobial efficacy of liposomes containing D-limonene and its effect on the storage life of blueberries. *Postharvest Biol. Technol.* **2017**, *128*, 130–137. [CrossRef]
24. Erasto, P.; Viljoen, A.M. Limonene—A review: Biosynthetic, ecological and pharmacological relevance. *Nat. Prod. Commun.* **2008**, *3*, 1193–1202. [CrossRef]
25. Roy, S.; Rhim, J.-W.; Jaiswal, L. Bioactive agar-based functional composite film incorporated with copper sulfide nanoparticles. *Food Hydrocoll.* **2019**, *93*, 156–166. [CrossRef]
26. Arrieta, M.P.; López, J.; Ferrándiz, S.; Peltzer, M.A. Characterization of PLA-limonene blends for food packaging applications. *Polym. Test.* **2013**, *32*, 760–768. [CrossRef]
27. Dias, M.V.; de Medeiros, H.S.; Soares, N.D.; de Melo, N.R.; Borges, S.V.; Carneiro, J.D.; de Assis Kluge, J.M. Development of low-density polyethylene films with lemon aroma. *LWT Food Sci. Technol.* **2013**, *50*, 167–171. [CrossRef]
28. Yao, Y.; Ding, D.; Shao, H.; Peng, Q.; Huang, Y. Antibacterial activity and physical properties of fish gelatin-chitosan edible films supplemented with D-Limonene. *Int. J. Polym. Sci.* **2017**, *2017*. [CrossRef]
29. Roy, S.; Rhim, J.-W. Carboxymethyl cellulose-based antioxidant and antimicrobial active packaging film incorporated with curcumin and zinc oxide. *Int. J. Biol. Macromol.* **2020**, *148*, 666–676. [CrossRef]
30. Roy, S.; Rhim, J.-W. Agar-based antioxidant composite films incorporated with melanin nanoparticles. *Food Hydrocoll.* **2019**, *94*, 391–398. [CrossRef]
31. Roy, S.; Rhim, J.-W. Curcumin incorporated poly(butylene adipate-co-terephthalate) film with improved water vapor barrier and antioxidant properties. *Materials* **2020**, *13*, 4369. [CrossRef]
32. Roy, S.; Van Hai, L.; Kim, H.C.; Zhai, L.; Kim, J. Preparation and characterization of synthetic melanin-like nanoparticles reinforced chitosan nanocomposite films. *Carbohydr. Polym.* **2020**, *231*, 115729. [CrossRef]
33. Roy, S.; Shankar, S.; Rhim, J.-W. Melanin-mediated synthesis of silver nanoparticle and its use for the preparation of carrageenan-based antibacterial films. *Food Hydrocoll.* **2019**, *88*, 237–246. [CrossRef]
34. Türkoğlu, G.C.; Sarışık, A.M.; Erkan, G.; Yıkılmaz, M.S.; Kontart, O. Micro-and nano-encapsulation of limonene and permethrin for mosquito repellent finishing of cotton textiles. *Iran Polym. J.* **2020**, *29*, 321–329. [CrossRef]
35. Dai, H.; Chang, P.R.; Geng, F.; Yu, J.; Ma, X. Preparation and properties of starch-based film using N, N-bis (2-hydroxyethyl) formamide as a new plasticizer. *Carbohydr. Polym.* **2010**, *79*, 306–311. [CrossRef]
36. Lee, J.H.; Jeong, D.; Kanmani, P. Study on physical and mechanical properties of the biopolymer/silver based active nanocomposite films with antimicrobial activity. *Carbohydr. Polym.* **2019**, 115159. [CrossRef] [PubMed]
37. Roy, S.; Rhim, J.-W. Preparation of carbohydrate-based functional composite films incorporated with curcumin. *Food Hydrocoll.* **2020**, *98*, 105302. [CrossRef]
38. Roy, S.; Rhim, J.-W. Preparation of carrageenan-based functional nanocomposite films incorporated with melanin nanoparticles. *Colloids Surfaces B Biointerfaces* **2019**, *176*, 317–324. [CrossRef]
39. Yeasmin, S.; Yeum, J.H.; Yang, S.B. Fabrication and characterization of pullulan-based nanocomposites reinforced with montmorillonite and tempo cellulose nanofibril. *Carbohydr. Polym.* **2020**, *240*, 116307. [CrossRef]
40. Vogler, E.A. Structure and reactivity of water at biomaterial surfaces. *Adv. Colloid Interface Sci.* **1998**, *74*, 69–117. [CrossRef]
41. Han, K.; Liu, Y.; Liu, Y.; Huang, X.; Sheng, L. Characterization and film-forming mechanism of egg white/pullulan blend film. *Food Chem.* **2020**, *315*, 126201. [CrossRef]

42. Ahmed, K.B.A.; Anbazhagan, V. Synthesis of copper sulfide nanoparticles and evaluation of in vitro antibacterial activity and in vivo therapeutic effect in bacteria-infected zebrafish. *RSC Adv.* **2017**, *7*, 36644–36652. [CrossRef]

Publisher's Note: MDPI stays neutral with regard to jurisdictional claims in published maps and institutional affiliations.

© 2020 by the authors. Licensee MDPI, Basel, Switzerland. This article is an open access article distributed under the terms and conditions of the Creative Commons Attribution (CC BY) license (http://creativecommons.org/licenses/by/4.0/).

Article

Curcumin and Diclofenac Therapeutic Efficacy Enhancement Applying Transdermal Hydrogel Polymer Films, Based on Carrageenan, Alginate and Poloxamer

Katarina S. Postolović [1], Milan D. Antonijević [2], Biljana Ljujić [3], Slavko Radenković [1], Marina Miletić Kovačević [4], Zoltan Hiezl [2], Svetlana Pavlović [2], Ivana Radojević [5] and Zorka Stanić [1,*]

1. Department of Chemistry, Faculty of Science, University of Kragujevac, 34000 Kragujevac, Serbia
2. Faculty of Engineering and Science, School of Science, Medway Campus, University of Greenwich, Chatham Maritime, Kent ME4 4TB, UK
3. Department of Genetics, Faculty of Medical Sciences, University of Kragujevac, 34000 Kragujevac, Serbia
4. Department of Histology and Embryology, Faculty of Medical Sciences, University of Kragujevac, 34000 Kragujevac, Serbia
5. Department of Biology and Ecology, Faculty of Science, University of Kragujevac, 34000 Kragujevac, Serbia
* Correspondence: zorka.stanic@pmf.kg.ac.rs; Tel.: +381-34336223; Fax: +381-34335040

Abstract: Films based on carrageenan, alginate and poloxamer 407 have been formulated with the main aim to apply prepared formulations in wound healing process. The formulated films were loaded with diclofenac, an anti-inflammatory drug, as well as diclofenac and curcumin, as multipurpose drug, in order to enhance encapsulation and achieve controlled release of these low-bioavailability compounds. The obtained data demonstrated improved drug bioavailability (encapsulation efficiency higher than 90%), with high, cumulative in vitro release percentages (90.10% for diclofenac, 89.85% for curcumin and 95.61% for diclofenac in mixture-incorporated films). The results obtained using theoretical models suggested that curcumin establishes stronger, primarily dispersion interactions with carrier, in comparison with diclofenac. Curcumin and diclofenac-loaded films showed great antibacterial activity against Gram-positive bacteria strains (*Bacillus subtilis* and *Staphylococcus aureus*, inhibition zone 16.67 and 13.67 mm, respectively), and in vitro and in vivo studies indicated that curcumin- and diclofenac-incorporated polymer films have great potential, as a new transdermal dressing, to heal wounds, because diclofenac can target the inflammatory phase and reduce pain, whereas curcumin can enhance and promote the wound healing process.

Keywords: curcumin; diclofenac; films; biopolymers; carrageenan/alginate/poloxamer; wound healing

Citation: Postolović, K.S.; Antonijević, M.D.; Ljujić, B.; Radenković, S.; Miletić Kovačević, M.; Hiezl, Z.; Pavlović, S.; Radojević, I.; Stanić, Z. Curcumin and Diclofenac Therapeutic Efficacy Enhancement Applying Transdermal Hydrogel Polymer Films, Based on Carrageenan, Alginate and Poloxamer. *Polymers* **2022**, *14*, 4091. https://doi.org/10.3390/polym14194091

Academic Editors: Swarup Roy and Jong-Whan Rhim

Received: 9 September 2022
Accepted: 24 September 2022
Published: 29 September 2022

Publisher's Note: MDPI stays neutral with regard to jurisdictional claims in published maps and institutional affiliations.

Copyright: © 2022 by the authors. Licensee MDPI, Basel, Switzerland. This article is an open access article distributed under the terms and conditions of the Creative Commons Attribution (CC BY) license (https://creativecommons.org/licenses/by/4.0/).

1. Introduction

As a specific biological process, wound healing refers to the growth and regeneration of tissues [1]. The wound healing process is considered to include five phases (hemostasis, inflammation, migration, proliferation, and maturation), where some of the phases may overlap [1–3]. The inflammatory phase is the first response to a skin injury, occurs immediately after the injury (together with the hemostatic phase) and lasts for approximately three days. During the inflammatory phase, various cellular and vascular processes stop further damage, eliminate pathogens and clean the wound [4]. After the inflammatory phase, the proliferation process takes place simultaneously with the migration process. During this phase, formation of granulation tissue, re-epithelization, and collagen synthesis by fibroblasts lead to wound damage repair [1,4].

A fluid called exudate is produced in the healing process and is present in almost all healing phases [5]. The produced exudate keeps the wound moist, which is an ideal environment for effective and efficient healing [6]. However, excess exudate can lead to complications in healing. Problems can also occur due to the appearance of pathogenic microorganisms (mainly bacteria and some strains of fungi) on the wound surface, which can cause severe infections and is often cited as the main reason for the prolonged healing process [1,7]. Furthermore, uncontrolled reproduction of pathogenic bacteria can lead to blood poisoning, sepsis, and even a fatal outcome [1,7]. From the above, it can be concluded that non-toxic and biocompatible formulations, with proper mechanical strength and capability to absorb excess exudate but prevent wound dehydration by maintaining a moist environment and effectively prevent or control infections, are among the key factors in wound treatment and healing and thus important tasks for the scientific community [8].

Bioactive or smart dressings are a class of dressings that are able to deliver bioactive compounds in the wound site and create an active and dynamic interaction with the wound's environment [8]. Some formulations incorporating various bioactive molecules (dicarboxylic acids, Ag-nanoparticles, chitosan/propolis nanoparticles) that can prevent infection and have a positive effect on different wound healing phases and accelerate healing have been dealt with in numerous studies [1,9–12]. The use of dry and solid formulations that allow controlled release of the bioactive component over a longer period can give better therapeutic results because the patient is exposed to a drug concentration that is optimal for treatment [13]. Carriers can achieve incorporation and later the controlled release of various low-bioavailability drugs and antibiotics [13]. Controlled release of the bioactive component on the wound is achieved by carrier swelling during its contact with the wound exudate [13]. During swelling, the distance between the polymer chains increases, thus creating a system that can release drugs in a controlled manner. In addition to polymer hydration and swelling, a significant role in drug release from the carrier can be played by crosslinking the polymer within the carrier and the rates of potential carrier degradation and drug diffusion through the polymer matrix [13,14]. Recently, a hydrogel in the form of thin, elastic films has been increasingly applied in the wound healing process. This allows the unhindered transfer of matter from the film to the wound, secreting a moderate exudate amount [1]. Various natural (polysaccharides, proteins, and lipids) and synthetic polymers (poly(ethylene glycol), PEG, poly(vinyl alcohol), PVA, poly(ethylene oxide), PEO, and poly(vinyl pyrrolidone), PVP) can be used as constituents of these formulations [1,15]. Wound dressings can be developed from a combination of bio and synthetic polymers. Biopolymers suffer from poor mechanical properties that can be overcome by combining them with synthetic polymers. Carrageenan and alginate are constituents of wound dressing materials [16,17], but can form a stable formulation in combination with synthetic polymer poloxamer 407 able to incorporate hydrophobic bioactive compounds [18].

Diclofenac (Dlf) is a non-steroidal, anti-inflammatory drug (NSAIDs) and has the greatest application in the treatment of the painful rheumatic process [19]. This drug is commercially available in the form of various formulations for oral, dermal, or intramuscular application [19]. Oral administration of diclofenac is limited due to its low solubility in acidic media and possible diclofenac intramolecular cyclization [19]. To improve diclofenac bioavailability and to avoid its side effects after oral intake, diclofenac dermal application is more common, where possible [20]. Diclofenac does not affect individual phases (except inflammatory) in the wound healing process but indirectly affects healing because of its antibacterial properties [21,22]. Although the use of antibiotics to prevent infections is an effective solution, due to the occurrence of resistant pathogenic microorganisms and slower synthesis/isolation of new antibiotics, there is a need for alternative solutions [9]. Previous studies [23–26] have shown that diclofenac-incorporated formulations have antimicrobial activity, to some extent, against various bacterial strains. Since diclofenac primarily acts as an anti-inflammatory drug, which reduces post-injury pain, its additional advantage in wound treatment is its antibacterial property.

Curcumin (Cur) is a hydrophobic polyphenolic compound with antioxidant, anti-cancer, anti-inflammatory, and antimicrobial properties [27,28]. However, despite its high efficacy, the use of curcumin is limited due to its very low solubility and thus bioavailability [29]. For this reason, increasing curcumin bioavailability and developing formulations that serve that purpose have been the subjects of numerous studies [29–32]. Curcumin has good potential for wound healing treatment due to its antimicrobial, anti-inflammatory, and antioxidant properties. The wound healing process also includes reactive oxygen species, which are part of the immune response to the appearance of microorganisms [33]. However, prolonged exposure to reactive oxygen species in higher concentrations leads to oxidative stress, inhibiting the maturation phase during wound healing. For this reason, reactive oxygen species are the leading cause of prolonged inflammation [34–36]. Since curcumin has excellent antioxidant properties, great attention is paid to developing formulations that can be dermally applied, thus achieving the maximum anti-inflammatory effect of curcumin [37]. In addition, it has been found that curcumin can improve wound healing by participating in granulation tissue formation, damaged tissue regeneration, collagen deposition, thus improving epithelial cell regeneration processes and increasing fibroblasts proliferation [38]. Due to the above, more recent studies and review articles are dedicated to developing and describing various curcumin-containing wound healing dressings [39–45]. Polymer-based wound dressings (hydrogels, films, membranes, nanoparticles, nanofibers, liposomes) loaded with curcumin exhibited great in vitro and in vivo therapeutic outcomes [44]. Innovative strategies include formulations based on combination of curcumin with other anti-bacterial or anti-inflammatory agents [45,46].

In our previous research [18], the films based on carrageenan (Car), alginate (Alg) and poloxamer 407 (Pol) were optimized. These optimized films were used in this work to examine the efficiency of encapsulation and release of diclofenac individually and in a mixture with curcumin. After film characterization, the results obtained by diclofenac and curcumin release were related to the interactions these drugs achieve with the carrier components studied using theoretical models and AIM (Atoms in Molecules) analysis. Furthermore, considering the anti-inflammatory effect of curcumin [37], the synergistic effect of curcumin and diclofenac during the treatment of inflammation [47], as well as the positive effect of curcumin in the proliferation phase [38], films based on carrageenan, alginate and poloxamer containing a combination of these two drugs were prepared with the ultimate purpose of their application for in vivo wound healing.

2. Materials and Methods

2.1. Materials

Sodium alginate and κ-carrageenan were obtained from Roth (Karlsruhe, Germany). Curcumin, diclofenac (sodium salt), poloxamer 407, calcium chloride dihydrate, potassium chloride, sodium chloride, glutamine, fetal bovine serum, penicillin, streptomycin, resazurin, amoxicillin, tetracycline, 3-(4,5-dimethylthiazol-2-yl)-2,5-diphenyltetrazolium bromide (MTT), trypan blue, ketamine, and xylazine were purchased from Sigma-Aldrich (Burlington, MA, USA). Glycerol and ethanol were obtained from Honeywell (Charlotte, NC, USA). Sodium hydrogen phosphate dihydrate was purchased from Poch (Gliwice, Poland) and potassium dihydrogen phosphate from Kemika (Zagreb, Croatia). Non-essential amino acids were obtained from Capricorn Scientific GmbH (Ebsdorfergrund, Germany).

2.2. Film Preparation

Polysaccharides and poloxamer 407-based films (Car/Alg/Pol) were prepared by the casting method, using the procedure described in our previous research [18]. Appropriate polysaccharide masses (0.4 g of carrageenan and 0.1 g of alginate, total saccharide concentration 2.0% w/w) and the aqueous solution of poloxamer 407 (0.15 g, 5.0% w/v) were added to the aqueous solution of glycerol as plasticizer (60% w/w relative to total mass of saccharides) [18]. The mixture was stirred on a magnetic stirrer at room temperature for 1 h, then heated to 70 °C, and a solution of calcium chloride (0.5% w/w) was gradually added dropwise to the mixture (1 mL/min). After CaCl$_2$ solution instillation, stirring was continued for 20 min under the same conditions, with further application of the ultrasonic bath. Then, the mixture was poured into Petri dishes (d = 9 cm) and dried for 20 h at a temperature of 40 °C. In the second phase, the dried semi-crosslinked films were immersed in a 10% glycerol and 3% calcium chloride solution for 10 min to achieve further crosslinking [18]. Finally, obtained crosslinked Car/Alg/Pol films were air-dried. To prepare films containing diclofenac (Car/Alg/Pol-Dlf), curcumin (Car/Alg/Pol-Cur), or a mixture of curcumin and diclofenac (Car/Alg/Pol-Cur+Dlf), an aqueous solution of diclofenac (1.0% w/v), i.e., a solution of curcumin in ethanol (1.0%, w/v), was added to the mixture of starting components (saccharides and poloxamer) after 30 min of initial stirring. Then, stirring was continued at room temperature for another 30 min, and the further work process was identical to that previously described.

2.3. Film Characterization

2.3.1. Infrared Spectroscopy

FTIR (Fourier-transform infrared) spectra of films (Car/Alg/Pol, Car/Alg/Pol-Dlf and Car/Alg/Pol-Cur+Dlf) and starting components were recorded using infrared (IR) spectroscopy (Perkin Elmer Spectrum Two spectrophotometer, Waltham, MA, USA) to characterize the prepared film's composition. The spectra were recorded in the range of 4000–500 cm^{-1}.

2.3.2. Texture Analysis

In order to investigate the mechanical properties of the prepared films, texture analysis was performed. The films were cut into a rectangular shape using a micrometer and a scalpel. The width of the samples was 10 mm and the gauge length was 30 mm, with a gripping length of 10 mm on each side. The thickness of each film was evaluated before tensile characterization, at five different points using the micrometer. The mechanical properties of the films were measured using a Texture Analyzer (TA.HD plus, Stable Micro Systems Ltd., Surrey, UK), equipped with a 5 kg load cell, using tensile grips A/TG. The test speed was 6 mm/s, with a trigger force of 0.09 N. The elongation at break (%EB) and tensile strength (TS) were estimated according to Equations (1) and (2), while Young's modulus (YM) was estimated from the linear part of the stress–strain curve, according to the Equation (3). The time needed for the sample to break was also investigated.

$$\text{Elongation at break} = \frac{\text{increase in length at break}}{\text{initial film length}} \times 100 \tag{1}$$

$$\text{Tensile strength} = \frac{\text{force at failure}}{\text{cross sectional area of the film}} \tag{2}$$

$$\text{Young's modulus} = \frac{\Delta \text{Stress}}{\Delta \text{Strain}} \tag{3}$$

The results of three replicates for each of the four films were expressed as the mean values ± SD.

2.3.3. Scanning Electron Microscopy

Scanning electron microscopy (SEM) and energy dispersive X-ray (EDX) microanalysis were completed on the films using a Hitachi SU8030 instrument (Tokyo, Japan) with a field emission electron gun. The SEM is coupled with a Thermo Scientific NORAN System 7 detector (Madison, WI, USA) for X-ray microanalysis. Strips of each film were secured onto alumina stubs. Surface characterization was completed using a 1.0 keV accelerating voltage (V_a) and a 10 µA emission current (I_e) in the low magnification mode. Elemental point analysis was carried out at 20.0 keV (V_a) and 10 µA (I_e).

2.3.4. XRD Analysis

X-ray diffraction (XRD) was used to evaluate the crystalline content of the films. Data were collected on a D8 Advance X-ray Diffractometer (Bruker, Germany) in theta–theta geometry in the transmission mode using Cu K$_\alpha$ radiation at 40 kV and 40 mA. A primary

Göbel mirror for parallel beam X-rays and removal of Cu K$_\beta$ radiation along with a primary 4° Soller slit, and a 0.2 mm exit slit was part of the setup. The sample rotation was set at 15 rpm, X-rays were collected using a LynxEye silicon strip position sensitive detector set with an opening of 3° with the LynxIris set at 6.5 mm and a secondary 2.5° Soller slit. Data collection was between 2 and 60° 2θ, step size of 0.02° and a counting time of 0.5 s per step. Two layers of the sample was secured between mylar film. Data were collected using DIFFRAC plus XRD Commander version 2.6.1 software (Bruker-AXS, Karlsruhe, Germany). Peak identification was completed using an EVA V6.0.0.7 (Bruker, Karlsruhe, Germany) software package.

2.3.5. Thermogravimetric Analysis

Thermogravimetric analysis (TGA) was conducted using the Discovery 5500 TGA (TA Instruments, Crawley, UK) in aluminum pans with sample size 3.0 ± 0.5 mg for starting materials and 7.0 ± 1.0 mg for all film formulations. Samples were heated from ambient temperature (20 °C) to 500 °C at 10 °C/min, under nitrogen (25 mL/min). Data were analyzed using TA Advantage Universal Analysis V4.5 software (Lukens Dr, New Castle, DE, USA).

2.3.6. Differential Scanning Calorimetry

Differential scanning calorimetry (DSC) was carried out using a Discovery 2500 DSC (TA Instruments, Crawley, UK) in hermetically sealed T zero aluminum pans with 3.0 ± 1.0 mg of sample. Sample was heated at 10 °C/min from −70 to 300 °C. Experiments were conducted in triplicate under nitrogen atmosphere (flow rate 50 mL/min). Data were analyzed using TA Advantage Universal Analysis V4.5 software (Lukens Dr, New Castle, DE, USA).

2.3.7. Encapsulation Efficiency of Drugs

The encapsulation efficiency of curcumin and diclofenac was determined by immersing the films with incorporated drugs (Car/Alg/Pol-Dlf and Car/Alg/Pol-Cur+Dlf) in phosphate buffer pH 7.40. After 24 h, aliquots were taken, and the concentration of encapsulated drugs was determined using UV/Vis spectrophotometry (Perkin Elmer UV/Vis, Lambda 365, Waltham, MA, USA), at a wavelength of 430 nm for curcumin and 276 nm for diclofenac. The ratio of the spectrophotometrically determined drug weight to the weight of drug added to films in the preparation process represents the encapsulation efficiency (Equation (4)). The measurements were performed in triplicate.

$$\text{EE } (\%) = \frac{\text{Spectrophotometrically determined amount of drug}}{\text{Added amount of drug}} \times 100 \qquad (4)$$

2.4. In Vitro Drug Release

The release of diclofenac from the Car/Alg/Pol-Dlf film as well as curcumin and diclofenac from the Car/Alg/Pol-Cur+Dlf film were monitored in vitro in conditions simulating wound exudate (PBS buffer, pH 7.4). For drug release testing, 2 × 2 cm films (diclofenac weight in the Car/Alg/Pol-Dlf film was 1.50 mg, while curcumin and diclofenac weight in the Car/Alg/Pol-Cur+Dlf film was 2.86 and 1.53 mg, respectively) were added to the buffer solution and incubated at 37 °C. Aliquots were taken at certain time intervals, and the concentrations of released diclofenac and curcumin were determined spectrophotometrically by measuring the absorbance at 276 and 430 nm, respectively. The measurements were performed in triplicate.

2.5. Drug Release Kinetics

Based on the results obtained during the in vitro release of drugs, the release kinetics were determined, indicating the mechanism of diclofenac and curcumin release from the films. The release kinetics were tested using various mathematical models, including zero-order kinetics, first-order kinetics, and the Higuchi, Hixon–Crowell, and Korsmeyer–

Peppas release models, where M_t/M_∞ represents the fraction of released drug at a given time (t) [48].

$$\text{Zero-order kinetic:} \quad M_t/M_\infty = kt \quad (5)$$

$$\text{First-order kinetic:} \quad ln\,(M_t/M_\infty) = kt \quad (6)$$

$$\text{Higuchi model:} \quad M_t/M_\infty = kt^{1/2} \quad (7)$$

$$\text{Hixon–Crowell model:} \quad (1 - M_t/M_\infty)^{1/3} = -kt \quad (8)$$

$$\text{Korsmeyer–Peppas model:} \quad M_t/M_\infty = kt^n \quad (9)$$

A mathematical model that best describes the release of drugs from films can be determined based on the correlation coefficient (R^2) value. Furthermore, the mechanism of drug release can be predicted based on the value of n (release exponent) [48].

2.6. Computational Details

Full geometry optimizations of the aggregate structures formed by attaching diclofenac and curcumin molecules to the drug carrier were performed at the semiempirical PM6 level of theory using the Gaussian 09 program package [49]. Structures of isolated curcumin and diclofenac molecules were optimized at the B3LYP/def2-SVP level of theory. All optimizations were done for six positions of two molecular systems (drug and carrier), which adopt face-to-face, side-to-side, and perpendicular arrangements, according to the scheme proposed in recent works [50]. Frequency calculations confirmed that the obtained optimized aggregate structures have no imaginary frequencies. Only the most stable structures were further examined.

In order to assess interactions between the drug molecule and its carrier, the binding energy (BE) was calculated through single point energy calculations at the B3LYP/def2-SVP level of theory. The BEs were computed as the difference between the B3LYP/def2-SVP electronic energy of the PM6 optimized aggregate structure and the sum of the B3LYP/def2-SVP electronic energies of the fragments whose geometries were extracted from the optimized aggregate structures. Van der Waals interactions in the studied complexes were estimated with Grimme's D3 scheme [51] The AIM analysis was carried out by the Multiwfn program [52] and the obtained electron density of the bond critical points ($\rho(r_{BCP})$) was used to calculate the hydrogen bond binding energy (HBBE) as proposed by Emamian et al. [53]. In particular, the following equations were used:

$$HBBE = -223.08 \times \rho(r_{BCP}) + 0.7423 \quad (10)$$

$$HBBE = -323.34 \times \rho(r_{BCP}) - 1.0661 \quad (11)$$

to calculate HBBEs for neutral and charged complexes, respectively.

2.7. Antibacterial Activity of Films

Antibacterial activity of the films (Car/Alg/Pol, Car/Alg/Pol-Cur, Car/Alg/Pol-Dlf, and Car/Alg/Pol-Cur+Dlf) was tested against four standard strains of bacteria. Antibiotic discs (A—amoxicillin 25 µg, T—tetracycline 30 µg, and S—streptomycin 10 µg) were used as positive controls. The experiment involved two Gram-positive bacteria (*Bacillus subtilis* ATCC 6633 and *Staphylococcus aureus* ATCC 25923) and two Gram-negative bacteria (*Pseudomonas aeruginosa* ATCC 27853 and *Escherichia coli* ATCC 25922).

Bacterial suspensions—preparation and standardization. Bacterial cultures were cultivated on nutrient agar before the experiment. The incubation period lasted 18–20 h at a temperature of 37 °C. The bacterial suspensions were prepared by the direct colony method. The procedure was performed under sterile conditions. First, 3–4 morphologically identical bacteria colonies were transferred to 5 mL of saline, mixed well to separate the cells and form a suspension. Then, the suspension turbidity was adjusted using a densitometer (DEN-1, BioSan, Latvia), McFarland 0.5 corresponding to 10^8 CFU/mL. Bacterial suspensions were prepared immediately before the experiment, as they should be used approximately within 30 min of preparation [54,55].

Disk diffusion method. The susceptibility of bacteria to the tested films and standard antibiotics was tested by the in vitro disk diffusion method. The disk diffusion test was performed in a Petri dish on Mueller–Hinton (MH) agar (25 mL of medium per plate). Films and antibiotics discs were cut into cylinders measuring 5 mm in diameter. Films with tested substances and discs with specific concentrations of antibiotics were placed on the surface of the medium (3 identical films/discs on 1 plate), on which a pure bacterial suspension with $1-2 \times 10^8$ CFU/mL was cultivated. After incubation (16–24 h), the inhibition zone diameter (the surface of the bacterial growth inhibition zone) was measured. The measured values were compared with the EUCAST standard [56], and the tested bacteria were classified as sensitive, moderately sensitive, and resistant [57]. All zones of inhibition were calculated in triplicates.

2.8. Cell Viability Study

Cell culture. In order to evaluate cell viability (proliferation) in the presence of the Car/Alg/Pol-Dlf and Car/Alg/Pol-Cur+Dlf films, a standard MTT test was applied [58]. A human fetal lung fibroblast cell line (MRC-5) was cultured in Dulbecco's modified eagle medium supplemented with 10% fetal bovine serum, 100 U/mL penicillin, 100 µg/mL streptomycin, 2 mM L-glutamine, and 1 mmol/L non-essential amino acids. Cells were cultivated at 37 °C in an atmosphere of 5% CO_2 and absolute humidity. The culture medium was completely replaced every 3 days, cell viability was determined using trypan blue staining, and only cell suspensions with viability greater than 95% were further used.

Cell viability assay. A viability study of the Car/Alg/Pol, Car/Alg/Pol-Dlf, Car/Alg/Pol-Cur+Dlf films was performed using an MTT assay. Firstly, the films were cut into cylinders of 9 mm in diameter. Secondly, the films were transferred into 96-well plates and irradiated by ultraviolet light for 30 min. Finally, the suspensions of MRC-5 cells (5000 cells per well, according to studies [40,59]) were dropped onto the sample surfaces. As a control, the same amounts of MRC-5 cells were dropped in the blank dishes. The plates were incubated for 24 and 48 h in an atmosphere of 5% CO_2 and absolute humidity, at 37 °C. Then, MTT solution was added to cell culture and incubated. After incubation, MTT solution was removed, DMSO was added, and absorbance was measured at 595 nm with a multiplate reader. Experiments were performed in triplicates and repeated in three independent series.

2.9. In Vivo Study

All the animal research studies were approved by the Animal Ethics Committee of the Faculty of Medicine, University of Kragujevac (Ethical Approval Number: 01-6121). The use of the prepared films with incorporated mixture of curcumin and diclofenac and films containing only diclofenac was investigated for in vivo healing of burn-caused wounds. For in vivo study, male Wistar albino rats (6 to 8 weeks old, average body weight 200–250 g) were used. One group of animals ($n = 3$) was exposed to burns and not further treated (control), the second group ($n = 3$) was treated with the Car/Alg/Pol films, the third ($n = 3$) with the Car/Alg/Pol-Dlf films, and the fourth ($n = 5$) with the Car/Alg/Pol-Cur+Dlf films. The process of causing burns to rats was performed following the protocols in the previously published study [39]. Before causing burns, the animals were anesthetized with intraperitoneal ketamine (10 mg/kg body weight) and xylazine (5 mg/kg body weight). Then, the backs of healthy rats were shaved using depilatory cream. On the shaved skin area, the burns were caused by applying a hot metal plate (measuring 2×2 cm) to the skin for 10 s. Wounds caused this way were covered with the prepared films (measuring 2×2 cm). The healing process was monitored for seven days, with the daily replacement of film samples.

Histopathological analysis. The intensity of the skin injury caused by the hot metal plate was estimated based on histopathological analysis of healthy skin and skin exposed to burns. The contribution of incorporated drugs (diclofenac, curcumin) to the healing process was determined by comparing histopathological analyzes of untreated burned skin (control) and burned film-treated skin. All rats were sacrificed by means of cervical dislocation on day 7 post-burning. The skin was aseptically removed and fixed in 10% buffered formalin fixative overnight. Paraffin wax-embedded skin sections (5 μm) were stained with hematoxylin and eosin (H&E), and stained slides were then examined under a light microscope to evaluate the extent of damage. The images were captured with a light microscope equipped with a digital camera.

3. Results and Discussion
3.1. Film Characterization
3.1.1. Basic Characteristics of Films

The average weights and thicknesses of the obtained films and the weight of drugs incorporated in films are shown in Table 1 ($n = 5$). It can be concluded from the obtained results that films of the same composition show uniformity in terms of both weight and thickness. According to the obtained results by statistical analysis, there were no significant differences between the mass and thickness of blank and drug-containing films ($p > 0.05$).

Table 1. Basic film characteristics ($n = 5$).

Film	Mass of Film (mg/cm^2)	Film Thickness (μm)	Mass of Drug (mg/cm^2 of Carrier)
Car/Alg/Pol	12.21 ± 0.65 [18]	104.27 ± 3.35 [18]	/
Car/Alg/Pol-Dlf	13.79 ± 0.65	121.12 ± 0.93	0.375 ± 0.012
Car/Alg/Pol-Cur+Dlf	14.29 ± 0.13	134.83 ± 2.17	0.718 ± 0.028 (Cur) 0.400 ± 0.017 (Dlf)

3.1.2. FTIR Spectroscopy

The FTIR spectra of Car/Alg/Pol, Car/Alg/Pol-Dlf and Car/Alg/Pol-Cur+Dlf, as well as pure curcumin and diclofenac, are shown in Figure 1.

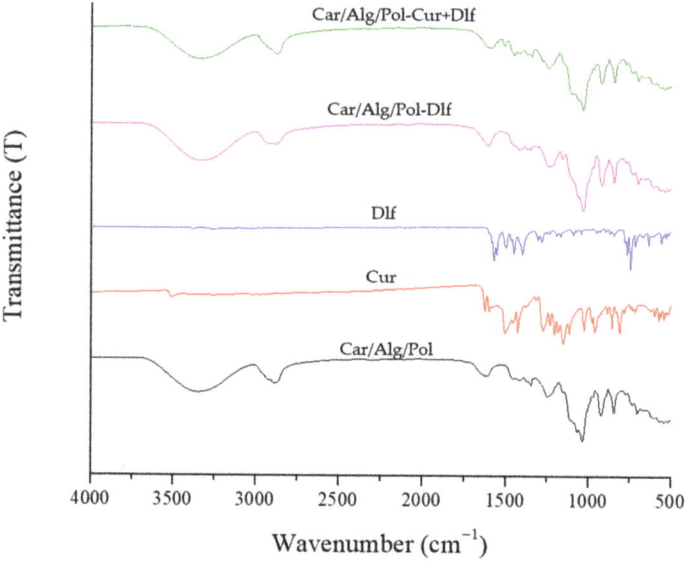

Figure 1. FTIR spectra of curcumin, diclofenac, and the Car/Alg/Pol, Car/Alg/Pol Dlf and Car/Alg/Pol-Cur+Dlf films.

The spectra of pure drugs (both curcumin and diclofenac) show vibrational characteristic of aromatic C−C and C−H bonds can be observed. In addition, on the curcumin spectrum, the sharp band at 3508 cm^{-1} originates from the vibrations of the phenolic O−H group, and a band at 1628 cm^{-1} is the result of the valence vibrations of the C=O bond [60]. On the other hand, in addition to the vibrations of the bonds in the aromatic ring, the diclofenac spectrum is also characterized by bands at 1574 and 745 cm^{-1}, which originate from the vibrations of the carboxylate anion and C−Cl bond, respectively [61].

The spectra of the Car/Alg/Pol, Car/Alg/Pol-Cur and Car/Alg/Pol-Cur+Dlf films, a wide absorption band in the range 3600–3000 cm^{-1} can be observed originating from the valence vibrations of the present saccharides −OH bonds. Additionally, valence vibrations of C−H bonds can be noticed in the range 3000–2840 cm^{-1}. The width of the bands corresponding to the vibrations of the O−H bonds is a consequence of established hydrogen bonds. The FTIR spectrum of the Car/Alg/Pol film contains all group vibrations that are characteristic of both carrageenan (sulfate group vibrations—band at 1245 cm^{-1}) and alginate (asymmetric and symmetric vibrations of carboxylate anion—bands at 1615 and 1417 cm^{-1}), as well as poloxamer (C−O bond vibrations—very intense band at 1033 cm^{-1}). From this, it can be concluded that a unique carrageenan/alginate/poloxamer hydrogel was formed. On the spectra of drug-containing films, bands characteristic of carrier constituents can also be observed. However, as a consequence of the addition and interactions of the drugs with the alginate from the carrier, there is a significant shift in the wavenumbers corresponding to the vibration of the carboxylate anion of diclofenac, 1574→1609 cm^{-1} (for the film Car/Alg/Pol-Dlf), or 1574→1602 cm^{-1} (for the film Car/Alg/Pol-Cur+Dlf). Due to the homogeneous drug distribution within the films, other characteristic bands of diclofenac and curcumin cannot be observed in the spectra of films containing these drugs.

3.1.3. Texture Analysis

A desirable wound dressing should have good mechanical properties and maintain integrity during use [16]. A wound dressing should be flexible, elastic, and not prone to tear or rupture upon application, whether applied topically to protect dermal wounds or when used as an internal wound support [62]. The mechanical properties of the prepared films were evaluated using Texture Analyzer, and the results are presented in Table 2 as the mean values of three replicates for each film ± SD.

Table 2. The mechanical properties of analyzed films (n = 3).

Sample Name	Elongation at Break (% ± SD)	Tensile Strength (MPa ± SD)	Young's Modulus (MPa ± SD)	Time to Break (s ± SD)
Car/Alg/Pol	32.41 ± 1.02	34.60 ± 1.31	4.00 ± 0.04	3.09 ± 0.13
Car/Alg/Pol-Cur	27.36 ± 4.20	27.62 ± 2.63	3.86 ± 0.20	2.69 ± 0.40
Car/Alg/Pol-Dlf	25.19 ± 3.40	28.14 ± 1.63	4.41 ± 0.43	2.66 ± 0.39
Car/Alg/Pol-Cur+Dlf	29.66 ± 3.38	32.66 ± 0.18	4.87 ± 0.43	2.77 ± 0.34

The mechanical strength of the prepared films was presented in terms of their tensile strength, the percentage of elongation to break (measure of extensibility), and Young's modulus, a parameter used to describe the rigidity and stiffness of the material, as well. Comparison of the films Car/Alg/Pol and Car/Alg/Pol-Cur indicates that the addition of curcumin led to the slight decrease in elongation at break, and therefore decrease in extensibility. Time to break, tensile strength and Young's Modulus values decreased, as well. Therefore, it can be concluded that addition of the curcumin to the films led to the slight decrease in material ductility and elasticity. The decrease in the strength and elongation of the break of the Car/Alg/Pol-Cur film can be a consequence of polymer–curcumin interactions on the films surface which results in crystals formation (can be also noticed by SEM analysis, Section 3.1.4). Similar results, in terms of %EB and TS reduction in the presence of curcumin, were obtained in the studies [63,64]. The results show that films containing diclofenac were stronger, stiffer, and less elastic than the Car/Alg/Pol-Cur films, as indicated by higher values of both TS and Young's modulus and lower values of %EB. Compared to the blank films, similar conclusion can be obtained, with the note that the tensile strength is smaller for the Car/Alg/Pol-Dlf film. The addition of both curcumin and diclofenac was responsible for a small decrease in the elongation at break and time to break, whereas Young's modulus increased in the comparison with blank films. Additionally, tensile strength was similar to blank films and higher in comparison with the Car/Alg/Pol-Cur and Car/Alg/Pol-Cur+Dlf films. The improvement in tensile strength of the transdermal Car/Alg/Pol-Cur+Dlf films might be attributed to the high aspect ratio and rigidity which results from the strong affinity between the polymers and drugs. For all drug containing films, the decrease in the extensibility could be attributed to the restriction of mobility of polymer chains in the presence of drugs due to strength of polymer–drug interactions (explained in the Section 3.4).

3.1.4. SEM Analysis

Secondary electron images (magnification ×500) captured from the four prepared films show clear difference in the surface morphology (Figure 2).

The Car/Alg/Pol film (Figure 2a) had a smooth, homogeneous, and uniform surface, indicating the excellent film formability of carrageenan, alginate and poloxamer. Meanwhile, the Car/Alg/Pol-Cur film presented somewhat uneven surfaces (Figure 2b). Some convex pieces and crystalline particles were observed on the surface when the curcumin was added to the film. Observation of the surface of Car/Alg/Pol-Cur (Figure 2b) reveals crystals that are embedded into the polymer matrix. This can be explained by the fact that hydroxyl groups presented in carrageenan, alginate and curcumin could dehydrate and condense with the carboxyl or sulfate group from polysaccharides. The complex reaction can result in the formation of a complex three-dimensional network structure. Similar results were obtained in the study of Xie et al. [65], where curcumin-loaded chitosan/pectin films were developed. The Car/Alg/Pol-Dlf film (Figure 2c) has a uniform and smooth surface, but contrast difference is visible because of the presence of diclofenac, in comparison with the Car/Alg/Pol film surface. Additionally, the surface of diclofenac-incorporated film also had some white spots and fine dust-like particles, which had been reported that the calcium ions could accumulate and form white patches in the polysaccharide film [65]. The morphological characteristics of the surface of the Car/Alg/Pol-Cur+Dlf film (Figure 2d) had more similarities with the Car/Alg/Pol-Cur film surface than with Car/Alg/Pol-Dlf, due to higher concentration of incorporated curcumin in comparison with diclofenac. The presence of crystalline particles and small dust-like particles covering the surface can be noticed at Figure 2d. Lower concentration of diclofenac and its higher solubility enabled more evenly distribution of diclofenac in the film network, in comparison to curcumin.

EDX analysis was carried out to identify the elemental composition of polymer matrix (Figure 2a), as well as drug-loaded films (Figure 2b–d). It was proved that the elemental structure of Car/Alg/Pol was mainly comprised of carbon (25.3 w%) and oxygen (61.2 w%), while the remaining mass is made up of sodium, sulfur, chlorine, calcium and potassium. The results are in concordance with chemical structure of polymers presented in the carrier (carbon and oxygen, as the main constituents of carbohydrates and poloxamer; sodium, potassium and chlorine, as usual impurities and counterions of alginate and carrageenan; sulfur, as a constituent of carrageenan and calcium as crosslinking ion). The EDX results of drug-loaded films showed similar results with the ones for the blank films. Addition of diclofenac could be confirmed by chlorine content increase, in comparison with the Car/Alg/Pol and Car/Alg/Pol-Cur films, whereas curcumin addition has no significant influence on elemental analysis due to its chemical composition.

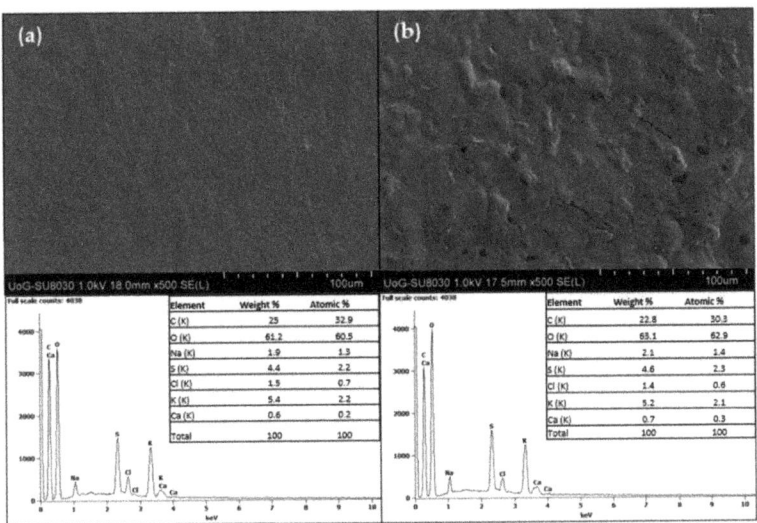
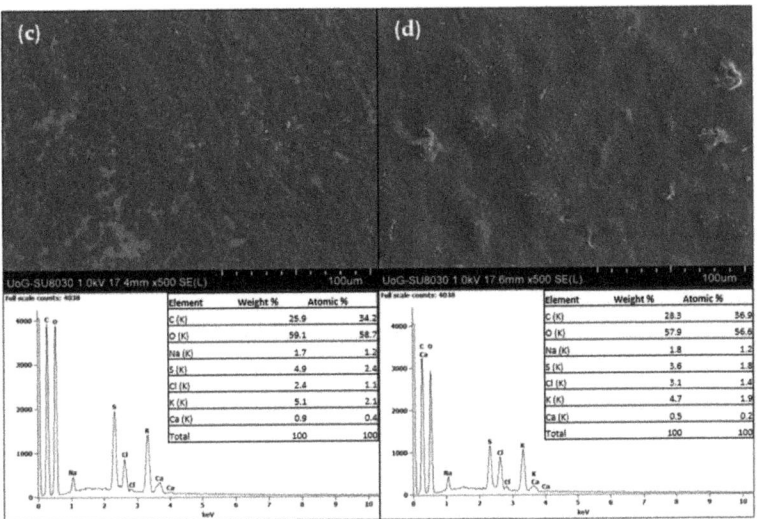

Figure 2. SEM images (×500) and EDX analysis of the (**a**) Car/Alg/Pol, (**b**) Car/Alg/Pol-Cur, (**c**) Car/Alg/Pol-Dlf, (**d**) Car/Alg/Pol-Cur+Dlf films.

3.1.5. XRD Analysis

The physical form of the films was determined using X-ray diffraction. The obtained diffractograms for the prepared films are shown in Figure 3.

Curcumin and diclofenac, as pure substances, are presented exclusively in crystalline form [18,23]. All the prepared films are predominantly amorphous, based on obtained XRD patterns, with the note that the Car/Alg/Pol-Cur and Car/Alg/Pol-Cur+Dlf films also have crystalline content. The diffractogram corresponding to the Car/Alg/Pol-Dlf film indicates the exclusively amorphous state of all present components, which means that diclofenac molecules have dispersed within the carrier. These results can be related to the results obtained by SEM analysis (Section 3.1.4), where crystalline particles are present on the surface of curcumin-containing films, while the Car/Alg/Pol-Dlf film exhibited smooth surface. In the case of the Car/Alg/Pol-Cur and Car/Alg/Pol-Cur+Dlf films, the presence of two crystalline peaks can be observed, which can be attributed to one of the curcumin polymorphs [66]. This can be explained by the conversion of one structure of curcumin to the other polymorph form (which differ from each other in the keto-enol orientation of curcumin molecules) during film preparation or its storage [66].

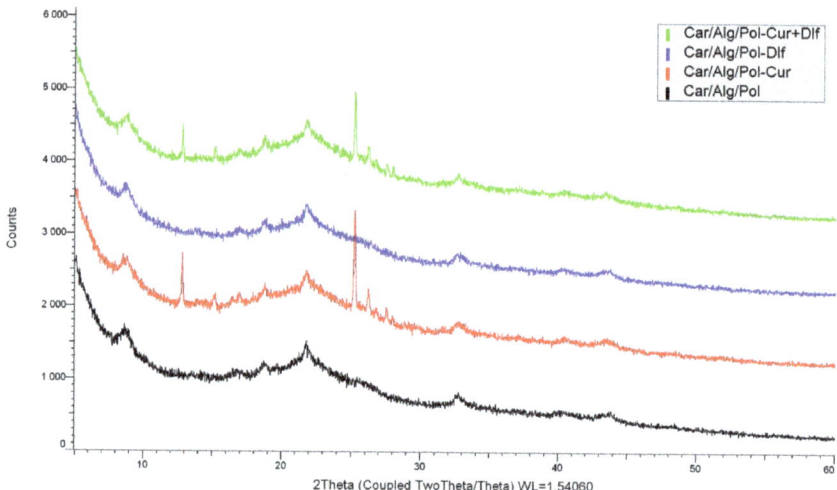

Figure 3. XRD patterns of the prepared films.

3.1.6. Thermogravimetric Analysis

The thermal stability of the films was studied using thermogravimetric analysis. Figure 4 shows the thermograms corresponding to the Car/Alg/Pol, Car/Alg/Pol-Dlf, Car/Alg/Pol-Cur and Car/Alg/Pol-Cur+Dlf films, as well as pure curcumin and diclofenac.

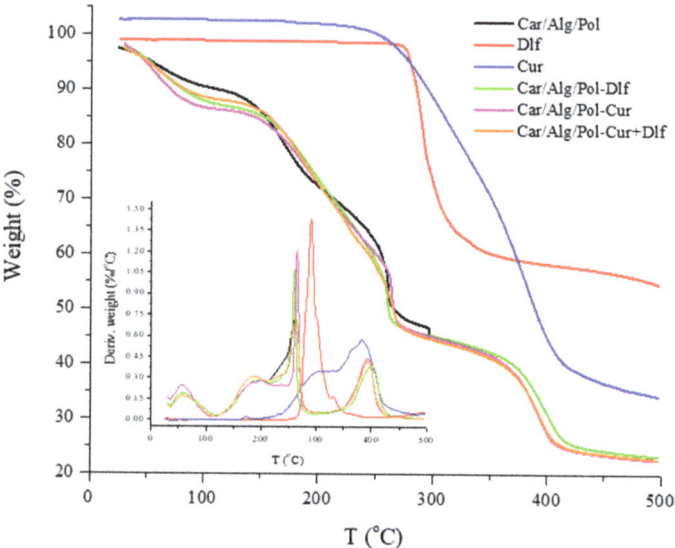

Figure 4. TGA and DTG curves of diclofenac, curcumin, and the Car/Alg/Pol, Car/Alg/Pol-Cur, Car/Alg/Pol-Dlf and Car/Alg/Pol-Cur+Dlf films.

Based on the obtained results, it can be noticed that thermograms corresponding to the films are similar, in terms of the weight loss stages. On the other hand, pure curcumin and diclofenac have one significant mass loss at 250 and 280 °C, respectively, related to their thermal decomposition. The initial mass loss in all films (around 10%), which occurs at temperatures below 100 °C, is caused by the evaporation of weakly bound water (remained after drying during film preparation) or water absorbed from the air, present in the sample. Therefore, the films have a tendency to absorb water.

The next significant mass losses occur in the range of temperatures from 150 to 280 °C, in two stages. The first weight loss at a temperature above 150 °C is the result of evaporation of glycerol, presented in films as a plasticizer [67]. This process is followed by alginate decomposition that starts at approximately 220 °C and continues up to 240 °C [68]. As a consequence of these processes, films lost approximately 26% of their mass. The second significant weight loss (calculated mass loss is approximately 15%) is related to the carrageenan decomposition that starts at approximately 250 °C and takes place up to 280 °C [69]. Finally, the last mass loss (approximately 22%) occurred due to poloxamer thermal decomposition in the temperature range of 380 to 400 °C [70]. Due to small concentrations of curcumin and diclofenac in the films, their decomposition is not clearly noticeable in thermograms; as it coincides with decomposition of carrageenan and alginate.

3.1.7. Differential Scanning Calorimetry

With the aim to investigate the drug physical state and polymers behavior in the films, DSC analysis of prepared formulations and starting materials was carried out at a temperature from −70 to 300 °C, as shown in Figure 5.

The DSC thermograms show a complex thermal behavior of starting materials and the film formulations. Carrageenan and alginate DSC thermograms did not show any thermal events at temperatures below 200 °C or sharp endothermic peaks above it, which confirms their predominantly amorphous nature. The presence of complex endothermic peak at 208 and 202 °C for carrageenan and alginate, respectively, can be linked to the thermal decomposition of the polymers. On the other hand, on DSC thermogram of poloxamer, an endothermic (melting) peak is present at 54 °C in addition to an exothermic peak at 163 °C, which could be attributed to poloxamer recrystallization from the melt [23,70]. Diclofenac, at higher temperatures, showed a significant exothermic peak, which was immediately followed by two endothermic peaks, which are result of diclofenac melting and its thermal decomposition [71,72]. The DSC thermogram of curcumin revealed a single sharp peak at 179 °C, which corresponds to the melting point of crystalline curcumin [70].

At higher temperatures, resemblance to the degradations processes in carrageenan and alginate is evident in all the prepared films, which can be seen in Figure 5b. The Car/Alg/Pol film showed an altered endothermic peak at 230 °C, suggesting that the thermal characteristics of polymers changed during film production, which is caused by polymer interaction, similar to the results obtained by Boateng et al. [23]. In the DSC thermogram which corresponds to the Car/Alg/Pol-Dlf film, only broad peaks can be observed, thus confirming its amorphous structure. Due to the interaction of polymers and diclofenac, which led to diclofenac transformation from crystalline to amorphous state, peaks attributed to its melting and decomposition cannot be noticed. On the other hand, endothermic peak which corresponds to biopolymers thermal decomposition can be observed at 220 and 225 °C, for the Car/Alg/Pol-Cur and Car/Alg/Pol-Cur+Dlf films, respectively. Additionally, curcumin-containing films showed two small endothermic peaks at 167 °C, which can be attributed to the melting point of curcumin polymorphs [66], similar to the results obtained by XRD analysis (Section 3.1.5).

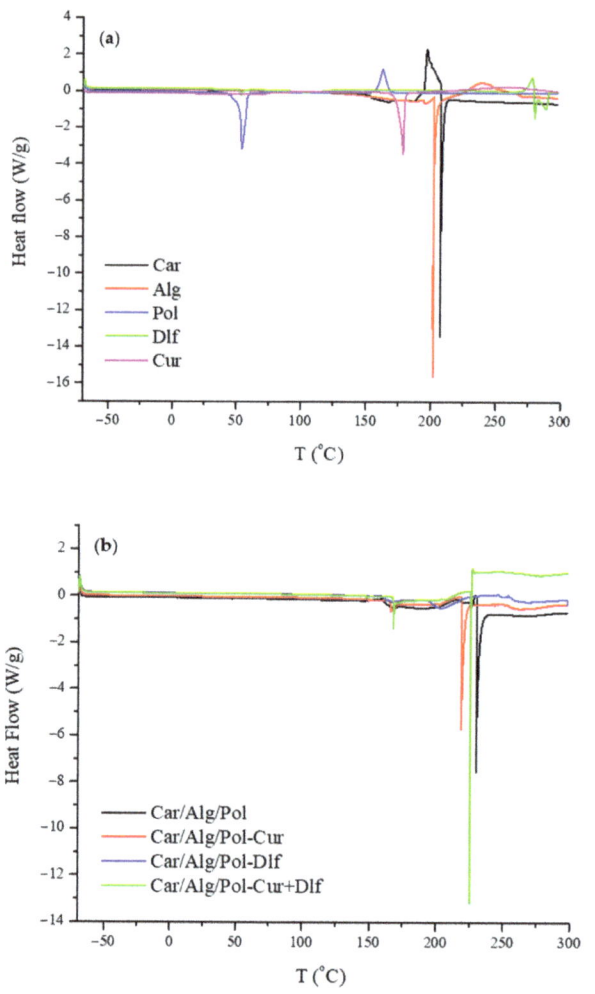

Figure 5. DSC thermograms of (**a**) starting components and (**b**) prepared films.

3.1.8. Drug Encapsulation Efficiency

In the previous research propylene glycol nanoliposomes containing curcumin were developed for burn wound healing with the encapsulation efficiency of 84.66% [73]. Additionally, natural (chitosan), synthetic (poly-lactic co-glycolic acid) and semi-synthetic (carboxymethylcellulose) polymer-based nanoparticles were used for curcumin delivery with high encapsulation efficiencies (higher than 90%) [74]. Previously published articles also studied polymer-based (chitosan and alginate/carboxymethyl chitosan/aminated chitosan) carriers with high diclofenac encapsulation efficiency (84 and 95%, respectively) [75,76]. The percentage of drug encapsulation in our study was defined by determining the weight of drugs (diclofenac and the mixture of curcumin and diclofenac) incorporated into the films. Polysaccharide- and poloxamer-based carriers easily interact with the added drugs, forming a unique, homogeneous film. The encapsulation efficiency of diclofenac in the Car/Alg/Pol-Dlf film is (92.65 ± 3.20)%, while the encapsulation efficiency of curcumin and diclofenac in the Car/Alg/Pol-Cur+Dlf film is (90.49 ± 3.90)% for Cur and (98.83 ± 4.25)% for Dlf. Results revealed that the tested drugs are incorporated within the films in a high percentage, which further leads to an increase in their bioavailability.

3.2. In Vitro Release Study

The in vitro study of diclofenac release from the films Car/Alg/Pol-Dlf (Figure 6) demonstrate that a high percentage of release was achieved at the beginning (initial burst in the first 15 min), and continues to grow gradually in a period of up to 3 h. Subsequently, it begins to stabilize within 24 h, with a final release percentage of (90.10 ± 4.89)%. By comparing the results obtained in this study with the results of other studies [23,24], where it was also monitored the release of diclofenac from polymer-based films, it can be concluded that a higher release percentage in 24 h is achieved in our work compared to the results of other works, where the release percentage after 72 h was 60% [23,24].

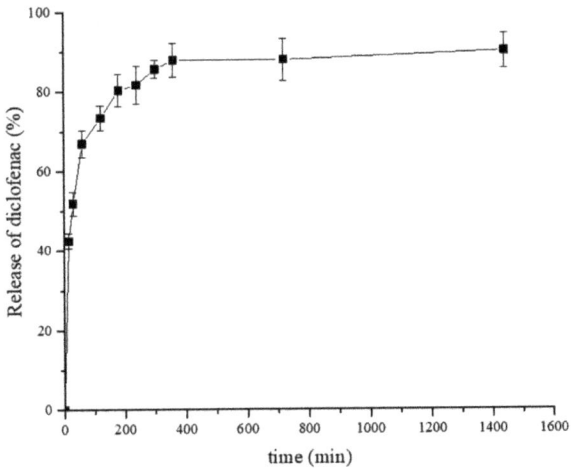

Figure 6. In vitro release of diclofenac from the Car/Alg/Pol-Dlf film (n = 3).

By studying the profiles of drug release from films containing a mixture of curcumin and diclofenac (Figure 7), it can be noticed that a slightly higher percentage of diclofenac release was obtained compared to the release results obtained from films of the same composition containing only diclofenac (Figure 6). Additionally, the final release percentage of diclofenac (95.61 ± 1.67)% after 24 h is greater than the curcumin release rate (90.48 ± 0.30)% over 24 h. The graphs that follow the drugs release process (Figure 7) are similar to the previously obtained graph corresponding to the release of individual drugs: diclofenac (Figure 6), or the graph obtained in the study monitoring the release of curcumin from films of the same composition [18]. Therefore, it can be concluded that curcumin and diclofenac generally retain their individual characteristics during the release process when mixed within the carrier, Car/Alg/Pol-Cur+Dlf.

Comparing the results obtained in our paper with the percentage of diclofenac release in the presence of *Curcuma longa* plant extract from the transdermal gel [77] reveals the advantages of prepared polysaccharide- and poloxamer-based films. The percentage of diclofenac release after 24 h was 84.19% [77], which is lower compared to the percentage achieved in our study (95.61%). Additionally, a study by Mendes et al. [78] investigated phospholipid nanofibers, based on polysaccharide chitosan, for transdermal delivery of individual drugs diclofenac and curcumin. The percentages of curcumin and diclofenac release after 24 h were approximately 20% and 60%, respectively [78], while after 7 days, the maximum release of 75% was achieved for curcumin and 80% for diclofenac [78]. Therefore, it can be concluded that our investigation gave significantly better results in terms of the efficiency of the prepared carriers.

Figure 7 shows that drug release is cumulative over 24 h. It can also be seen that curcumin is released from the carrier at a much slower rate in the initial hours compared to diclofenac, which indicates stronger curcumin and carrier interactions. In addition to drug interactions with the carrier, swelling of the carrier itself, diffusion of the solute, and carrier degradation are crucial factors influencing the release of drugs from polymeric carriers [79]. Since the solubility of curcumin in buffer is significantly lower than the solubility of diclofenac, its diffusion rate into buffer solution will also be lower. Therefore, the solubility of curcumin directly causes its slower release compared to the release of diclofenac.

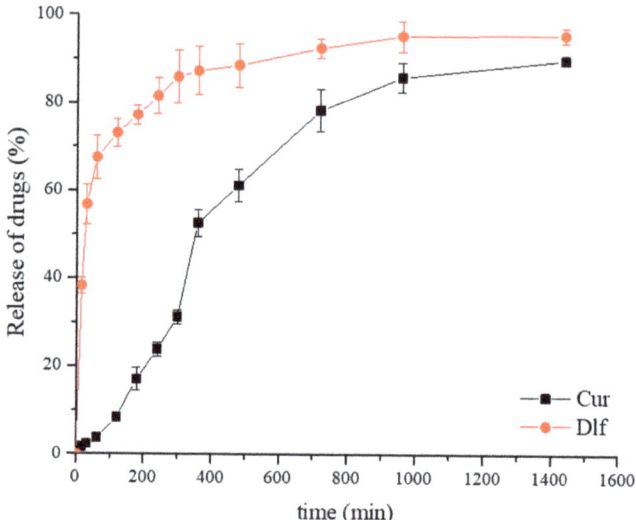

Figure 7. Curcumin and diclofenac in vitro release from the Car/Alg/Pol-Cur+Dlf film (n = 3).

In vitro release results are in accordance with the fact that the local reduction in the inflammation response is advocated in burn management. Fast diclofenac release is preferable because anti-inflammatory drugs suppress a persistent inflammatory response, leading to improved wound healing [80,81]. Biphasic pattern of diclofenac release involving the two stages is targeted to control the local inflammation and pain associated with a burn injury. Thus, the burst anti-inflammatory drug release effect ensures both a rapid reduction in painful sensation and the management of the pro-inflammatory mediators' cascade released at the burn level and is needed immediately after lesion occurrence [82]. After mentioned burst release, the gradual drug delivery phase offers an anti-inflammatory and analgesic local effect over the longer period needed for burn healing. Diclofenac release profile obtained in this work (especially from the Car/Alg/Pol-Cur+Dlf film) is desirable for burn treatment as the first 12 h are critical and correspond to the peak of the inflammatory phase [83]. On the other hand, release of curcumin, as a drug which accelerate different phases of wound healing, should be in a sustained manner, similar to the release profiles obtained in previous studies [46,73]. Release mechanism which includes a burst release of antibacterial drug diclofenac followed by a sustained release of curcumin with stronger antibacterial activity is expected to be effective in controlling and preventing infection in the very early stages of wound infliction. Prolonged curcumin release might indicate a long scale antimicrobial potency fabricated biocomposite dressings [84].

3.3. Drug Release Kinetics

The equations corresponding to zero-order kinetics, first-order kinetics, as well as the Higuchi, Hixon–Crowell and Korsmeyer–Peppas release models were applied to the results obtained by in vitro diclofenac and curcumin release study to investigate the mechanism of drug release from films. The correlation coefficient values obtained by fitting the results in accordance with the equations corresponding to the models are shown in Table 3. Additionally, the values of the release rate constants are shown and the value of n in the Korsmeyer–Peppas model.

Table 3. Values of correlation coefficients, release rate constants and release exponent.

Film	Zero-Order Kinetics		First-Order Kinetics		Highuchi Model	
	k_0	R^2	k_I	R^2	k_H	R^2
Car/Alg/Pol-Dlf	0.0156	0.3896	0.0202	0.3357	0.1083	0.6372
Car/Alg/Pol-Cur+Dlf (Cur)	0.0483	0.8529	0.1567	0.6112	0.2716	0.9370
Car/Alg/Pol-Cur+Dlf (Dlf)	0.0182	0.5347	0.0241	0.4304	0.1159	0.7541
	Hixon−Crowell Model		Korsmeyer−Peppas Model			
	k_{HC}	R^2	k_{KP}	n		R^2
Car/Alg/Pol-Dlf	0.0328	0.7955	0.7412	0.1400		0.8466
Car/Alg/Pol-Cur+Dlf (Cur)	0.0431	0.9930	0.1427	0.6807		0.9213
Car/Alg/Pol-Cur+Dlf (Dlf)	0.0321	0.9283	0.6861	0.1529		0.9069

The release of curcumin, as well as diclofenac from the Car/Alg/Pol-Dlf and Car/Alg/Pol-Cur+Dlf films, is best described with the Hixon–Crowell model, which is otherwise characteristic of systems in which the release rate is largely controlled by drug solubility in buffer rather than by diffusion of particles through the matrix [48]. Still, it should be considered that curcumin solubility is significantly lower than that of diclofenac, which further explains the slower curcumin release from the carrier. Curcumin release can also be described with the Highuchi model, which is characteristic of drugs with particles dispersed within a uniform, solid matrix, which acts as a diffusion medium, where drug release is largely controlled by Fick's law of diffusion [48]. These mechanisms of drug release from the tested formulations were further confirmed using the Korsmeyer–Peppas model, which describes very well the release of curcumin and diclofenac from all tested films (high correlation coefficient values were obtained).

The Korsmeyer–Peppas model is significant for estimating the mechanism of drug release and is mainly used to determine the parameter that has the greatest impact on the release rate (polymer swelling, diffusion of incorporated substance, polymer erosion) [48]. The value obtained for the release exponent n of 0.5 directly indicates the release controlled by drug diffusion, while the value of $n = 1$ indicates that drug release occurs primarily due to polymer swelling [48]. If the values of n differ from the above, then the release mechanism is influenced by several factors. In general, values of the release exponent below 0.5 correspond to Fick's law-controlled diffusion, above 0.5 to diffusion that does not obey this law, where release is also caused by polymers erosion, and values above 1 correspond to the super-transport case [48].

Considering the results obtained using the Korsmeyer–Peppas model, and following the values of n, it can be concluded that the diclofenac mechanism of release is the same during its release from the film containing only diclofenac ($n = 0.14$) and from the film containing a mixture of curcumin and diclofenac ($n = 0.15$). The obtained low value of the release exponent indicates that diclofenac release is primarily controlled by its diffusion from the carrier into the buffer, explaining the high release rate. Additionally, this result is in agreement with the Hixon–Crowell model, which describes the release of diclofenac from these formulations. The value of the curcumin release exponent from the film containing a mixture of curcumin and diclofenac is 0.68, which indicates that the release mechanism is influenced by the diffusion rate and swelling of the polymer. The obtained results are in agreement with the description of curcumin release using Higuchi and Hixon–Crowell models.

3.4. Theoretical Study of Component Interaction in Developed Films

In this section, the experimentally obtained results were further rationalized by means of quantum chemical computations. The optimized, most stable structures of the complexes formed between the diclofenac and curcumin molecules and the corresponding drug carriers are displayed in Figures 8 and 9. The most stable structures of the complexes were obtained by curcumin binding to carrageenan, as well as diclofenac binding to alginate from drug carrier. It should be noted that the curcumin-based structure presented in Figure 9 is somewhat different from that found in our previous study [18]. In the present work a more stable aggregate structure was obtained in which the curcumin molecule better adopts the shape of the drug carrier, thus maximizing bonding interactions.

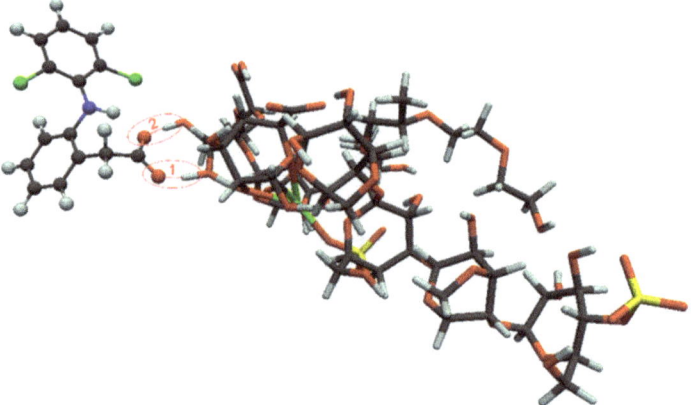

Figure 8. Optimized structure of diclofenac-carrier complex. Hydrogen bonds are denoted by red dashed lines (1 and 2). For the sake of clarity, the ball-and-stick and licorice visualization models were used for molecular structures of the drug and drug carrier, respectively.

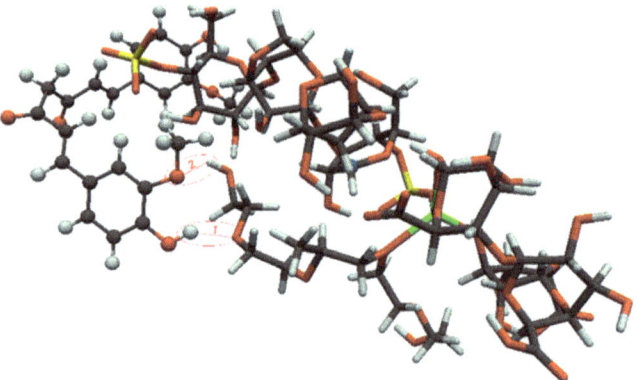

Figure 9. Optimized structure of curcumin-carrier complex. Hydrogen bonds are denoted by red dashed lines (1 and 2). For the sake of clarity, the ball-and-stick and licorice visualization models were used for molecular structures of the drug and drug carrier, respectively.

It was found that the bonding interactions between carboxylate anion of diclofenac and hydroxyl groups of alginate from carrier are dominated by two hydrogen bonds, of which the hydrogen bond 1 is found to be very strong. Based on Equation 11 the binding energies of hydrogen bonds 1 and 2 in the diclofenac containing complex are found to be -38.9 and -9.5 kcal/mol, respectively. These results are also in concordance with results obtained by FTIR analysis (Section 3.1.2) and wavenumber shift of carboxylate anion in films containing diclofenac, as a consequence of formation of hydrogen bonds between diclofenac carboxylate anion and alginate. On the other hand, in the case of curcumin there are two rather weak hydrogen bonds (formed between phenolic group/oxygen from ether group of the drug and carrageenan from carrier). According to Equation 10 the HBBE for bonds 1 and 2 in the curcumin-based complex are -7.3 and -1.7 kcal/mol, respectively.

The BEs calculated at the B3LYP/def2-SVP level of theory are given in Table 4. The obtained BEs predict that the bonding interactions between diclofenac and the drug carrier are more pronounced than curcumin-carrier interactions. From the optimized aggregate structures shown in Figure 8, it can be anticipated that bonding interactions between diclofenac and the drug carrier are manly determined by the strength of the formed hydrogen bonds, whereas curcumin binds through much more pronounced dipol-dipol and van der Waals interactions (Figure 9). Relevance of the van der Waals interactions in the case of curcumin comes from the fact that this molecule, in comparison with diclofenac, has much wider molecular surface and much more flexible geometry which enables an efficient adsorption on the drug carrier surface.

Table 4. Binding energies (BEs in kcal/mol) for diclofenac and curcumin in the respective complexes calculated the B3LYP/def2-SVP and B3LYP-D3/def2-SVP levels of theory.

Method	Diclofenac	Curcumin
B3LYP/def2-SVP	−36.5	−28.5
B3LYP-D3/def2-SVP	−44.9	−56.9

The BEs calculated with a more appropriate theoretical treatment, which accounts dispersion interactions (characteristic for curcumin) through Grimme's D3 method, show that curcumin is much stronger bonded to the carrier than diclofenac (Table 4). It should be pointed out that these results are in agreement with the experimentally obtained in vitro release data, which shows that diclofenac can be easier released from carrageenan/alginate/poloxamer carrier than curcumin.

3.5. Antibacterial Activity of Films

All tested bacteria showed sensitivity to the tested antibiotic disks (amoxicillin, tetracycline, streptomycin) prescribed in the manufacturer's instructions and the valid EU-CAST standard [56]. Gram-negative bacteria: *Pseudomonas aeruginosa* ATCC 27853 and *Escherichia coli* ATCC 25922 did not show sensitivity to any of the tested films. The resistance of Gram-negative bacteria to the effect of films containing components with antimicrobial potential can be explained by the carrier's structure. Carrageenan is an anionic polysaccharide, and since it is present in a large percentage in the carrier, it can be assumed that the carrier also will carry a negative charge. In addition, Gram-negative bacteria contain an additional outer layer composed of negatively charged lipopolysaccharides. Therefore, it is believed that more positively charged carriers will have a greater antimicrobial effect against these bacteria strains because they can achieve more favorable electrostatic interactions. On the other hand, films prepared in our work are expected to have a more pronounced tendency against Gram-positive bacteria.

The Car/Alg/Pol carrier and the Car/Alg/Pol-Cur film did not show antibacterial activity against the tested bacterial strains. In contrast, the Car/Alg/Pol-Dlf and Car/Alg/Pol-Cur+Dlf films were active against the Gram-positive bacteria strains. Table 5 shows the zones of bacterial inhibition obtained by the effect of the prepared Car/Alg/Pol-Dlf and Car/Alg/Pol-Cur+Dlf films, as well as antibiotics (amoxicillin, tetracycline and streptomycin) as controls, on strains of *Bacillus subtilis* ATCC 6633 and *Staphylococcus aureus* ATCC 25923.

As can be seen from the presented results, films containing a mixture of curcumin and diclofenac give significantly better results compared to films containing only diclofenac as the active substance. In the case of the Car/Alg/Pol-Cur+Dlf film, the antibacterial activity of both diclofenac and curcumin is pronounced. The obtained results can be related to the results achieved by monitoring the in vitro release of the mixture of drugs. Films containing only curcumin did not exhibit antimicrobial activity against the tested bacteria, despite curcumin's favorable antibacterial properties. This can be explained by the fact that the present bacteria reproduce much faster compared to the rate of antibacterial agent curcumin release from the carrier. Experimental results indicate that diclofenac is rapidly released from formulations, approximately 50% in the first 30 min, while only 2% of curcumin is released during this time [18]. Even though diclofenac shows low antibacterial activity (compared to commercially available and used antibiotics – amoxicillin, tetracycline, and streptomycin), its concentration in the initial phase of release is high enough to slow the growth of bacteria. Thanks to the fast action of diclofenac, conditions for the further antibacterial action of curcumin were created despite its slow release. As can be seen from the attached results (Table 5), the inhibition zone of the Car/Alg/Pol-Cur+Dlf film for bacteria *Bacillus subtilis* ATCC 6633 corresponds to the inhibition zone induced by the antibiotic amoxicillin and is similar to the inhibition zone provided by the antibiotic streptomycin. For strain *Staphylococcus aureus* ATCC 25923, the zone of inhibition of the applied film is only slightly smaller than the zone of inhibition caused by the antibiotic streptomycin.

Table 5. Results of susceptibility of tested Gram-positive bacteria to films and antibiotics discs.

Bacteria	*Bacillus subtilis* ATCC 6633	*Staphylococcus aureus* ATCC 25923
Tested films	Inhibition zone (mm) $\sum$	
Car/Alg/Pol-Dlf	9.33	6.33
Car/Alg/Pol-Cur+Dlf	16.67	13.67
Antibiotic disks	Inhibition zone (mm) $\sum$ and sensitivity category (S [1])	
A	16.67—S	26.67—S
T	29.67—S	23.33—S
S	18.33—S	16.67—S

[1] S—sensitive.

The antimicrobial activity of the film with a mixture of drugs against Gram-positive bacteria can also be related to the results of the study by Adamczak et al. [85] that investigated the antimicrobial activity of curcumin on more than 100 strains of pathogens. The results indicated that the susceptibility of Gram-positive bacteria was significantly higher than that of Gram-negative bacteria. The susceptibility of the species was not related to its genus. It was concluded that curcumin has great potential as a very selective antibacterial agent [85]. Other studies [86–89] confirmed the synergistic antibacterial activity of curcumin with different antibiotics (cefaclor, cefodizime, cefotaxime, gentamicin, amikacin, and ciprofloxacin) against different strains of bacteria. The antimicrobial effect of diclofenac in the presence of antibiotics was also examined [23]. Films with incorporated streptomycin (30%, v/v) and diclofenac (10%, v/v) were prepared to be used for faster healing of chronic wounds. The application of films enabled the controlled release of streptomycin and diclofenac for 72 h. Films with incorporated drugs gave higher inhibition zones against *Staphylococcus aureus*, *Pseudomonas aeruginosa*, and *Escherichia coli* compared to zones of inhibition provided by pure drugs. Still, the concentration of drugs used in the films was very high [23].

Our work studied the synergistic effect that occurs in the combination of curcumin with the non-steroidal anti-inflammatory drug diclofenac. Based on all the above, it can be concluded that the film Car/Alg/Pol-Cur+Dlf has great potential for treating infections caused by strains of *Bacillus subtilis* ATCC 6633 and *Staphylococcus aureus* ATCC 25923 as its action is similar to the effect of antibiotics (amoxicillin, streptomycin). Additionally, the side effects of the Car/Alg/Pol-Cur+Dlf film (resistance and undesired effects in the gastrointestinal tract) are expected to be significantly lower compared to that of commercially available antibiotics.

3.6. Cell Viability Assay and In Vivo Wound Healing Study

The effect of Car/Alg/Pol, Car/Alg/Pol-Dlf, and Car/Alg/Pol-Cur+Dlf on MRC-5 cell line viability was examined by MTT test after cultivation for 24 and 48 h (Figure 10).

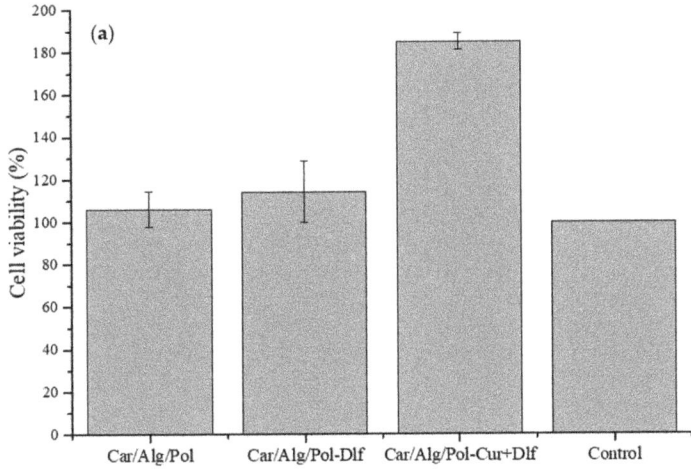

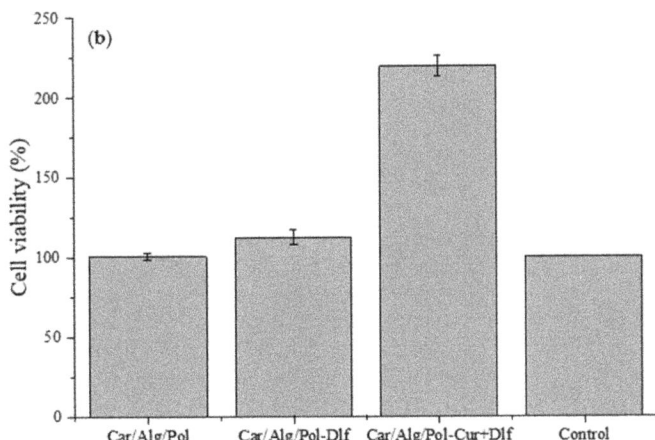

Figure 10. Influence of the Car/Alg/Pol, Car/Alg/Pol-Dlf i Car/Alg/Pol-Cur+Dlf films on MRC-5 cell viability after (**a**) 24 h and (**b**) 48 h.

The obtained results indicate that the film Car/Alg/Pol has no effect on the healing process as the percentage of examined cells viability is the same as in the control sample. Additionally, after 24 h of incubation, in the presence of films containing diclofenac, it is observed that the viability of MRC-5 cells is higher compared to the control, but without a statistically significant difference. However, cell viability is significantly higher in the presence of films containing a mixture of curcumin and diclofenac during the incubation period of 24 and 48 h. Comparing the obtained results with the results obtained in the study of curcumin films [18], it can be concluded that, after 24 h, the increase in cell viability occurs exclusively due to the presence of curcumin in films since similar viability percentages were obtained for films containing only curcumin and a mixture of diclofenac and curcumin. Based on the percentage of viable cells, it can also be concluded that a film containing a mixture of curcumin and diclofenac may have a potential for application in the wound healing process. The presence of diclofenac, although it does not contribute to the healing process, does not interfere with the positive effect of curcumin. On the other hand, the film containing a mixture of drugs shows significant antibacterial activity. This additionally demonstrates its suitability for use since, in addition to affecting the inflammation and proliferation phases in the wound healing process, it also prevents infections. The obtained positive results of in vitro assay have directed further research, and the prepared films were tested as formulations potentially applicable for wound healing of the skin of rats.

Histopathological analysis. A representative photomicrograph of rats' skin sections from all groups (control, treated, and non-treated) stained with H&E is shown in Figure 11.

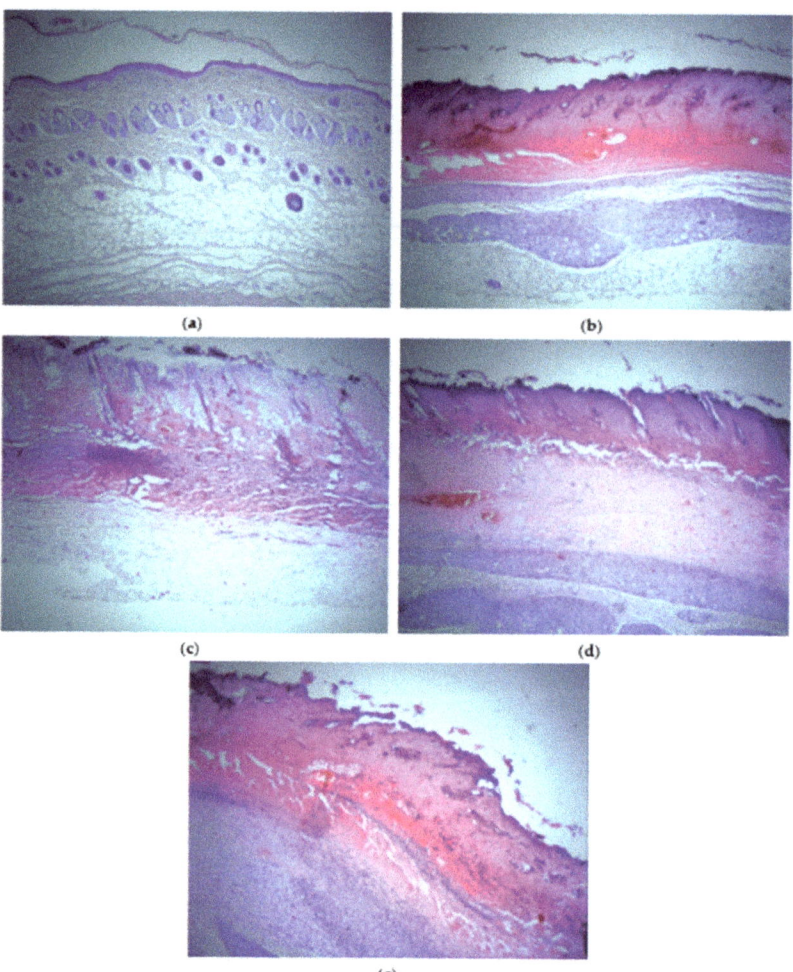

Figure 11. Histopathological observation of H&E stained skin sections (200×) of (**a**) healthy skin, (**b**) burned skin, (**c**) burned skin treated with the Car/Alg/Pol film, (**d**) burned skin treated with the Car/Alg/Pol-Dlf film, and (**e**) burned skin treated with the Car/Alg/Pol-Cur+Dlf film.

Analysis of paraffin sections of skin tissue of untreated animals, stained by hematoxylin-eosin technique, showed a healthy skin structure. Epidermis, dermis, and subcutaneous adipose tissue had normal histological structure, with intact sweat and sebaceous glands. In addition, the structure of hair follicles is preserved. A sample of burned skin tissue showed clear signs of epidermis and dermis damage compared to the untreated group of animals. The analysis showed damage to all epidermis layers—infiltration of inflammatory cells is observed in the dermis, indicating inflammatory skin changes. Additionally, heavy bleeding in the dermis was recorded. In the group treated with the Car/Alg/Pol films, disorganization of all epidermis layers and infiltration of inflammatory cells with minimal signs of re-epithelialization were observed. Histopathological analysis of burned skin tissue showed minimal tissue regeneration signs in animals treated with the Car/Alg/Pol-Dlf films, including reduced infiltration of inflammatory cells compared to the group treated with only the Car/Alg/Pol films. In contrast to the above films, a notable degree of skin regeneration was observed in the group treated with the Car/Alg/Pol-Cur+Dlf films. On the sections of burned skin tissue treated the with Car/Alg/Pol-Cur+Dlf films, significant regeneration of the epidermis is observed, with well-organized layers and minimal infiltration of inflammatory cells. Results of histopathological analysis can be related to the cell viability study because the obtained results indicated that the Car/Alg/Pol-Cur+Dlf film enhances cell proliferation. Considering both in vitro and in vivo data, our findings clearly demonstrate that the application of films, containing both curcumin and diclofenac, improves the healing of burns remarkably. Available data from the literature related to the dermatological effects of curcumin, mentioned above, indicate its anti-inflammatory and antioxidant effect achieved through enhanced synthesis of hyaluronic acid and the effect of increasing skin moisture [35–38,44], along with the known anti-inflammatory effect of diclofenac [19]. The presented drug characteristics and the results obtained through our in vivo study led to a conclusion that the use of curcumin and diclofenac films achieves effective healing of burn-caused dermal wounds.

4. Conclusions

In biocompatible films based on polymers κ-carrageenan, alginate and poloxamer, diclofenac (an anti-inflammatory drug that has antibacterial properties) as well as a mixture of curcumin (a drug that exhibits antioxidant, anti-inflammatory and antibacterial properties) and diclofenac were incorporated. The characterization of the films showed that the prepared films have a smooth, homogeneous surface, while XRD analysis indicated a decrease in the crystallinity degree of curcumin and diclofenac after their incorporation into the films, while diclofenac transforms to an amorphous state. The in vitro release study showed that the bioavailability of curcumin and diclofenac was significantly improved by using developed carriers. Based on the results obtained by drug release kinetics, it was concluded that the polymer swelling degree has the greatest influence on curcumin release, while the release of diclofenac is largely controlled by diffusion. Theoretical examination of interactions that carriers establish with curcumin and diclofenac indicated that diclofenac formed strong hydrogen bonds with alginate from the carrier, while curcumin established stronger, primarily dispersion interactions with carrageenan. The antibacterial study of the prepared films showed that films with diclofenac and a mixture of curcumin and diclofenac inhibit the growth and development of Gram-positive bacteria *Bacillus subtilis* and *Staphylococcus aureus*. Additionally, it was determined that drug-loaded films are not cytotoxic, whereas films containing a mixture of curcumin and diclofenac can increase cell viability and thus have a favorable effect on cell proliferation, which is a phase during wound healing. Based on the results of in vivo study, it can be concluded that the produced films have great potential for healing wounds caused by burns.

Author Contributions: Conceptualization, Z.S.; methodology, K.S.P. and Z.S.; formal analysis, S.R.; investigation, K.S.P., B.L., M.D.A., M.M.K., Z.H., S.P. and I.R.; resources, M.D.A., B.L. and Z.S.; writing—original draft preparation, K.S.P.; writing—review and editing, Z.S.; supervision, Z.S. All authors have read and agreed to the published version of the manuscript.

Funding: This work was supported by the Serbian Ministry of Education, Science and Technological Development (Agreement No. 451-03-68/2022-14/200122).

Institutional Review Board Statement: All the animal research studies were approved by the Animal Ethics Committee of the Faculty of Medicine, University of Kragujevac (Ethical Approval Number: 01-6121).

Informed Consent Statement: Not applicable.

Data Availability Statement: The data are contained within this article.

Conflicts of Interest: The authors declare no conflict of interest. The funders had no role in the design of the study; in the collection, analyses, or interpretation of data; in the writing of the manuscript; or in the decision to publish the results.

References

1. Boateng, J.S.; Matthews, K.H.; Stevens, H.N.; Eccleston, G.M. Wound healing dressings and drug delivery systems: A review. *J. Pharm. Sci.* **2008**, *97*, 2892–2923. [CrossRef]
2. Schultz, G.S. Molecular regulation of wound healing. In *Acute and Chronic Wounds: Nursing Management*, 2nd ed.; Mosby: St. Loius, MO, USA, 1999; pp. 413–429.
3. Martin, P. Wound healing—Aiming for perfect skin regeneration. *Science* **1997**, *276*, 75–81. [CrossRef]
4. Cañedo-Dorantes, L.; Cañedo-Ayala, M. Skin acute wound healing: A comprehensive review. *Int. J. Inflamm.* **2019**, *2019*, 3706315. [CrossRef] [PubMed]
5. Sweeney, I.R.; Miraftab, M.; Collyer, G. A critical review of modern and emerging absorbent dressings used to treat exuding wounds. *Int. Wound J.* **2012**, *9*, 601–612. [CrossRef] [PubMed]
6. Eaglstein, W.H.; Davis, S.C.; Mehle, A.L.; Mertz, P.M. Optimal use of an occlusive dressing to enhance healing: Effect of delayed application and early removal on wound healing. *Arch. Dermatol.* **1988**, *124*, 392–395. [CrossRef] [PubMed]
7. Bowler, P.G.; Duerden, B.I.; Armstrong, D.G. Wound microbiology and associated approaches to wound management. *Clin. Microbiol. Rev.* **2001**, *14*, 244–269. [CrossRef]
8. Farahani, M.; Shafiee, A. Wound healing: From passive to smart dressings. *Adv. Healthc. Mater.* **2021**, *10*, 2100477. [CrossRef]
9. Boateng, J.; Catanzano, O. Advanced therapeutic dressings for effective wound healing—A review. *J. Pharm. Sci.* **2015**, *104*, 3653–3680. [CrossRef]
10. Ávila-Salas, F.; Marican, A.; Pinochet, S.; Carreño, G.; Valdés, O.; Venegas, B.; Donoso, W.; Cabrera-Barjas, G.; Vijayakumar, S.; Durán-Lara, E.F. Film dressings based on hydrogels: Simultaneous and sustained-release of bioactive compounds with wound healing properties. *Pharmaceutics* **2019**, *11*, 447. [CrossRef]
11. Song, S.; Liu, Z.; Abubaker, M.A.; Ding, L.; Zhang, J.; Yang, S.; Fan, Z. Antibacterial polyvinyl alcohol/bacterial cellulose/nano-silver hydrogels that effectively promote wound healing. *Mater. Sci. Eng. C* **2021**, *126*, 112171. [CrossRef]
12. Sharaf, S.M.; Al-Mofty, S.E.-D.; El-Sayed, E.-S.M.; Omar, A.; Dena, A.S.A.; El-Sherbiny, I.M. Deacetylated cellulose acetate nanofibrous dressing loaded with chitosan/propolis nanoparticles for the effective treatment of burn wounds. *Int. J. Biol. Macromol.* **2021**, *193*, 2029–2037. [CrossRef]
13. Langer, R. Polymeric delivery systems for controlled drug release. *Chem. Eng. Commun.* **1980**, *6*, 1–48. [CrossRef]
14. Slaughter, B.V.; Khurshid, S.S.; Fisher, O.Z.; Khademhosseini, A.; Peppas, N.A. Hydrogels in regenerative medicine. *Adv. Mater.* **2009**, *21*, 3307–3329. [CrossRef]
15. Mogosanu, G.D.; Grumezescu, A.M. Natural and synthetic polymers for wounds and burns dressing. *Int. J. Pharm.* **2014**, *463*, 127–136. [CrossRef]
16. Pereira, R.; Carvalho, A.; Vaz, D.C.; Gil, M.H.; Mendes, A.; Bártolo, P. Development of novel alginate based hydrogel films for wound healing applications. *Int. J. Biol. Macromol.* **2013**, *52*, 221–230. [CrossRef]
17. Jaiswal, L.; Shankar, S.; Rhim, J.W. Carrageenan-based functional hydrogel film reinforced with sulfur nanoparticles and grapefruit seed extract for wound heal-ing application. *Carbohydr. Polym.* **2019**, *224*, 115191. [CrossRef]
18. Postolović, K.; Ljujić, B.; Kovačević, M.M.; Đorđević, S.; Nikolić, S.; Živanović, S.; Stanić, Z. Optimization, characterization, and evaluation of carrageenan/alginate/poloxamer/curcumin hydrogel film as a functional wound dressing material. *Mater. Today Commun.* **2022**, *31*, 103528. [CrossRef]
19. Todd, P.A.; Sorkin, E.M. Diclofenac sodium—A reappraisal of its pharmacodynamic and pharmacokinetic properties, and therapeutic efficacy. *Drugs* **1988**, *35*, 244–285. [CrossRef]
20. Goh, C.F.; Lane, M.E. Formulation of diclofenac for dermal delivery. *Int. J. Pharm.* **2014**, *473*, 607–616. [CrossRef]
21. Salem-Milani, A.; Balaei-Gajan, E.; Rahimi, S.; Moosavi, Z.; Abdollahi, A.; Zakeri-Milani, P.; Bolourian, M. Antibacterial effect of diclofenac sodium on Enterococcus faecalis. *J. Dent.* **2013**, *10*, 16.
22. Dutta, N.K.; Dastidar, S.G.; Kumar, A.; Mazumdar, K.; Ray, R.; Chakrabarty, A.N. Antimycobacterial activity of the antiinflammatory agent diclofenac sodium, and its synergism with streptomycin. *Braz. J. Microbiol.* **2004**, *35*, 316–323. [CrossRef]
23. Boateng, J.S.; Pawar, H.V.; Tetteh, J. Polyox and carrageenan based composite film dressing containing anti-microbial and anti-inflammatory drugs for effective wound healing. *Int. J. Pharm.* **2013**, *441*, 181–191. [CrossRef] [PubMed]
24. Pawar, H.V.; Tetteh, J.; Boateng, J.S. Preparation, optimisation and characterisation of novel wound healing film dressings loaded with streptomycin and diclofenac. *Colloids Surf. B* **2013**, *102*, 102–110. [CrossRef] [PubMed]
25. Alqahtani, F.Y.; Aleanizy, F.S.; El Tahir, E.; Alquadeib, B.T.; Alsarra, I.A.; Alanazi, J.S.; Abdelhady, H.G. Preparation, characterization, and antibacterial activity of diclofenac-loaded chitosan nanoparticles. *Saudi Pharm. J.* **2019**, *27*, 82–87. [CrossRef] [PubMed]
26. Sarwar, M.N.; Ullah, A.; Haider, M.K.; Hussain, N.; Ullah, S.; Hashmi, M.; Khan, M.Q.; Kim, I.S. Evaluating antibacterial efficacy and biocompatibility of PAN nanofibers loaded with diclofenac sodium salt. *Polymers* **2021**, *13*, 510. [CrossRef] [PubMed]

27. Sharma, R.A.; Gescher, A.J.; Steward, W.P. Curcumin: The story so far. *Eur. J. Cancer* **2005**, *41*, 1955–1968. [CrossRef] [PubMed]
28. Stanić, Z. Curcumin, a compound from natural sources, a true scientific challenge—A review. *Plant Foods Hum. Nutr.* **2017**, *72*, 1–12. [CrossRef]
29. Anand, P.; Kunnumakkara, A.B.; Newman, R.A.; Aggarwal, B.B. Bioavailability of curcumin: Problems and promises. *Mol. Pharm.* **2007**, *4*, 807–818. [CrossRef]
30. Strimpakos, A.S.; Sharma, R.A. Curcumin: Preventive and therapeutic properties in laboratory studies and clinical trials. *Antioxid. Redox Signal.* **2008**, *10*, 511–546. [CrossRef]
31. Roy, S.; Rhim, J.W. Preparation of carbohydrate-based functional composite films incorporated with curcumin. *Food Hydrocoll.* **2020**, *98*, 105302. [CrossRef]
32. Roy, S.; Rhim, J.W. Antioxidant and antimicrobial poly (vinyl alcohol)-based films incorporated with grapefruit seed extract and curcumin. *J. Environ. Chem. Eng.* **2021**, *9*, 104694. [CrossRef]
33. Imlay, J. Pathways of oxidative damage. *Annu. Rev. Microbiol.* **2003**, *57*, 395–418. [CrossRef]
34. Mohanty, C.; Das, M.; Sahoo, S.K. Sustained wound healing activity of curcumin loaded oleic acid based polymeric bandage in a rat model. *Mol. Pharm.* **2012**, *9*, 2801–2811. [CrossRef]
35. Panchatcharam, M.; Miriyala, S.; Gayathri, V.S.; Suguna, L. Curcumin improves wound healing by modulating collagen and decreasing reactive oxygen species. *Mol. Cell. Biochem.* **2006**, *290*, 87–96. [CrossRef]
36. Thangapazham, R.L.; Sharad, S.; Maheshwari, R.K. Skin regenerative potentials of curcumin. *Biofactors* **2013**, *39*, 141–149. [CrossRef]
37. Mohanty, C.; Sahoo, S.K. Curcumin and its topical formulations for wound healing applications. *Drug Discov. Today* **2017**, *22*, 1582–1592. [CrossRef]
38. Joe, B.; Vijaykumar, M.; Lokesh, B.R. Biological properties of curcumin-cellular and molecular mechanisms of action. *Crit. Rev. Food Sci. Nutr.* **2004**, *44*, 97–111. [CrossRef]
39. Qu, J.; Zhao, X.; Liang, Y.; Zhang, T.; Ma, P.X.; Guo, B. Antibacterial adhesive injectable hydrogels with rapid self-healing, extensibility and compressibility as wound dressing for joints skin wound healing. *Biomaterials* **2018**, *183*, 185–199. [CrossRef]
40. Duan, Y.; Li, K.; Wang, H.; Wu, T.; Zhao, Y.; Li, H.; Tang, H.; Yang, W. Preparation and evaluation of curcumin grafted hyaluronic acid modified pullulan polymers as a functional wound dressing material. *Carbohydr. Polym.* **2020**, *238*, 116195. [CrossRef]
41. Wathoni, N.; Motoyama, K.; Higashi, T.; Okajima, M.; Kaneko, T.; Arima, H. Enhancement of curcumin wound healing ability by complexation with 2-hydroxypropyl-γ-cyclodextrin in sacran hydrogel film. *Int. J. Biol. Macromol.* **2017**, *98*, 268–276. [CrossRef]
42. Li, X.; Nan, K.; Li, L.; Zhang, Z.; Chen, H. In vivo evaluation of curcumin nanoformulation loaded methoxy poly (ethylene glycol)-graft-chitosan composite film for wound healing application. *Carbohydr. Polym.* **2012**, *88*, 84–90. [CrossRef]
43. Sajjad, W.; He, F.; Ullah, M.W.; Ikram, M.; Shah, S.M.; Khan, R.; Khan, T.; Khan, A.; Khalid, A.; Yang, G.; et al. Fabrication of bacterial cellulose-curcumin nanocomposite as a novel dressing for partial thickness skin burn. *Front. Bioeng. Biotechnol.* **2020**, *8*, 553037. [CrossRef]
44. Alven, S.; Nqoro, X.; Aderibigbe, B.A. Polymer-based materials loaded with curcumin for wound healing applications. *Polymers* **2020**, *12*, 2286. [CrossRef]
45. Sharifi, S.; Fathi, N.; Memar, M.Y.; Khatibi, S.M.H.; Khalilov, R.; Negahdari, R.; Vahed, S.Z.; Dizaj, S.M. Anti-microbial activity of curcumin nanoformulations: New trends and future perspectives. *Phytother. Res.* **2020**, *34*, 1926–1946. [CrossRef]
46. Hadizadeh, M.; Naeimi, M.; Rafienia, M.; Karkhaneh, A. A bifunctional electrospun nanocomposite wound dressing containing surfactin and curcumin: In vitro and in vivo studies. *Mater. Sci. Eng. C* **2021**, *129*, 112362. [CrossRef]
47. De Paz-Campos, M.A.; Ortiz, M.I.; Piña, A.E.C.; Zazueta-Beltrán, L.; Castañeda-Hernández, G. Synergistic effect of the interaction between curcumin and diclofenac on the formalin test in rats. *Phytomedicine* **2014**, *21*, 1543–1548. [CrossRef]
48. Costa, P.; Lobo, J.M.S. Modeling and comparison of dissolution profiles. *Eur. J. Pharm. Sci.* **2001**, *13*, 123–133. [CrossRef]
49. Frisch, M.J.; Trucks, G.W.; Schlegel, H.B.; Scuseria, G.E.; Robb, M.A.; Cheeseman, J.R.; Scalmani, G.; Barone, V.; Mennucci, B.; Petersson, G.A.; et al. *Gaussian 09*; Gaussian Inc.: Wallingford, CT, USA, 2009.
50. Costa, M.P.; Prates, L.M.; Baptista, L.; Cruz, M.T.; Ferreira, I.L. Interaction of polyelectrolyte complex between sodium alginate and chitosan dimers with a single glyphosate molecule: A DFT and NBO study. *Carbohydr. Polym.* **2018**, *198*, 51–60. [CrossRef] [PubMed]
51. Grimme, S. Semiempirical GGA-type density functional constructed with a long-range dispersion correction. *J. Comput. Chem.* **2006**, *27*, 1787–1799. [CrossRef] [PubMed]
52. Lu, T.; Chen, F. Multiwfn: A multifunctional wavefunction analyzer. *J. Comput. Chem.* **2012**, *33*, 580–592. [CrossRef]
53. Emamian, S.; Lu, T.; Kruse, H.; Emamian, H. Exploring nature and predicting strength of hydrogen bonds: A correlation analysis between atoms-in-molecules descriptors, binding energies, and energy components of symmetry-adapted perturbation theory. *J. Comput. Chem.* **2019**, *40*, 2868. [CrossRef]
54. Andrews, J.M. Determination of minimum inhibitory concentrations. *J. Antimicrob. Chemother.* **2001**, *48*, 5–16. [CrossRef]
55. Andrews, J.M. BSAC standardized disc susceptibility testing method (version 4). *J. Antimicrob. Chemother.* **2005**, *56*, 60–76. [CrossRef]
56. EUCAST—The European Committee on Antimicrobial Susceptibility Testing Breakpoint Tables for Interpretation of MICs and Zone Diameters. Version 9.0. 2019. Available online: http://www.eucast.org (accessed on 14 June 2022).

57. Jorgensen, J.H.; Turnidge, J.D. Susceptibility test methods: Dilution and disc diffusion methods. In *Manual of Clinical Microbiology*, 8th ed.; American Society of Microbiology: Washington, DC, USA, 2003; pp. 1119–1125.
58. Mosmann, T. Rapid colorimetric assay for cellular growth and survival: Application to proliferation and cytotoxicity assays. *J. Immunol. Methods* **1983**, *65*, 55–63. [CrossRef]
59. Li, H.; Xue, Y.; Jia, B.; Bai, Y.; Zuo, Y.; Wang, S.; Zhao, Y.; Yang, W.; Tang, H. The preparation of hyaluronic acid grafted pullulan polymers and their use in the formation of novel biocompatible wound healing film. *Carbohydr. Polym.* **2018**, *188*, 92–100. [CrossRef]
60. Mohan, P.K.; Sreelakshmi, G.; Muraleedharan, C.V.; Joseph, R. Water soluble complexes of curcumin with cyclodextrins: Characterization by FT-Raman spectros-copy. *Vib. Spectrosc.* **2012**, *62*, 77–84. [CrossRef]
61. Aielo, P.B.; Borges, F.A.; Romeira, K.M.; Miranda, M.C.R.; Arruda, L.B.D.; Filho, P.N.L.; Drago, B.C.; Herculano, R.D. Evaluation of sodium diclofenac release using natural rubber latex as carrier. *Mater. Res.* **2014**, *17*, 146–152. [CrossRef]
62. Rezvanian, M.; Ahmad, N.; Amin, M.C.I.M.; Ng, S.F. Optimization, charac-terization, and in vitro assessment of alginate-pectin ionic cross-linked hydrogel film for wound dressing applications. *Int. J. Biol. Macromol.* **2017**, *97*, 131–140. [CrossRef]
63. Roy, S.; Rhim, J.W. Preparation of bioactive functional poly (lactic acid)/curcumin composite film for food packaging application. *Int. J. Biol. Macromol.* **2020**, *162*, 1780–1789. [CrossRef]
64. Luo, N.; Varaprasad, K.; Reddy, G.V.S.; Rajulu, A.V.; Zhang, J. Preparation and characterization of cellulose/curcumin composite films. *RSC Adv.* **2012**, *2*, 8483–8488. [CrossRef]
65. Xie, Q.; Zheng, X.; Li, L.; Ma, L.; Zhao, Q.; Chang, S.; You, L. Effect of curcumin addition on the properties of biodegradable pectin/chitosan films. *Molecules* **2021**, *26*, 215. [CrossRef]
66. Thorat, A.A.; Dalvi, S.V. Solid-state phase transformations and storage sta-bility of curcumin polymorphs. *Cryst. Growth Des.* **2015**, *15*, 1757–1770. [CrossRef]
67. Ezati, P.; Rhim, J.W. pH-responsive pectin-based multifunctional films incorporated with curcumin and sulfur nanoparticles. *Carbohydr. Polym.* **2020**, *23*, 115638. [CrossRef]
68. Xiao, C.; Liu, H.; Lu, Y.; Zhang, L. Blend films from sodium alginate and gela-tin solutions. *J. Macromol. Sci. A* **2001**, *38*, 317–328. [CrossRef]
69. Liu, Y.; Qin, Y.; Bai, R.; Zhang, X.; Yuan, L.; Liu, J. Preparation of pH-sensitive and antioxidant packaging films based on κ-carrageenan and mulberry polyphenolic extract. *Int. J. Biol. Macromol.* **2019**, *134*, 993–1001. [CrossRef]
70. Postolović, K.S.; Antonijević, M.D.; Ljujić, B.; Miletić Kovačević, M.; Gazdić Janković, M.; Stanić, Z.D. pH-Responsive hydrogel beads based on alginate, κ-carrageenan and poloxamer for enhanced curcumin, natural bioactive compound, encapsulation and controlled release efficiency. *Molecules* **2022**, *27*, 4045. [CrossRef]
71. Pasquali, I.; Bettini, R.; Giordano, F. Thermal behaviour of diclofenac, diclofenac sodium and sodium bicarbonate compositions. *J. Therm. Anal. Calorim.* **2007**, *90*, 903–907. [CrossRef]
72. Tudja, P.; Khan, M.Z.I.; Meštrovic, E.; Horvat, M.; Golja, P. Thermal behaviour of diclofenac sodium: Decomposition and melting characteristics. *Chem. Pharm. Bull.* **2001**, *49*, 1245–1250. [CrossRef]
73. Kianvash, N.; Bahador, A.; Pourhajibagher, M.; Ghafari, H.; Nikoui, V.; Rezayat, S.M.; Dehpour, A.R.; Partoazar, A. Evaluation of propylene glycol nanoliposomes containing curcumin on burn wound model in rat: Biocompatibility, wound healing, and anti-bacterial effects. *Drug Deliv. Transl. Res.* **2017**, *7*, 654–663. [CrossRef]
74. Shende, P.; Gupta, H. Formulation and comparative characterization of nanoparticles of curcumin using natural, synthetic and semi-synthetic polymers for wound healing. *Life Sci.* **2020**, *253*, 117588. [CrossRef]
75. Gull, N.; Khan, S.M.; Butt, O.M.; Islam, A.; Shah, A.; Jabeen, S.; Khan, S.U.; Khan, A.; Khan, R.U.; Butt, M.T.Z. Inflammation targeted chitosan-based hydrogel for controlled release of diclofenac sodium. *Int. J. Biol. Macromol.* **2020**, *162*, 175–187. [CrossRef] [PubMed]
76. Omer, A.M.; Ahmed, M.S.; El-Subruiti, G.M.; Khalifa, R.E.; Eltaweil, A.S. pH-Sensitive alginate/carboxymethyl chitosan/aminated chitosan microcapsules for efficient encapsulation and delivery of diclofenac sodium. *Pharmaceutics* **2021**, *13*, 338. [CrossRef] [PubMed]
77. Khan, M.R.U.; Raza, S.M.; Hussain, M. Formulation and in-vitro evaluation of cream containing diclofenac sodium and curcuma longa for the management of rheumatoid arthritis. *Int. J. Pharma Sci.* **2014**, *4*, 654–660.
78. Mendes, A.C.; Gorzelanny, C.; Halter, N.; Schneider, S.W.; Chronakis, I.S. Hybrid electrospun chitosan-phospholipids nanofibers for transdermal drug delivery. *Int. J. Pharm.* **2016**, *510*, 48–56. [CrossRef] [PubMed]
79. Momoh, F.U.; Boateng, J.S.; Richardson, S.C.; Chowdhry, B.Z.; Mitchell, J.C. Development and functional characterization of alginate dressing as potential protein delivery system for wound healing. *Int. J. Biol. Macromol.* **2015**, *81*, 137–150. [CrossRef]
80. Hajiali, H.; Summa, M.; Russo, D.; Armirotti, A.; Brunetti, V.; Bertorelli, R.; Athanassiou, A.; Mele, E. Alginate–lavender nanofibers with antibacterial and anti-inflammatory activity to effectively promote burn healing. *J. Mater. Chem. B* **2016**, *4*, 1686–1695. [CrossRef]
81. Ferreira, H.; Matamá, T.; Silva, R.; Silva, C.; Gomes, A.C.; Cavaco-Paulo, A. Functionalization of gauzes with liposomes entrapping an anti-inflammatory drug: A strategy to improve wound healing. *React. Funct. Polym.* **2013**, *73*, 1328–1334. [CrossRef]
82. Ghica, M.V.; Albu Kaya, M.G.; Dinu-Pîrvu, C.-E.; Lupuleasa, D.; Udeanu, D.I. Development, Optimization and in vitro/in vivo characterization of collagen-dextran spongious wound dressings loaded with flufenamic acid. *Molecules* **2017**, *22*, 1552. [CrossRef]

83. Morgado, P.I.; Lisboa, P.F.; Ribeiro, M.P.; Miguel, S.P.; Simões, P.C.; Correia, I.J.; Aguiar-Ricardo, A. Poly(vinyl alcohol)/chitosan asymmetrical membranes: Highly controlled morphology toward the ideal wound dressing. *J. Membr. Sci. Technol.* **2017**, *159*, 262–271. [CrossRef]
84. Tummalapalli, M.; Berthet, M.; Verrier, B.; Deopura, B.L.; Alam, M.S.; Gupta, B. Composite wound dressings of pectin and gelatin with aloe vera and curcumin as bioactive agents. *Int. J. Biol. Macromol.* **2016**, *82*, 104–113. [CrossRef]
85. Adamczak, A.; Ożarowski, M.; Karpiński, T.M. Curcumin, a natural antimicrobial agent with strain-specific activity. *Pharmaceuticals* **2020**, *13*, 153. [CrossRef]
86. Moghaddam, K.M.; Iranshahi, M.; Yazdi, M.C.; Shahverdi, A.R. The combination effect of curcumin with different antibiotics against *Staphylococcus aureus*. *Int. J. Green Pharm.* **2009**, *3*, 141–143.
87. Amrouche, T.; Noll, K.S.; Wang, Y.; Huang, Q.; Chikindas, M.L. Antibacterial activity of subtilosin alone and combined with curcumin, poly-lysine and zinc lactate against listeriamonocytogenes strains. *Probiotics Antimicrob. Proteins* **2010**, *2*, 250–257. [CrossRef]
88. Marathe, S.A.; Kumar, R.; Ajitkumar, P.; Nagaraja, V.; Chakravortty, D. Curcumin reduces the antimicrobial activity of ciprofloxacin against Salmonella Typhimurium and Salmonella Typhi. *J. Antimicrob. Chemother.* **2013**, *68*, 139–152. [CrossRef]
89. Teow, S.Y.; Ali, S.A. Synergistic antibacterial activity of Curcumin with antibiotics against *Staphylococcus aureus*. *Pak. J. Pharm. Sci.* **2015**, *28*, 2109–2114.

Article

Bio-Based Polymeric Substrates for Printed Hybrid Electronics

Enni Luoma [1,*], Marja Välimäki [2], Jyrki Ollila [2], Kyösti Heikkinen [2] and Kirsi Immonen [1]

[1] Sustainable Products and Materials, VTT Technical Research Centre of Finland, Visiokatu 4, 33720 Tampere, Finland; kirsi.immonen@vtt.fi
[2] Digital Technologies, VTT Technical Research Centre of Finland, Kaitoväylä 1, 90570 Oulu, Finland; marja.valimaki@vtt.fi (M.V.); jyrki.ollila@vtt.fi (J.O.); kyosti.heikkinen@vtt.fi (K.H.)
* Correspondence: enni.luoma@vtt.fi

Abstract: Printed flexible hybrid electronics (FHE) is finding an increasing number of applications in the fields of displays, sensors, actuators and in energy harvesting and storage. The technology involves the printing of conductive and insulating patterns as well as mounting electronic devices and circuits on flexible substrate materials. Typical plastic substrates in use are, for example, non-renewable-based poly(ethylene terephthalate) (PET) or poly(imides) (PI) with high thermal and dimensional stability, solvent resistance and mechanical strength. The aim of this study was to assess whether renewable-based plastic materials can be applied on sheet-to-sheet (S2S) screen-printing of conductive silver patterns. The selected materials were biaxially oriented (BO) bio-based PET (Bio-PET BO), poly(lactic acid) (PLA BO), cellulose acetate propionate (CAP BO) and regenerated cellulose film, NatureFlex™ (Natureflex). The biaxial orientation and annealing improved the mechanical strength of Bio-PET and PLA to the same level as the reference PET (Ref-PET). All renewable-based substrates showed a transparency comparable to the Ref-PET. The printability of silver ink was good with all renewable-based substrates and printed pattern resistance on the same level as Ref-PET. The formation of the printed pattern to the cellulose-based substrates, CAP BO and Natureflex, was very good, showing 10% to 18% lower resistance compared to Ref-PET and obtained among the bio-based substrates the smallest machine and transverse direction deviation in the S2S printing process. The results will open new application possibilities for renewable-based substrates, and also potentially biodegradable solutions enabled by the regenerated cellulose film and PLA.

Keywords: flexible printed electronics; flexible hybrid electronics; biopolymer films; renewable-based substrate; screen-printing; surface energy

Citation: Luoma, E.; Välimäki, M.; Ollila, J.; Heikkinen, K.; Immonen, K. Bio-Based Polymeric Substrates for Printed Hybrid Electronics. *Polymers* **2022**, *14*, 1863. https://doi.org/10.3390/polym14091863

Academic Editors: Swarup Roy and Jong-Whan Rhim

Received: 31 March 2022
Accepted: 28 April 2022
Published: 2 May 2022

Publisher's Note: MDPI stays neutral with regard to jurisdictional claims in published maps and institutional affiliations.

Copyright: © 2022 by the authors. Licensee MDPI, Basel, Switzerland. This article is an open access article distributed under the terms and conditions of the Creative Commons Attribution (CC BY) license (https://creativecommons.org/licenses/by/4.0/).

1. Introduction

Flexible thin film electronics have increasing innovative potential in the fields of displays, sensors, actuators, and in energy harvesting and storage. Flexible hybrid electronics' technology can be characterised as a combination of printing processes and ink chemistry, as well as substrate material engineering for the manufacturing of lightweight electronic components. The technology involves mounting electronic devices and circuits on flexible substrate materials, which are typically polymer films. The target sectors for flexible electronics are especially in health care, automotive, wearables, mobile communications, and human–machine interfaces. Internet-of-Things-based sensing applications are comprised of a large variety of potential user interfaces in sensing, e.g., for logistic operations, diagnostics, and environmental conditions. Additionally, the expectations for the operational conditions and for the lifespan can vary a lot [1–3].

Screen-printing is one of the most utilised technologies for manufacturing flexible electronics. It is a fast and versatile process, which can be carried out cost- and material-effectively. Screen-printing pastes have a typical viscosity range of 500–5000 mPas and, compared to other printing techniques, allow higher material thickness, thus being suitable for printing interconnects and passive circuit elements [3].

Screen-printing can be carried out in sheet-to-sheet (S2S) and roll-to-toll (R2R) processes by using a flat-bed or rotary type of screen-printer. In flat-bed screen-printing the substrates are typically sheets and are printed one by one with a planar printing screen, whereas in rotary screen-printing, the printing screen is a rotating cylinder with a fixed position and the printing is usually executed in continuous R2R processing [3–5]. It should also be noted that in R2R processing, both the substrate and the printed patterns need to have flexibility.

Metallic particle inks are typically used in screen-printing due to their high conductivity. To attain a high level of conductivity, they typically require curing at elevated temperatures (150–250 °C). This heat treatment removes insulating components, such as stabilising agents and other additives. Silver is the most widely used metal in inks because of its high bulk conductivity and resistance to oxidation, even though silver is an expensive material, resulting in high ink prices [2]. Ordinarily, a high curing temperature of silver inks also puts a limit on plastic substrates suitable for printed electronics, which has also led to developments for conductive ink materials with lower curing temperatures [6,7].

Nowadays, most of the substrate materials in printed electronics are derived from non-renewable fossil raw materials. Examples of substrate materials are poly(imides) (PI), poly(ether ether ketone) (PEEK), poly(ether sulfone) (PES), poly(ether imide) (PEI), poly(ethylene naphthalate) (PEN), and poly(ethylene terephthalate) (PET). Requirements for substrate materials include dimensional stability, thermal stability, solvent resistance, and mechanical strength [1,8]. Furthermore, the thermal management forms a challenge for the devices, and for the polymer substrate. For instance, the thermal management of the printed and hybrid-integrated light-emitting diode (LED) foil can be improved by implementing a special heat management structure, including thermal vias and slugs, to conduct the excess heat through the substrate to the heat sink, as presented in the paper by Keränen et al. [9].

Materials chosen for this study were bio-based or partially bio-based. Bio-based materials decrease dependency on fossil raw materials, as they can be derived from plant-based renewable resources. Plant-based materials also have a lower carbon footprint than materials derived from fossil resources. Their use is in line with the United Nations (UN) Sustainable Goals 12 and 13 and meets the requirements of the European Union (EU) Bioeconomy and EU Plastic strategies [10–12]. Globally, increasing use of electronic equipment will also create environmental concerns at the end-of-life, especially if integrated in packaging, textiles, or small gadgets. Biodegradable or compostable biopolymers can offer an alternative end-of-life solution [13]. Besides having an eco-friendly label, the material also has to exhibit sufficient properties, such as heat stability, mechanical strength, and flexibility, to function well in printed electronics applications.

Poly(ethylene terephthalate) (PET) is already used in printed electronics applications. PET is inexpensive and it has good processability for various products. Nowadays, PET is also produced as a partially bio-based resin (bio-PET), containing 30% bio-based ethylene glycol, which reduces its carbon footprint in comparison to fully fossil-based grades. However, there are several potential routes for the production of bio-PET entirely from renewable resources [14]. At the molecular level, the bio-PET is similar to fossil-based PET and is thus categorised as drop-in polymer and as an easy replacement for it. PET is a semi-crystalline clear polymer with a high melting temperature (T_m) (250–260 °C). Typically, PET film processing involves orientation, which generates a semi-crystalline structure via strain-induced crystallisation, improving the mechanical strength and heat resistance properties. The glass transition temperature (T_g) for amorphous PET is 67 °C and 81 °C for crystalline PET [15–17].

Poly(lactic acid) (PLA) currently has the biggest production volume of bioplastics and has already found its place in various applications [18]. PLA is manufactured entirely from renewable feedstocks, and it is compostable in industrial composting conditions. It has good mechanical strength and stiffness; however, its brittleness limits some applications. Furthermore, PLA has good processability and a transparent appearance. The properties of PLA are affected by crystallinity. Similar to PET, the crystallinity of PLA can be improved through orientation and annealing. Crystallisation is primarily defined by two stereo-isomeric forms: poly(D-lactide) (PDLA) and poly(L-lactide) (PLLA). To be crystalline, PLA has to be optically pure, containing >90% of PLLA isomer. The melting temperature and glass transition temperature are dependent on the molecular structure being typically between 150–200 °C and 45–65 °C, respectively [8,19]. Castro-Aguirre et al. and Luoma et al. have studied the suitability of cast-extruded PLA films in printed hybrid electronics processing, with promising results [5,20,21].

Cellulose esters have been commercially important polymers since the 18th century, given that they are some of the oldest biopolymers. They are produced using highly purified cellulose and contain several manufacturing steps which make the cost structure of cellulose esters more expensive than oil-based polymers, even though they are very good materials from a performance perspective. Cellulose esters such as cellulose acetate propionate (CAP) or cellulose acetate butyrate (CAB), are the thermoplastic polymers used in film, coatings, inks, moulding, and fibre applications. They typically have a high glass transition temperature and high modulus [22]. Thermoplastic cellulose esters such as CAP typically contain plasticisers to modify the processing properties that also affect other properties such as T_m and T_g. Another way to modify the properties is to control the substitution degree of cellulose hydrogen groups. A typical processing temperature of CAP materials is between 180–230 °C. CAP has T_g in range of 140–150 °C. CAP has high clarity and it has properties similar to cellulose acetate (CA) [23–25]. CAP is typically used as a film former in printing inks and overprint varnishes; being soluble in various ink and coating solvents that show compatibility with other resins in these applications [26].

Another cellulose-based substrate with a transparent plastic-like outlook, but without thermoplasticity, is regenerated cellulose, also called Cellophane™ according to the tradename of its first manufacturer, DuPont, in the early stages of the 20th century [27,28]. During the last 20 years, the manufacturing process for regenerated cellulose has improved and the common viscose process is changing to more environmental processes, e.g., those using ionic liquids and other cellulose dissolving processes, which has increased interest in novel uses for regenerated cellulose [29,30]. As such, the regenerated cellulose films are biodegradable, transparent, thermally stable in typical printing process conditions up to 200 °C, and a good oxygen barrier. Cellophane™ has high mechanical strength properties such as tensile strength of 120 MPa and modulus 12 GPa [31]. Its drawbacks include a low water-vapour barrier and it is not heat-sealable, which has led to the development of different coating methods for improving the barrier properties and heat-seal layers [32–34]. The use of regenerated cellulose and other cellulose-based materials in printed electronics was reviewed by Brunetti et al. [35] Another cellulose-based, home compostable, and suitable for marine degradation film is NatureFlex™ from Futamura [36]. Its strength and biodegradation properties were studied, for example, by Rapisarda et al. [37].

In this study, different bio-based materials and their performance in the screen-printing process were examined. The main objective was to find more sustainable substrate film materials for printed electronics applications without compromising their performance.

2. Materials and Methods

2.1. Materials

A commercially manufactured 125 μm thick PET MELINEX® ST506 from DuPont Teijin Films™ (Dumfries, United Kingdom) film was used as reference (Ref-PET), since PET is a commonly used substrate material in printed electronics applications. MELINEX® ST506 is a clear and heat-stabilised polyester film and has pre-treatment on both sides of its film. The manufacturer claims that film has excellent dimensional stability at temperatures up to 150 °C [38].

Partially bio-based poly(ethylene terephthalate) (Bio-PET) Eastlon PET CB-602AB was purchased from FKuR Kunstoff GmbH (Willich, Germany). It is a multipurpose PET grade comprised of 20% bio-based carbon content, since the second monomer of PET, monoethylene glycol (MEG), is derived from bio-based ethanol. The grade is especially suitable for the production of bottles and films and exhibits good optical clarity. The density of Bio-PET is 1.3–1.4 g/cm^3.

Poly(lactic acid) Luminy® L175 (PLA) was purchased from Total Corbion Ltd. (Gorinchem, Netherlands). PLA L175 is specified as high-heat grade, and it is 99% optically pure poly(L-lactic acid). It is suitable for film extrusion, thermoforming and fibre spinning. The density of PLA L175 is 1.24 g/cm^3 and it has a melt flow index (MFI) of 8 g/10 min at 210 °C. Melting of this PLA grade occurs at 175 °C.

Cellulose acetate propionate (CAP), Cellidor® CP 300-13, was purchased from Albis Plastics (Hamburg, Germany). The cellulose raw materials for Cellidor® CAP are drawn from sustainable and natural resources. According to its manufacturer, Cellidor is an amorphous polymer and has a high degree of light transmission. It also has high impact strength even at temperatures below freezing. The material contains 13% phthalate-free plasticiser as an additive. CAP has a heat deflection temperature of 85 °C and density of 1.2 g/cm^3.

NatureFlex™ 30 NVO is a commercial regenerated cellulose film with a thickness of 30 μm, from Futamura Chemical Co. Ltd (Nagoya, Japan). It is derived from renewable resources (bio-based carbon content 96%) and is certified as compostable in industrial and home composting environments. The manufacturer states that the NatureFlex™ is suitable for printing and for lamination to other biopolymers. The principal raw material for the NatureFlex™ is cellulose derived from wood pulp. Natureflex film has heat-seal coating on both sides of the film. The coating material was not disclosed by the manufacturer [34].

2.2. Cast Film Extrusion

Materials were dried prior to processing. CAP and PLA were dried overnight at 65 °C in a vacuum oven. Bio-PET was first dried at 150 °C in an oven, after which the granules were put into a vacuum chamber for three hours. Films were manufactured with a laboratory-scale single-screw extruder Brabender Plastograph EC plus 19/25 D (Brabender GmbH & Co. KG, Duisburg, Germany). The sheet die was 120 mm wide, and the films were cast on heated roll stack to avoid heat-shock. To avoid excess moisture absorption, a nitrogen stream of 5 L/min was fed into the feeder. Screw geometry was conical with a compression ratio of 3:1. A sieve of 40/80 mesh was used in the extruder. Table 1 shows processing parameters and film thicknesses for Bio-PET, PLA, and CAP.

Table 1. Processing parameters and film thicknesses for Bio-PET, PLA, and CAP.

Material	Screw	RPM	Temperature Profile from Die to Feeder	Chill Roll Temperature (°C)	Film Thickness (μm)
Bio-PET	3:1	50	270–270–270–270–270–270 °C	80	700
PLA	3:1	50	205–205–200–200–200–200 °C	60	650–700
CAP	3:1	50	210–210–210–200–200–200 °C	80	650–700

2.3. Orientation and Annealing

Cast films were oriented to make thinner films and to attain oriented microstructure. In the case of the semi-crystalline polyesters, PET and PLA, orientation improves crystallinity via strain-induced crystallisation [39]. Orientation was carried out biaxially (BO) with a Brückner Karo IV laboratory-scale stretcher (Brückner Maschinenbau GmbH & Co. KG, Siegsdorf, Germany). Orientation parameters were tailored for each material so that the maximum orientation ratio could be achieved without film tearing. Biaxial stretching was executed in simultaneous mode. Samples were cut into 8.5 cm × 8.5 cm squares and inserted into stretcher clips. The applied clip pressure was 40–50 bar depending on how thick the films were. During the orientation process, the films were first pre-heated for a defined time, after which stretching occurred at a pre-defined stretch rate.

In case of CAP, the orientation parameters were experimentally defined, as there was not much information about its orientation behaviour. For polyesters such as PET and

PLA a suitable orientation temperature can usually be found 15–20 °C above their glass transition temperature. Table 2 shows the applied orientation parameters for each material.

Table 2. Biaxial orientation (BO) parameters for Bio-PET, PLA and CAP films.

Material	Temperature (°C)	Pre-Heating Time (s)	BO Ratio	Stretch Rate (%/s)	Thickness after (µm)
Bio-PET	88	120	2.9 × 2.9	100	80–100
PLA	77	120	2.9 × 2.9	100	80–100
CAP	137	90	2.1 × 2.1	100	150–170

Annealing was performed in a custom-made annealing frame. The purpose of the frame was to keep the oriented film in tension and thereby prevent shrinkage during heat treatment. The annealing frame was adjustable so that the free area in the middle of the frame was as large as possible for films having different orientation ratios. Annealing temperatures were chosen, based on the orientation temperatures and each material's melting temperature, so that the annealing was carried out below T_m. The annealing time for each material was four minutes and it was carried out in an oven. The annealing Temperature for Bio-PET was 225 °C, for PLA 140 °C, and for CAP 110 °C.

2.4. Thermal Analysis

Cast films were examined with differential scanning calorimetry (DSC) to observe the transition temperatures and crystallinities of materials. Transitions such as glass transition (T_g) and melting temperature (T_m) are important in evaluating the heat resistance of materials. Further, the development of crystallinity due to orientation and annealing were studied with differential scanning calorimetry in the case of semi-crystalline materials.

During DSC measurements, heating and cooling runs were carried out twice at a rate of 10 K/min. The temperature range was slightly different for some of the materials depending on their expected melting temperature and glass transition temperature.

The crystallinity of semi-crystalline polymers can be calculated as follows.

$$X_c = \frac{\Delta H_m - \Delta H_{cc}}{\Delta H_m^\circ} \times 100\% \quad (1)$$

where ΔH_m is the melting enthalpy (J/g), ΔH_{cc} is the enthalpy of cold crystallisation (J/g) and ΔH_m° is the melting enthalpy of 100% crystalline polymer. Enthalpies of melting for 100% crystalline material can be found in the literature for PET and PLA and they are listed in Table 3 below.

Table 3. Enthalpies of melting (ΔH_m) of 100% crystalline polymers found in literature.

Material	ΔH_m° (J/g)
PET [40]	140.1
PLA [41]	93.6

2.5. Tensile Properties

Tensile testing for films was performed according to SFS-EN ISO 527-1 and SFS-EN ISO 527-3 with the Instron 4505 universal material tester (Instron Corp., Norwood, MA, USA) with 1000 N load cell and the Instron 2665 Series High Resolution Digital Automatic Extensometer (Instron Corp., Norwood, MA, USA). Tests were performed with a 5 mm/min extension rate. Tests were performed at standard conditions (23 °C and 50% RH). The specimen geometry was type 2. The initial distance between grips was 100 mm with gauge length of 50 mm for un-oriented samples. Some of the oriented and annealed samples were too small for a 100 mm gripping distance, thus the distance was adjusted so that those specimens could be tested.

2.6. Screen-Printing of Silver

Asahi LS 411 AW silver paste (Asahi Chemical Research Laboratory Co., Ltd., Tokyo, Japan) was printed with sheet-based flat-bed screen-printing equipment (EKRA XH STS, ASYS Group GmbH, Boennigheim, Germany) using 325 L mesh screen with 15 mm emulsion and 28 mm wire (Murakami, Tokyo, Japan). The silver was printed using a 1.2 mm snap off distance, a 0.2 mm down stop, 70 N pressure, and a 30 mm/s printing speed and dried in an oven (circulating heat) under 120 °C (Ref-PET), 100 °C (Bio-PET) and 80 °C (CAP, Natureflex, and PLA) for 30 min.

2.7. Hybrid-Integration of LEDs

Red, green and yellow WL-SMCC SMT Mono-color Chip LED Compact (Würth Electronics eiSos GmBH Co. & KG, Waldenburg, Germany) and white ASMT-CW20 SMT 0.2 mm Top Fire Chip LED (AVAGO Technologies Ltd., San Jose, CA, USA) were bonded on silver printed PET, PLA, and CAP films with EpoTek H20E (Epoxy Technology, Inc., Billerica, MA, USA) and cured at 80 °C for three hours. UV-curing adhesive Dymax 9008 (Dymax Corporation, Torrington, CT, USA) was dispensed around the LEDs to improve the mechanical support.

2.8. Film Characterisation

The film thickness was measured with an optical profilometer (Vantage Controller 50, cyberTECHNOLOGIES GmbH, Eching-Dietersheim, Germany) and optical transmission with a UV–Vis–NIR spectrophotometer (Cary 5000, Agilent, Santa Clara, CA, USA). The surface roughness was measured with a white light interferometer (Wyko NT3300, Veeco, Plainview, NY, USA) in vertical scanning interferometry (VSI) mode with 20.8× magnification and plane fit (tilt). Arithmetic roughness (Ra was calculated by fitting the absolute surface profile data with the mean surface level, root-mean-squared roughness (Rq) by fitting the root-mean-squared data from the surface profile with the mean surface level, and total roughness height (Rt) from the peak-to-valley differences. Prior to the measurement, a 60 nm silver layer was thermally vacuum evaporated (MB 200B, MBRAUN, Munich, Germany).

The accuracy of printed patterns was measured with a microscope (Smartscope OGP250, Optical Gaging Products, Rochester, NY, USA), the printed silver thickness with an optical profilometer (Vantage Controller 50, cyberTECHNOLOGIES GmbH, Eching-Dietersheim, Germany) and the resistance with a Multimeter (189 True RMS Multimeter, Fluke, Everett, WA, USA). Surface energy measurements were obtained using a contact angle meter (CAM200, Biolin Scientific, Stockholm, Sweden), and measured using water, ethylene glycol, and diiodomethane. Surface energy calculations were performed according to the Fowkes' Geometric Mean method.

2.9. Scanning Electron Microscopy (SEM)

Scanning electron microscopy (SEM) images were taken from cross-sections of the S2S screen-printed polymer films to study adhesion between the silver paste and film surface. The equipment was JEOL JSM-6360LV SEM (JEOL Ltd., Tokyo, Japan) with a tungsten hairpin filament as an electron source. Imaging was performed with secondary electron signals (SE). Prior to SEM imaging, cross-sections were prepared by breaking samples in liquid nitrogen immersion to ensure a brittle fracture without plastic deformation. Samples were gold sputtered for 130 s with a Baltec Balzers (Balzers Union, Vaduz, Liechtenstein) sputter coater. Images were taken with acceleration voltage of 8 kV and a spot size of 34. Working distance was between 10 and 16 mm.

3. Results and Discussion

3.1. Thermal Properties and Crystallinity

Thermal properties, including glass transition (T_g), cold crystallisation (T_{cc}), melting (T_m), and melt crystallisation (T_{mc}) temperatures, of the analysed film samples are presented in Table 4.

Table 4. Glass transition (T_g), cold crystallisation (T_{cc}), melting (T_m), and melt crystallisation temperatures (T_{mc}) of different film materials as un-oriented, biaxially oriented (BO) and biaxially oriented and annealed.

Material	T_g (°C)	T_{cc} (°C)	T_m (°C)	T_{mc} (°C)
Ref-PET	77.7 ± 0.0	–	254.1 ± 1.5	195.9 ± 1.2
Bio-PET	72.3 ± 0.0	131.3 ± 0.20	248.8 ± 0.7	186.3 ± 0.4
Bio-PET BO	83.5 ± 0.0	–	249.3 ± 0.1	188.8 ± 0.6
Bio-PET BO annealed	80.5 ± 0.2	–	248.7 ± 0.1	186.8 ± 0.1
PLA	60.3 ± 0.1	107.9 ± 0.1	176.9 ± 0.3	–
PLA BO	66.9 ± 0.1	93.5 ± 3.3	176.5 ± 0.0	–
PLA BO annealed	65.7 ± 0.1	–	175.6 ± 0.3	–
CAP	97.0 ± 1.1	–	155.8 ± 0.1	–
CAP BO	102.1 ± 0.2	–	158.4 ± 1.2	–
CAP BO annealed	109.5 ± 3.5	–	162.7 ± 1.4	–
Natureflex	68.2 ± 0.0	–	178.9 ± 0.0	–

From Table 4, Ref-PET shows the highest melting temperature at 254 °C of all of the tested materials. Bio-PET has a slightly lower melting temperature at 249 °C than the commercial Ref-PET. Orientation and annealing increased the glass transition temperature of Bio-PET from 72.3 °C to 80.5 °C, which is most likely a result of increased crystallinity and thereby restricted chain mobility in an amorphous phase. Both PET grades show melt-crystallisation behaviour upon cooling. Melt crystallisation of Ref-PET occurs at 196 °C while Bio-PET shows melt-crystallisation at temperatures below 190 °C.

Similar to Bio-PET, the glass transition of PLA also increased as a consequence of orientation and annealing (Table 4). Unoriented PLA has a T_g of 60.3 °C, while the oriented and annealed PLA is 65.7 °C. Melting of PLA occurs around 176 °C and is not affected by orientation or annealing. However, annealing causes the disappearance of the cold crystallization peak due to highly crystalline structure, as presented in Table 5.

Table 5. Melting enthalpies (ΔH_m), crystallisation enthalpies (ΔH_{cc}) and calculated crystallinities (X_C) of Ref-PET, Bio-PET and PLA films.

Sample	ΔH_m (J/g)	ΔH_{cc} (J/g)	X_C (%)
Ref-PET	62.7 ± 3.7	–	44.7 ± 2.6
Bio-PET	40.6 ± 1.2	31.4 ± 0.6	6.5 ± 0.4
Bio-PET BO	50.7 ± 0.2	–	36.2 ± 0.1
Bio-PET BO annealed	46.7 ± 0.6	–	33.3 ± 0.4
PLA	45.7 ± 1.4	46.3 ± 0.10	0.0 ± 1.33
PLA BO	61.0 ± 2.4	31.6 ± 1.2	31.4 ± 3.9
PLA BO annealed	48.8 ± 1.0	2.9 ± 1.2	49.0 ± 2.3

Presented in Table 4, CAP has a glass transition around 100 °C before annealing. After annealing, the glass transition seems to increase to 109.5 °C. There is also a minor melting peak at 155–163 °C, however, it does not have well-defined shape such as semi-crystalline polymers, and the integrated enthalpies are less than 16 J/g.

Natureflex has a T_g of 68 °C and clear melting peak of 179 °C. Natureflex is a regenerated cellulose film which is not a thermoplastic material, thereby the melting peak is most likely connected to the thermoplastic heat-seal coating on both sides of the Natureflex film.

The transition enthalpies and crystallinities of semi-crystalline polymers Ref-PET, Bio-PET, and PLA are presented in Table 5. Natureflex is excluded from this table, since the observed melting enthalpy describes the thermal behaviour of the heat-seal coating, not the Natureflex film itself.

Table 5 shows that Ref-PET has higher crystallinity at 44.7% than oriented and annealed Bio-PET, which has crystallinity of 33.3%. PLA has crystallinity of 49% after orientation and annealing. Biaxial orientation of the Bio-PET and PLA cast films is an important step toward achieving crystalline structure. Unoriented PLA exhibits amorphous morphology and unoriented Bio-PET has low crystallinity of 6.5%. Crystalline structure is known to improve mechanical performance and thermal stability of semi-crystalline polymers [41].

3.2. Tensile Properties

The tensile strength properties, including the Young's modulus, yield strength, and elongation at break values, for the films are presented in Table 6. The results are presented graphically in Figures 1–3. Figure 1 shows the Young's modulus values of tested materials, Figure 2 presents a visual comparison of yield strengths and Figure 3 compares measured elongation at break values.

Table 6. Tensile strength properties of films.

Sample	Young's Modulus (GPa)	Yield Strength (MPa)	Strain at Break (%)
Ref-PET	4.75 ± 0.13	94.2 ± 2.5	100.5 ± 8.5
Bio-PET	2.43 ± 0.09	55.2 ± 5.4	3.55 ± 0.17
Bio-PET BO annealed	4.66 ± 0.23	91.2 ± 4.8	95.4 ± 23.6
PLA	3.39 ± 0.13	50.3 ± 9.3	2.5 ± 2.0
PLA BO annealed	4.29 ± 0.44	80.7 ± 3.7	32.2 ± 29.2
CAP	1.30 ± 0.04	25.9 ± 1.1	59.0 ± 16.2
CAP BO annealed	1.69 ± 0.08	27.35 ± 1.3	9.1 ± 2.2
Natureflex	5.49 ± 0.50	44.97 ± 1.7	8.5 ± 1.5

Presented in Table 6 and Figure 1, the highest modulus value (5.49 GPa) is achieved with the commercial Natureflex film. Bio-PET and commercial reference PET also exhibit high modulus values. As a result of biaxial orientation, an annealed Bio-PET has 92% higher modulus than unoriented Bio-PET. The modulus of the oriented Bio-PET is 4.66 GPa, whereas the un-oriented Bio-PET has a modulus of 2.43 GPa. Stiffness of the oriented and annealed Bio-PET is similar to the commercial Ref-PET, which has modulus of 4.75 GPa. Similar results have been reported by Bandla et al., in their study the unoriented PET film has a modulus of 4.3 GPa while biaxial orientation results in modulus of 4.7 GPa [16].

As presented in Table 6 and Figure 1, orientation and annealing also improved the PLA stiffness: the unoriented PLA has modulus of 3.39 GPa while the biaxially oriented and annealed PLA has 4.29 GPa. When comparing stiffness properties, CAP exhibits lower modulus values than other films. Unoriented CAP film has a modulus of 1.30 GPa, while the biaxially oriented and annealed CAP has 1.69 GPa.

Presented in Table 6 and Figure 2, the highest yield strengths are obtained with the oriented Bio-PET and commercial Ref-PET. Biaxially oriented Bio-PET has a yield strength of 91 MPa and Ref-PET has 94 MPa. Orientation of Bio-PET results in a 65% higher yield strength. A similar trend is observed with the PLA, wherein orientation improves yield strength by 60%, resulting in a yield strength of 80 MPa. In the case of CAP, orientation and annealing has less dramatic impact on the yield strength; the yield strength of both unoriented and oriented CAP is in the range of 25–27 MPa. Natureflex has a lower yield strength than the PET grades or PLA at 45 MPa.

As presented in Table 6 and Figure 3, the unoriented PET and PLA are brittle with elongation at break values less than 5%. Orientation and annealing improved their ductility in comparison to the untreated films. The oriented and annealed Bio-PET has elongation of 95% which is similar to the Ref-PET, which has elongation at break of 100%. The oriented and annealed PLA has elongation at break of 32%. In the case of the CAP, orientation and annealing decreased elongation at break values by 85%. The Natureflex film is relatively brittle, with elongation at break value of 8.5%.

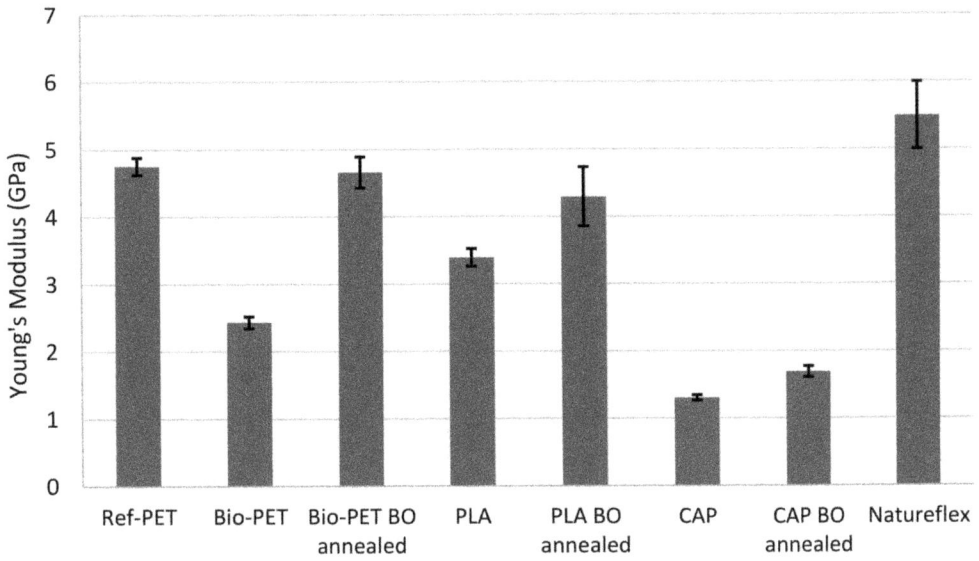

Figure 1. Young's modulus values for film materials.

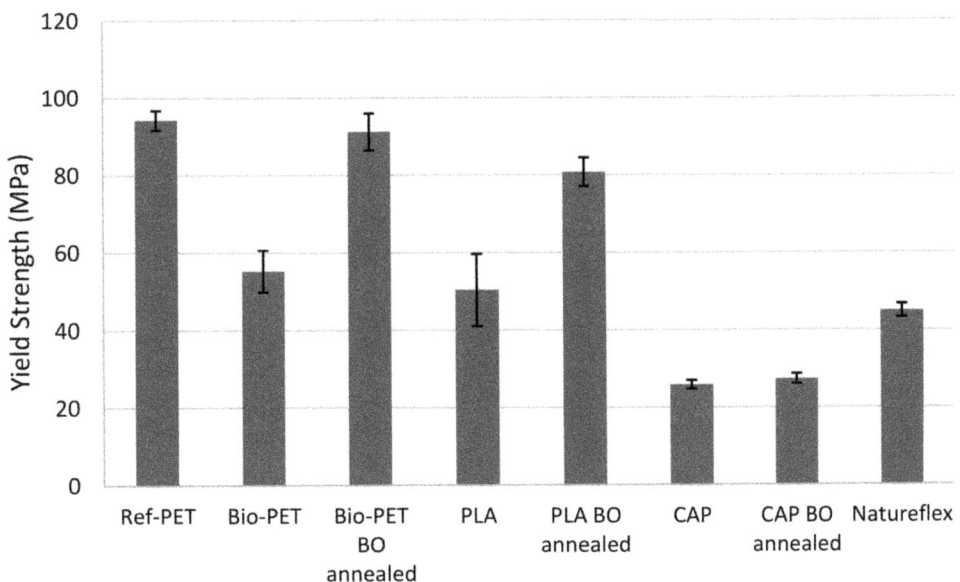

Figure 2. Yield strength of analysed film materials.

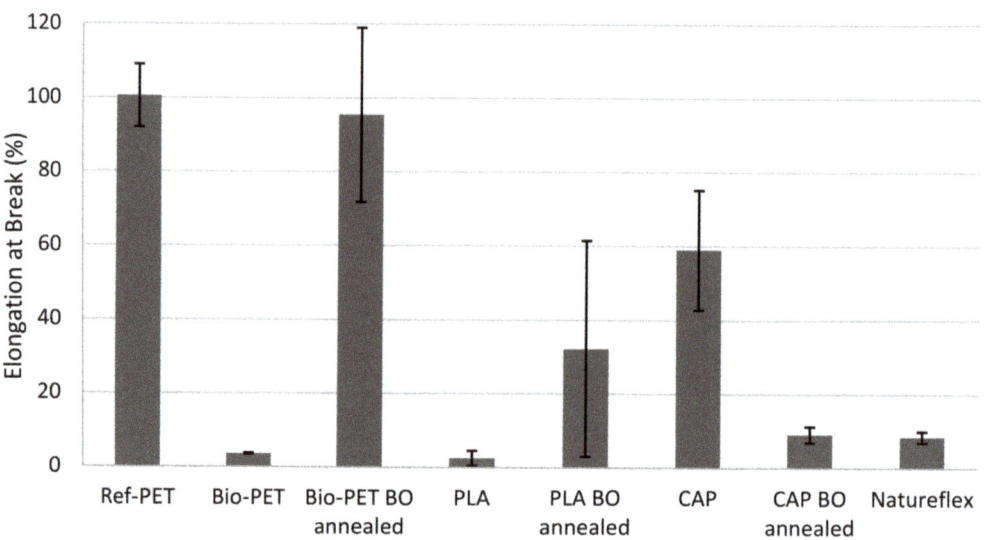

Figure 3. Elongation at break (%) values.

3.3. Transparency and Surface Properties

The optical transmission, surface roughness, surface topography, and surface energy of the Ref-PET, Bio-PET, PLA, CAP, and Natureflex are presented in Figures 4 and 5, and Tables 7 and 8.

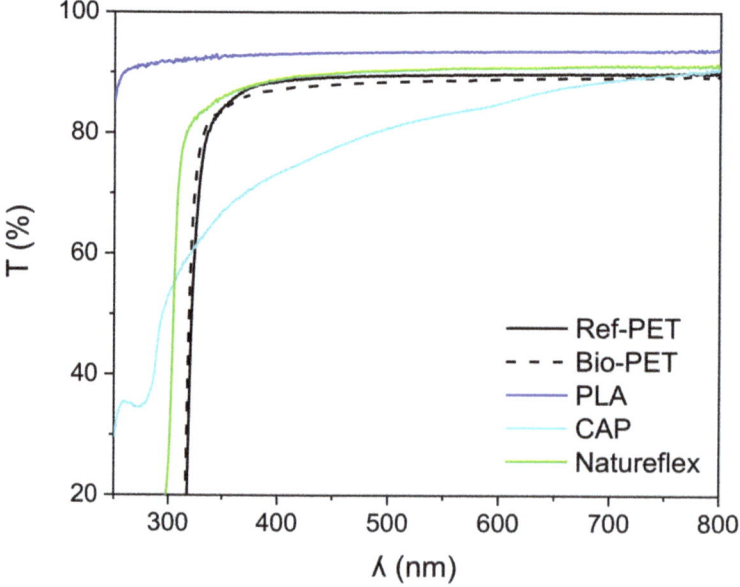

Figure 4. Optical transmission (T) of the Ref-PET, Bio-PET, PLA, CAP and Natureflex films versus wavelength (λ).

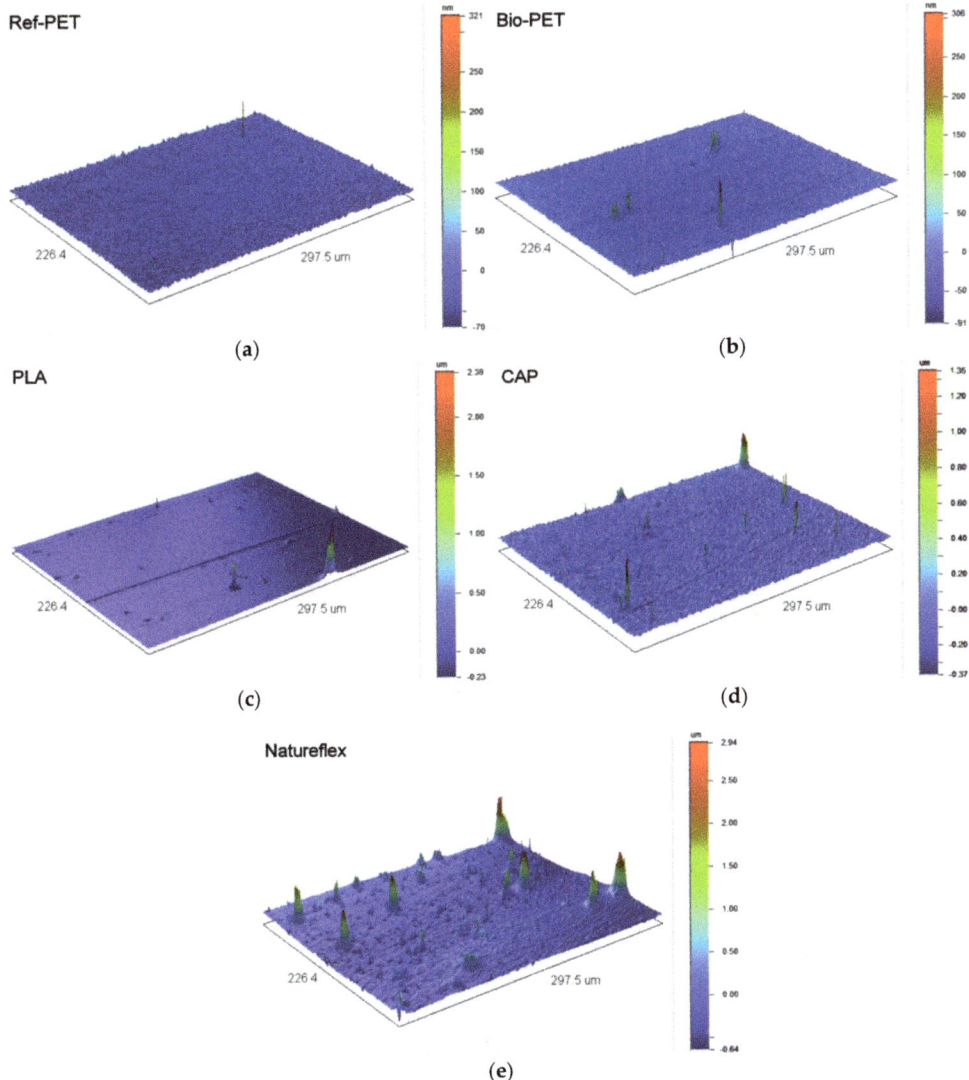

Figure 5. Surface profile of (**a**) Ref-PET; (**b**) Bio-PET; (**c**) PLA; (**d**) CAP; and (**e**) Natureflex film using white light interferometer with a magnitude of 20.8×.

Figure 4 and Table 7 show that, in comparison to the Ref-PET, the optical transmission of PLA and Natureflex was higher, being between 90 and 94% at 500 nm and 700 nm, respectively, and the optical transmission of the Bio-PET film was similar to the Ref-PET, being between 88 and 89%. The CAP film exhibited the lowest transmission, due to the milky appearance of the film, and the transmission was between 81 and 89% at 500 nm and 700 nm. However, it should be noted that the film thickness values were not the same and the film thickness ranged between 30 μm and 170 μm, thus the small differences in the transmission at a visible region can be explained by the film thickness.

Figure 4 plots the optical transmission versus the wavelength for the films in this study. The transmission of the Bio-PET and Natureflex films were comparable to the Ref-PET film between 350 and 800 nm wavelengths, and higher with PLA. The transmission of the

PET, Bio-PET, and Natureflex films rapidly decreased at 310–330 nm, and at 250 nm, the transmission was 1%, whereas the transmission of PLA film was 85–92% between 250 and 350 nm.

Table 7. Mean arithmetic roughness (Ra), root-mean-squared roughness (Rq), total height of roughness (Rt) and optical transmission (T), of substrate films.

Substrate	Ra (nm)	Rq (nm)	Rt (µm)	T (%) @500 nm	T (%) @700 nm
Ref-PET	7 ± 2	9 ± 2	0.3 ± 0.1	89	90
Bio-PET BO Annealed	4 ± 0	6 ± 1	0.4 ± 0.1	88	89
PLA BO Annealed	9 ± 1	37 ± 7	2.2 ± 0.5	93	94
CAP BO Annealed	23 ± 7	36 ± 5	1.6 ± 0.1	81	89
Natureflex	105 ± 12	184 ± 20	3.7 ± 0.1	90	91

Table 8. Polar and dispersive part of surface energy (SFE) of different substrates according to Fowkes.

Substrate	SFE Dispersive Part (mN/m)	SFE Polar Part (mN/m)	SFE Total (mN/m)
Ref-PET	34.42	9.60	44.02
Bio-PET BO Annealed	38.64	1.76	40.39
PLA BO Annealed	31.34	6.78	38.12
CAP BO Annealed	32.06	7.90	39.96
Natureflex	36.40	0.80	37.20

Film thickness values were comparable to Ref-PET with an exception of the 30 µm thin Natureflex, which made it more challenging to handle. From Table 7 and Figure 5, the surface of the Bio-PET film was comparable to Ref-PET in terms of the surface roughness (R_a, R_q and R_t) and topology. The surface roughness of PLA and CAP was an order of magnitude higher, while Natureflex was two orders of magnitude higher. Heat-seal coating on the surface of the Natureflex film provides the main explanation for the higher surface roughness (Figure 5). Here, the thickness of screen-printed silver is 20 µm, thus, the roughness values presented in Table 7 are acceptable for this work. If the electronic structures comprise thinner layers and the material thickness would range from a few tens of nm to 1 µm, the roughness of Natureflex would be too high.

Table 8 shows results from surface energy measurements for the analysed materials.

In printing and coating, the wetting and adhesion is associated with the surface tension of the ink and the surface energy of the substrate. In general, a proper bonding can be achieved when the surface energy is between 2 to 10 mN/m higher than the surface tension although that is not always the only determinant factor. Problems in the adhesion or wetting are less common if the solvent-based inks or pastes comprise low surface tension solvents, such as ethyl alcohol (22.1 mN/m) and ethyl acetate (24.4 mN/m) [42].

According to the Fowkes model, the total surface energy of the PET, Bio-PET, PLA, CAP, and Natureflex substrates ranged from 38 to 44 mN/m (Table 8), with the dispersive part of the surface tension varying from 31 to 38 mN/m and the polar part from 1 to 10 mN/m. Asahi LS 411 AW is ethyl carbitol acetate-based screen-printing paste that has a surface tension of 31.1 mN/m. In that respect, the surface tension of the printing paste solvent is 6–13 mN/m lower than that of the surface energy of the substrate, and supports proper wetting and adhesion of the ink [42].

Good adhesion is mainly associated with the polar interactions between the substrate and the ink, the polar part of the surface energy can be increased e.g., with plasma treatment [42,43]. Corona treatment and adhesion primers are also possible [42]. In this work, reported in Table 8, the polar part of the surface energy for CAP and PLA substrates was comparable with the Ref-PET, and significantly lower with the Bio-PET and Natureflex. The surface energy data predicted good adhesion strength for the Ref-PET, CAP, and PLA substrates, and lower adhesion strength for Bio-PET and Natureflex. However, the heat-seal coating on Natureflex film is expected to enhance the adhesion [34].

3.4. Dimensional Stability and Resistance of S2S Screen-Printed Films

Figure 6 presents the test layout for S2S screen-printing of silver paste, and Table 9 and Figure 7 present the dimensional accuracies of S2S screen-printed LS 411 AW silver on substrate film.

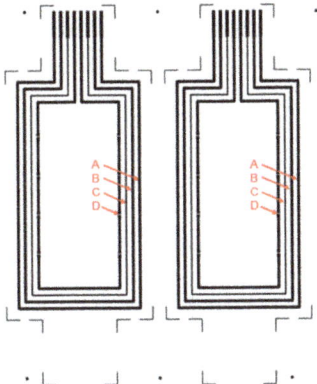

Figure 6. Test layout for S2S screen-printing of silver for comprising two test structures with 1.0 mm line width patterns (outer loop A and B) and 0.5 mm line width (loop C) for characterisation and loop D (inner loop) for LED integration.

Table 9. Machine direction (MD) and transverse direction (TD) accuracy of S2S screen-printed LS 411 AW silver patterns compared to the dimensions of the layout.

Substrate	Process T (°C)	Mean Dev MD (%)	Max; Min Dev MD (%)	Mean Dev TD (%)	Max; Min Dev TD (%)
Ref-PET	120	0.00	0.02; −0.01	0.12	0.17; 0.07
Bio-PET BO Annealed	100	−0.77	−0.49; −1.18	−0.71	−0.48; −1.10
PLA BO Annealed	80	−0.74	0.07; −1.12	−0.52	−0.34; −0.72
CAP BO Annealed	80	−0.65	−0.50; −0.79	−0.64	−0.10; −1.06
Natureflex	80	−0.63	−0.41; −0.70	−0.54	−0.31; −0.74

Temperatures for the thermal curing of the printed silver were evaluated with pre-tests. Higher temperatures can provide better electrical properties; thus, the aim was to have as high a processing temperature as possible. Based on pre-test results, the Ref-PET was processed at 120 °C, Bio-PET at 100 °C and PLA, CAP, and Natureflex at 80 °C for 30 min. Table 9 shows that the screen-printed LS 411 AW pattern obtained the dimensions of the layout in MD on the Ref-PET substrate, and were reduced between 0.6 and 0.8% on the Bio-PET, PLA, CAP and Natureflex. The printed LS 411 AW pattern was increased in TD on Ref-PET 0.1% and decreased between 0.5 and 0.8% on Bio-PET, PLA, CAP and Natureflex, respectively. Within biofilms, the best dimensional accuracy and smallest min–max deviation was obtained when LS 411 AW was printed on Natureflex, thus enabling the scale-up of the dimensional changes and improved reproducibility. Furthermore, the processing at 80 °C would likely increase the dimensional accuracy of the Bio-PET. It should also be noted that, in comparison to R2R processing, the unrestrained film was prone to deform during the S2S thermal treatment, whereas R2R processing provided the possibility of controlling the web and improving dimensional accuracy [5].

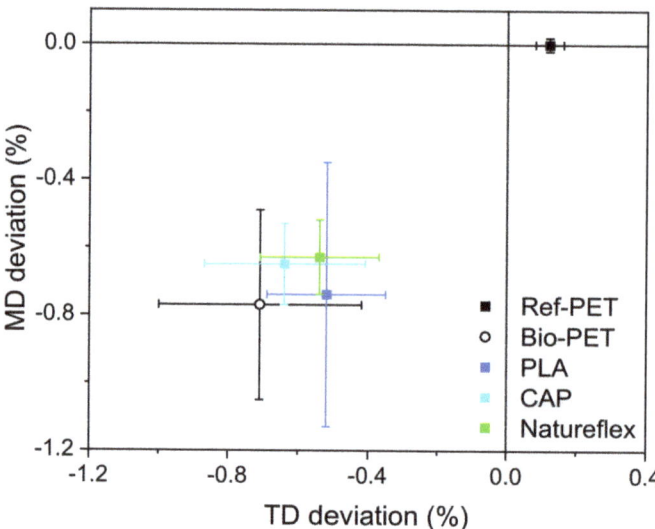

Figure 7. Machine direction (MD) and transverse direction (TD) accuracy of S2S screen-printed LS 411 AW silver patterns compared to the dimensions of the layout.

Table 10 presents the resistance measurements of S2S screen-printed LS 411 AW silver on substrate films using the test pattern presented in Figure 6, and the functionality of the printed structures was demonstrated with the hybrid-integration of LEDs presented in Figure 8.

Table 10. Resistance of S2S screen-printed LS 411 AW silver patterns with 1.0 mm (A, B) and 0.5 mm (C) line width on different substrates.

Substrate	Ag Thickness (μm)	Process T (°C)	Pattern-A Mean R (Ω)	Pattern-B Mean R (Ω)	Pattern-C Mean R (Ω)
Ref-PET	20 ± 2	120	7.2 ± 0.2	7.0 ± 0.2	12.4 ± 0.3
Bio-PET BO Annealed	21 ± 1	100	8.4 ± 0.2	8.3 ± 0.2	14.8 ± 0.6
PLA BO Annealed	23 ± 3	80	8.4 ± 0.6	8.1 ± 0.6	15.2 ± 1.3
CAP BO Annealed	18 ± 2	80	5.9 ± 0.3	5.7 ± 0.2	10.6 ± 0.4
Natureflex	21 ± 1	80	6.4 ± 0.1	6.3 ± 0.1	11.5 ± 0.3

Figure 8. Screen-printed hybrid-integrated LED foil.

In comparison to the 7.0 Ω resistance of the 1.0 mm-wide screen-printed silver pattern on Ref-PET, the silver printing on CAP and Natureflex resulted in resistance of 5.7 Ω and 6.3 Ω which was a significantly lower value and notably it was obtained by using 80 °C for the thermal curing of the silver. The resistance of silver on the Bio-PET and PLA comprised higher resistance. To understand the printed structure, the silver-substrate interface, and the possible correlation with the resistance values, the printed films were analysed with SEM.

3.5. SEM Analysis for S2S Screen-Printed Films

Cross-sectional SEM images of screen-printed LS 411 AW silver and substrate film interfaces are presented in Figure 9.

The cross-sectional image from the Ref-PET film depicts the reference point. As presented in Section 3.3, the higher polar part in the surface energy of the Ref-PET (Table 8) increases the polar interactions between the printed silver and the substrate promoting the adhesion. The cross-sectional image from the Bio-PET (Figure 9b) shows inadequate contact between the printed silver and the Bio-PET surface. In addition, the printed silver contains cracks that weaken the uniformity of the printed pattern, which can explain the higher resistance presented in Table 10. The poor contact between the printed silver and the substrate can be attributed to the lower polar part of the surface energy (Table 8), as well as to the limited thermal stability of the substrate (Table 9).

The cross-sectional SEM image from the PLA (Figure 9c) reveals that there is a slight gap between the printed silver and the PLA. although the higher polar part in the surface energy of the PLA (Table 8) enhances the adhesion properties. In comparison to the silver-PET, the higher printed silver resistance can be explained with the inadequate silver–PLA contact (Table 10). A cross-sectional SEM image from the CAP (Figure 9d) shows that the adhesion between the printed silver and the CAP is excellent; there are no cracks or cavities in the interface. In comparison to the Ref-PET, the silver-CAP contact is significantly better, and the printed silver resistance is the lowest. In addition to the high polar part of the surface energy, the compatibility is supported by the chemical similarity of CAP and acetate-based printing paste. Similar to the CAP, the cross-sectional image from the Natureflex (Figure 9e) exhibits a good printed silver–Natureflex contact and low printed silver resistance, although the polar part of the surface energy is low. Notably, the heat-seal coating on top of the Natureflex film will likely melt during the thermal curing and improve the printed silver–Natureflex adhesion.

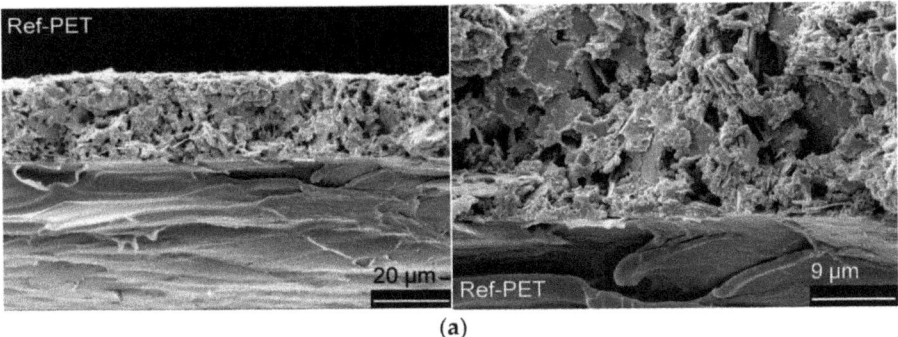

(a)

Figure 9. *Cont.*

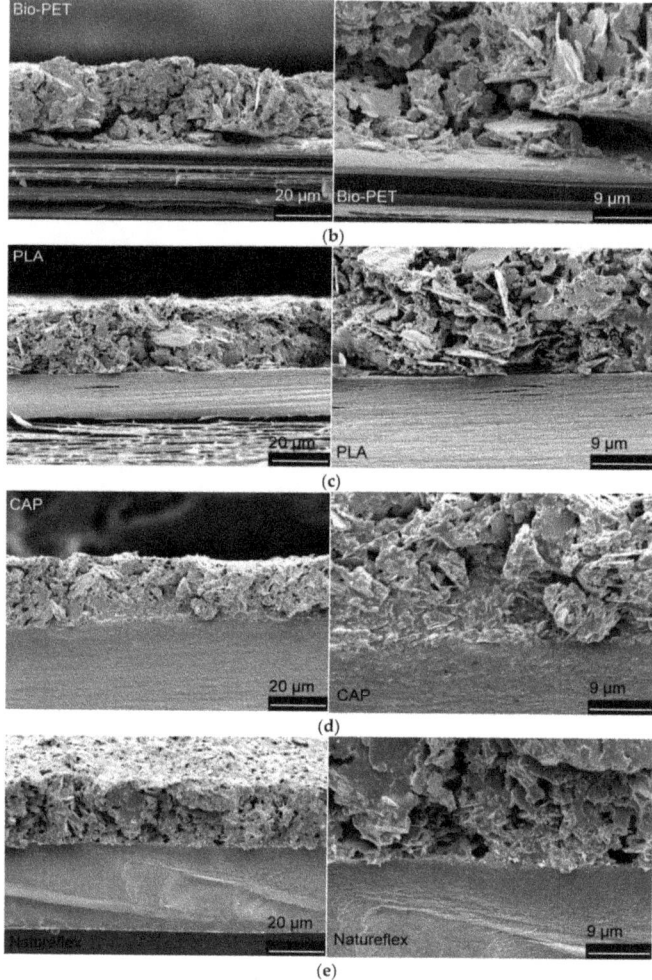

Figure 9. Cross-sectional SEM images of screen-printed LS 411 AW silver on (**a**) Ref-PET; (**b**) Bio-PET; (**c**) PLA; (**d**) CAP; and (**e**) Natureflex films. The images on top are magnified 1000× and the images on below are magnified 2500×.

4. Conclusions

The aim of this study was to assess the compatibility of various types of bio-based polymer substrates for printed hybrid electronics applications. The main objective was to find more sustainable substrate film materials to replace the fossil-based substrates such as poly(ethylene terephthalate) (PET), without compromising the performance. Very few research articles exist concerning the use of bio-based materials in thin film flexible electronics applications. This study fills a gap in the knowledge and combines screen-printing processing and material development in a novel way.

Three commercial bio-based polymer resins, bio-based poly(ethylene terephthalate) (Bio-PET), poly(lactic acid) (PLA) and cellulose acetate propionate (CAP), were used to manufacture cast-extruded substrate films on a laboratory-scale. In addition, the commercially manufactured bio-based regenerated cellulose film, Natureflex, was investigated. A commercial fossil-based PET film was used as a reference material. Laboratory- scale manufactured films were oriented and annealed to obtain oriented microstructure and improved heat stability via strain-induced crystallization.

Results show that bio-based materials can exhibit similar or even higher performance over the commercial fossil PET film (Ref-PET), even under lower printing process temperatures than the typically used 120 °C. Furthermore, the optical transmission and surface properties were comparable to the reference material. For example, the PLA films had even higher optical transmission than the commercial fossil-based PET-film. The compatibility in surface chemistries also supports the use of bio-based substrates in printed and hybrid electronics, although the limited thermal stability caused some challenges with the PLA and Bio-PET films in selected optimal printing process temperatures.

Screen-printed silver patterns on the CAP and Natureflex films had 18% and 11% lower resistance values than on the commercial PET film, providing better electrical properties due to more uniform and compact structures in the printed patterns. Scanning electron microscopy images showed that the contact between the screen-printed silver and substrate was better with the CAP and Natureflex films. The CAP and Natureflex films had lower dimensional changes during the printing process than the PLA and Bio-PET films, indicating better heat stability.

Surface energies of the substrate films were analysed, assuming correlation to printability and better ink wetting and adhesion. However, thermal stability during the printing process had a significant effect on final formation of ink layer structure. This proves that the surface energies are not enough alone to predict the electrical performance of novel substrates.

Optimization of the bio-based substrates through film orientation and annealing can improve the thermal stability and the surface properties, with limitations caused by the material properties being taken into account in electronics' processing. Importantly, good electrical performance endorses the use of bio-based substrates. In terms of mechanical performance, the biaxially oriented PLA and Bio-PET films have similar stiffness and tensile strength as the commercial PET film.

This study promotes the use of bio-based substrate materials, especially biaxially oriented CAP, in flexible electronics without compromising the quality and performance of the film and the printed electronics structures. In applications where stiffness, tensile strength, and optical clarity are emphasized, the biaxially oriented Bio-PET and PLA films have excellent performance. The regenerated cellulose-based film, Natureflex, can also open novel solutions that require biodegradability. For further studies, a detailed surface analysis could be employed to identify the substrate-ink adhesion properties of different material combinations.

Author Contributions: E.L.: Validation, Formal analysis, Investigation and Writing—Original Draft; M.V.: Validation, Formal analysis, Investigation and Writing—Original Draft; J.O.: Reviewing and Investigation; K.H.: Reviewing and Investigation; K.I.: Writing—Reviewing and Editing. All authors have read and agreed to the published version of the manuscript.

Funding: Part of the facilities used were provided by the Academy of Finland Research Infrastructure Printed Intelligence Infrastructure (PII-FIRI, grant No. 320020). The work is part of the Academy of Finland Flagship Programme, Photonics Research, and Innovation (PREIN), decision 320168. VTT internal funding was used to finalize the manuscript.

Institutional Review Board Statement: Not applicable.

Informed Consent Statement: Not applicable.

Data Availability Statement: The authors confirm that the data supporting the findings of this study are available within the article.

Acknowledgments: The authors thank Riikka Hedman, Emmi Herukka, Anne Peltoniemi, and Antti Veijola for participating in the laboratory-scale preparation and characterisation of LED foils. The Authors also thank Sini-Tuuli Rauta and Timo Flyktman for help in preparation and testing of laboratory-scale extrusion films.

Conflicts of Interest: The authors declare no conflict of interest.

References

1. Nathan, A.; Ahnood, A.; Cole, M.T.; Lee, S.; Suzuki, Y.; Hiralal, P.; Bonaccorso, F.; Hasan, T.; Garcia-Gancedo, L.; Dyadyusha, A.; et al. Flexible electronics: The next ubiquitous platform. *Proc. IEEE* **2012**, *100*, 1486–1517, Special Centennial Issue. [CrossRef]
2. Hoeng, F.; Denneulin, A.; Bras, J. Use of nanocellulose in printed electronics: A review. *Nanoscale* **2016**, *8*, 13131–13154. [CrossRef] [PubMed]
3. Khan, Y.; Thielens, A.; Muin, S.; Ting, J.; Baumbauer, C.; Arias, A.C. A New Frontier of Printed Electronics: Flexible Hybrid Electronics. *Adv. Mater.* **2020**, *32*, 1905279. [CrossRef] [PubMed]
4. Välimäki, M.; Apilo, P.; Po, R.; Jansson, E.; Bernardi, A.; Ylikunnari, M.; Vilkman, M.; Corso, G.; Puustinen, J.; Tuominen, J.; et al. R2R-printed inverted OPV modules-Towards arbitrary patterned designs. *Nanoscale* **2015**, *7*, 9570–9580. [CrossRef] [PubMed]
5. Luoma, E.; Välimäki, M.; Rokkonen, T.; Sääskilahti, H.; Ollila, J.; Rekilä, J.; Immonen, K.E. Oriented and annealed poly(lactic acid) films and their performance in flexible printed and hybrid electronics. *J. Plast. Film Sheeting* **2021**, *37*, 429–462. [CrossRef]
6. Liu, P.; He, W.; Lu, A. Preparation of low-temperature sintered high conductivity inks based on nanosilver self-assembled on surface of graphene. *J. Cent. South Univ.* **2019**, *26*, 2953–2960. [CrossRef]
7. DuPont de Nemours Inc. Printed Electronic Materials-Low Temperature Electronic Inks. Available online: https://www.dupont.com/products/low-temperature-electronic-inks.html (accessed on 31 March 2022).
8. Khan, S.; Lorenzelli, L.; Dahiya, R.S. Technologies for printing sensors and electronics over large flexible substrates: A review. *IEEE Sens. J.* **2015**, *15*, 3164–3185. [CrossRef]
9. Keränen, K.; Korhonen, P.; Rekilä, J.; Tapaninen, O.; Happonen, T.; Makkonen, P.; Rönkä, K. Roll-to-roll printed and assembled large area LED lighting element. *Int. J. Adv. Manuf. Technol.* **2015**, *81*, 529–536. [CrossRef]
10. UN Sustainable Development Goals. Available online: https://sdgs.un.org/goals (accessed on 19 April 2022).
11. EU Bioeconomy Strategy. Available online: https://ec.europa.eu/info/research-and-innovation/research-area/environment/bioeconomy/bioeconomy-strategy_en (accessed on 19 April 2022).
12. EU Plastic Strategy. Available online: https://ec.europa.eu/environment/strategy/plastics-strategy_en (accessed on 19 April 2022).
13. Reddy, M.M.; Vivekanandhan, S.; Misra, M.; Bhatia, S.K.; Mohanty, A.K. Biobased plastics and bionanocomposites: Current status and future opportunities. *Prog. Polym. Sci.* **2013**, *38*, 10–11. [CrossRef]
14. Andreeßen, C.; Steinbüchel, A. Recent developments in non-biodegradable biopolymers: Precursors, production processes, and future perspectives. *Appl. Microbiol. Biotechnol.* **2019**, *103*, 143–157. [CrossRef]
15. Jabarin, S.A. Orientation studies of polyethylene terephthalate. *Polym. Eng. Sci.* **1984**, *24*, 376–384. [CrossRef]
16. Bandla, S.; Allahkarami, M.; Hanan, J.C. Out-of-plane orientation and crystallinity of biaxially stretched polyethylene terephthalate. *Powder Diffr.* **2014**, *29*, 123–126. [CrossRef]
17. Groeninckx, G.; Berghmans, H.; Overbergh, N.; Smets, G. Crystallization of poly(ethylene terephthalate) induced by inorganic compounds. I. Crystallization behaviour from the glassy state in a low-temperature region. *J. Polym. Sci. B Polym. Phys. Polymer Phys. Ed. Abbr.* **1974**, *12*, 303–316. [CrossRef]
18. Bioplastics Market Data. Available online: https://www.european-bioplastics.org/market/ (accessed on 31 March 2022).
19. Farah, S.; Anderson, D.G.; Langer, R. Physical and mechanical properties of PLA, and their functions in widespread applications—A comprehensive review. *Adv. Drug Deliv. Rev.* **2016**, *107*, 367–392. [CrossRef] [PubMed]
20. Castro-Aguirre, E.; Iñiguez-Franco, F.; Samsudin, H.; Fang, X.; Auras, R. Poly(lactic acid)—Mass production, processing, industrial applications, and end of life. *Adv. Drug Deliv. Rev.* **2016**, *107*, 333–366. [CrossRef] [PubMed]
21. Välimäki, M.K.; Sokka, L.I.; Peltola, H.B.; Ihme, S.S.; Rokkonen, T.M.J.; Kurkela, T.J.; Ollila, J.T.; Korhonen, A.T.; Hast, J.T. Printed and hybrid integrated electronics using bio-based and recycled materials—increasing sustainability with greener materials and technologies. *Int. J. Adv. Manuf. Technol.* **2020**, *111*, 325–339. [CrossRef]
22. White, A.W.; Buschanan, C.M.; Pearcy, B.G.; Wood, M.D. Mechanical properties of cellulose acetate propionate/aliphatic polyester blends. *J. Appl. Polym. Sci.* **1994**, *52*, 525–530. [CrossRef]
23. Albis Plastics Data for Cellidor®. Available online: www.albis.com (accessed on 31 March 2022).
24. Abd Manaf, M.E.; Nitta, K.H.; Yamaguchi, M. Mechanical properties of plasticized cellulose ester films at room and high temperatures. *ARPN J. Eng. Appl. Sci.* **2006**, *11*, 2354–2358.
25. Eastman Eastman™ Cellulose-Based Specialty Polymers. Available online: https://cms.chempoint.com/ChemPoint/media/ChemPointSiteMedia/General-CE-Brochure.pdf (accessed on 31 March 2022).
26. Edgar, K.J.; Buchanan, C.M.; Debenham, J.S.; Rundquist, P.A.; Seiler, B.D.; Shelton, M.C.; Tindall, D. Advances in cellulose ester performance and application. *Prog. Polym. Sci.* **2001**, *26*, 1605–1688. [CrossRef]
27. Hyden, W.L. Manufacture and Properties of Regenerated Cellulose Films. *Ind. Eng. Chem.* **1929**, *21*, 405–410. [CrossRef]
28. APG Global Data for Cellophane. Available online: http://www.apgglobe.com/en/cellophane/ (accessed on 31 March 2022).
29. Bajpai, P. *Biermann's Handbook of Pulp and Paper*, 3rd ed.; Elsevier Ltd.: Amsterdam, The Netherlands, 2018; pp. 233–247.
30. El Seoud, O.A.; Kostag, M.; Jedvert, K.; Malek, N.I. Cellulose Regeneration and Chemical Recycling: Closing the 'Cellulose Gap' Using Environmentally Benign Solvents. *Macromol. Mater. Eng.* **2020**, *305*, 1900832. [CrossRef]
31. Ge, X. Regenerated Cellulose Films from Binary Solvent Systems. Master's Thesis, Aalto University, Espoo, Finland, July 2018. Available online: https://aaltodoc.aalto.fi/handle/123456789/33678 (accessed on 31 March 2022).

32. Zhang, L.; Zhou, J.; Huang, J.; Gong, P.; Zhou, Q.; Zheng, L.; Du, Y. Biodegradability of regenerated cellulose films coated with polyurethane/natural polymers interpenetrating polymer networks. *Ind. Eng. Chem. Res.* **1999**, *38*, 4284–4289. [CrossRef]
33. Zhang, Y.; Ji, Q.; Qi, H.; Liu, Z. Structure and Properties of Regenerated Cellulose Film with Fluorocarbon Coating. *Phys. Procedia* **2012**, *32*, 706–713. [CrossRef]
34. Futamura Group Futamura NatureFlex™ Info. Available online: http://www.futamuragroup.com/en/divisions/cellulose-films/products/natureflex/heat-resistant/ (accessed on 31 March 2022).
35. Brunetti, F.; Operamolla, A.; Castro-Hermosa, S.; Lucarelli, G.; Manca, V.; Farinola, G.M.; Brown, T.M. Printed Solar Cells and Energy Storage Devices on Paper Substrates. *Adv. Funct. Mater.* **2019**, *29*, 1806798. [CrossRef]
36. Biomasspackaging. Available online: http://www.biomasspackaging.com/brands/natureflex/ (accessed on 19 April 2022).
37. Rapisarda, M.; Patanè, C.; Pellegrino, A.; Malvuccio, A.; Rizzo, V.; Muratore, M.; Rizzarelli, P. Compostable Polylactide and Cellulose Based Packaging for Fresh-Cut Cherry Tomatoes: Performance Evaluation and Influence of Sterilization Treatment. *Materials* **2020**, *13*, 3432. [CrossRef] [PubMed]
38. DuPontTeijinFilms MELINEX® ST506 Datasheet. Available online: https://usa.dupontteijinfilms.com/wp-content/uploads/2017/01/ST506-Datasheet.pdf (accessed on 31 March 2022).
39. Jabarin, S.A. Strain-induced crystallization of poly(ethylene terephthalate). *Polym. Eng. Sci.* **1992**, *32*, 1341–1349. [CrossRef]
40. Sichina, W.J. DSC as Problem Solving Tool: Measurement of Percent Crystallinity of Thermoplastics, Application Note. Available online: https://www.perkinelmer.com/Content/applicationnotes/app_thermalcrystallinitythermoplastics.pdf (accessed on 31 March 2022).
41. Wu, J.H.; Yen, M.S.; Wu, C.P.; Li, C.H.; Kuo, M.C. Effect of Biaxial Stretching on Thermal Properties, Shrinkage and Mechanical Properties of Poly (Lactic Acid) Films. *J. Polym. Environ.* **2013**, *21*, 303–311. [CrossRef]
42. Aydemir, C.; Altay, B.N.; Akyol, M. Surface analysis of polymer films for wettability and ink adhesion. *Color Res. Appl.* **2021**, *46*, 489–499. [CrossRef]
43. Petasch, W.; Räuchle, E.; Walker, M.; Elsner, P. Improvement of the adhesion of low-energy polymers by a short-time plasma treatment. *Surf. Coatings Technol.* **1995**, *74–75*, 682–688. [CrossRef]

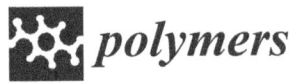

Article

Solvent Effect to the Uniformity of Surfactant-Free Salmon-DNA Thin Films

Jake Richter, Moses Nnaji and Heungman Park *

Department of Physics and Astronomy, Texas A&M University-Commerce, Commerce, TX 75428, USA; jrichter2@leomail.tamuc.edu (J.R.); mnnaji95@gmail.com (M.N.)
* Correspondence: heungman.park@tamuc.edu

Abstract: Fabrication of surfactant-modified DNA thin films with high uniformity, specifically DNA–CTMA, has been well considered via drop-casting and spin-coating techniques. However, the fabrication of thin films with pure DNA has not been sufficiently studied. We characterize the uniformity of thin films from aqueous salmon DNA solutions mixed with ethanol, methanol, isopropanol, and acetone. Measurements of thickness and macroscopic uniformity are made via a focused-beam ellipsometer. We discuss important parameters for optimum uniformity and note what the effects of solvent modifications are. We find that methanol- and ethanol-added solutions provide optimal fabrication methods, which more consistently produce high degrees of uniformity with film thickness ranging from 20 to 200 nm adjusted by DNA concentration and the physical parameters of spin-coating methods.

Keywords: DNA films; spin coating; film uniformity; solvent effect

Citation: Richter, J.; Nnaji, M.; Park, H. Solvent Effect to the Uniformity of Surfactant-Free Salmon-DNA Films. *Polymers* **2021**, *13*, 1606. https://doi.org/10.3390/polym13101606

Academic Editor: Swarup Roy

Received: 15 April 2021
Accepted: 11 May 2021
Published: 16 May 2021

Publisher's Note: MDPI stays neutral with regard to jurisdictional claims in published maps and institutional affiliations.

Copyright: © 2021 by the authors. Licensee MDPI, Basel, Switzerland. This article is an open access article distributed under the terms and conditions of the Creative Commons Attribution (CC BY) license (https://creativecommons.org/licenses/by/4.0/).

1. Introduction

Deoxyribonucleic acid (DNA) has always been an interest in the biological sciences but has recently become an interest to researchers in materials science, physics, and chemistry for applications as a biodegradable component in optical and electronic devices [1–6]. Specifically, the band gap of DNA makes it favorable to function as a hole injection/electron blocking layer in organic-based electronic devices [7–9]. Previous authors have demonstrated the ability of DNA to improve LED performance [7–13] and as a component in the fabrication of photovoltaics [14–17]. Moreover, DNA films can be used as a matrix and doped with functional materials that can alter the optical and electronic properties of the thin film, which may be useful for different types of organic devices [18–20]. Due to the discovery of these promising physical properties, DNA thin-film devices have become a notable research interest in materials science.

Despite the potential of DNA-based thin-film devices, not much study has been reported to systematize the fabrication of DNA thin films from aqueous and alcohol solutions [14]. Previous research has focused on thin films fabricated from surfactant modifications, primarily with cetyltrimethylammonium (CTMA) modified DNA thin films with consistent results [13,15–17,21–25]. Thus, research has focused on controlling the electronic and optical properties of surfactant-modified thin films, and applications have relied on characterizing results from DNA–CTMA complexes. Recently, groups have been making progress towards efficient surfactant-free thin films because the toxicity of CTMA makes DNA–CTMA solutions ineligible for many biological applications and becomes insoluble in water [14,19,26]. However, research on surfactant-free thin DNA films has focused on how to affect the physical properties of the films, such as optical dispersion and refractive indices [19,25,27–30] and not on improving the efficiency of thin-film fabrication or standardizing the production of films with tunable thickness within a controlled range of uniformity. Developing a process for unmodified DNA is important for consistent research results and enabling future researchers to develop their own thin films with or without modification.

We comment on some complications in the current methodologies for reporting thin-film fabrication. First, there are varying reports on how solutions are prepared. It has been noted that sonication times of pure DNA solutions will affect the size of polymer chains and, as a result, will affect both the viscosity and characteristics of the resulting thin film [24]. This, along with other significant variations in the solution preparation phase, such as mixing practices and minor changes in the deposition phase of spin coating, can cause serious issues when reproducing thin films, especially when solutions are spin-coated onto a hydrophobic surface. For viscous solutions, the type of dispense (i.e., static or dynamic dispense) can have considerable effects if there are bubbles in the solution before or after deposition, and the area of initial deposition can have a tremendous impact on the resulting uniformity of the film even when one controls for other parameters [31–35]. Additionally, previous results rely on microscopic measurements of uniformity using AFM and SEM techniques, although the usefulness of these measurements are undeniable at the smallest scales, there is little evidence of expectations for macroscopic uniformity or thickness. Thus, it is plausible that researchers can be fabricating films promising microscopic measurements, but the existence of uniformity over millimeter scales could compromise the results of the film and its applications.

We present a systematic analysis of spin-coating procedures to fabricate DNA thin films from solvent added aqueous solutions with tunable thickness and desired macroscopic uniformity (millimeters to centimeters). Since there are a continuous number of changes that may be made to the fabrication process, such as DNA concentration, solvent/water ratio, spin rate, and so on, we focus on a range of DNA concentrations and alcohol choices; we then modify spin rate for each one. We present analysis on quantifying uniformity and it is assumed that a reasonable interpolation of the data between concentration and solvent choice can be made upon data analysis.

2. Materials and Methods

2.1. Aqueous DNA Solution Preparation with Solvents

Deoxyribonucleic acid salts extracted from salmon testes (salmon DNA salt) was purchased from Sigma-Aldrich. DNA solutions were made by first dissolving a determined mass of salmon DNA in a pre-determined amount of distilled water for concentrations of DNA that were 4 mg/mL (~0.4 wt.%), 8 mg/mL (~0.8 wt.%), or 12 mg/mL (~1.2 wt.%). For consistency, all DNA concentrations were dissolved in 15 mL of water; thus only the total mass of DNA was changed. Once the DNA and water were added to the vial, a quick vigorous shake was given to ensure all DNA salts were integrated into the solution. The aqueous solutions of DNA were stirred at room temperature with magnetic bars for 24 h to develop a homogeneous mixture. Upon completion of the initial mixing process, each solution of DNA was then separated into two smaller vials to be diluted by a chosen solvent with one of two ratios, either 0.5 mL of solvent to 2 mL of DNA solution or 2 mL of solvent to 2 mL of DNA solution. The resulting solvent–water–DNA mixtures were then stirred with magnetic bars for 24 h to ensure proper mixture with the solvents, thus a total mixing time of 48 h. The high-concentration DNA solution at 12 mg/mL became highly viscous like gel.

2.2. Thin-Film Fabrication by Spin Coating

Silicon wafers were cut by 1 cm × 1 cm and treated with sonication baths in clean acetone, water, and methanol and dried with nitrogen between each bath. Once the sonication bath of methanol was complete and the wafer dried, it was subject to UV–ozone treatment for 25 min. The thickness of the native oxide of the silicon wafers was measured by an ellipsometer, which was about 1.9 nm. Within 1 h of being cleaned, the wafers were used as a substrate for a thin film.

A clean wafer was then centered on a spin coater, and thin films were created by depositing the same volume, ~125 µL, of DNA solution onto the center of each wafer until the solution naturally covered the entirety of the wafer (i.e., no forced spreading with a pipette tip), a process which generally requires less than 5 s. For high-viscosity solutions this is imperative since rearrangement of deposited solution with a pipette tip, as is commonly practiced with inks and low-viscosity solutions during static dispense, can lead to unwanted inconsistencies in the development process and create many confounding variables that render experimental analysis impractical. Immediately after dispense the spin process was initiated and lasted for 300 s. This process was repeated on new wafers controlling only for the RPM and solvent-DNA solution. After the film was deposited onto the wafer it was vacuum-dried for 1 h at ~20 mTorr. The films were subsequently measured with an ellipsometer.

2.3. Macroscopic Uniformity Measurement by a Focused-Beam Spectroscopic Ellipsometer

Ellipsometry measurements were taken on a 5 × 5 grid of points using an alpha-SE Ellipsometer (J.A. Woollam, Lincoln, NE, USA). Each wafer with a DNA thin film was centered on the measuring platform, and measurements were taken at 1.25 mm intervals as shown in Figure 1. The focused-beam size is about 500 µm. A Cauchy model was used in the transparent visible-to-near-IR regions to determine the thickness of the films.

Figure 1. (a) A schematic diagram of focused-beam ellipsometry. (b) A 5 × 5 measurement grid on DNA films on silicon wafers. The spacing between the measurement points is 1.25 mm.

2.4. UV–VIS Absorbance Measurements

DNA solutions even at 1 mg/mL produce absorbance peaks that are ill-defined with less than 0.1% transmittance in the expected UV range for DNA for the spectrometer. To solve this issue, a procedure similar to the one discussed Section 2.1 was used to produce solutions of DNA at 0.04 mg/mL (~0.004 wt.%) and diluted with the same volume of solvents to make 1:1 ratio solvent-to-water DNA solutions. The solutions were placed into 1 cm path length fused quartz cuvettes for absorbance measurements. For each measurement, a baseline of solvent–water solution was taken prior to the UV–VIS measurement for solvent–water DNA solution to isolate the DNA absorbance peaks and determine any effect alcohol choice might have on DNA.

3. Results

3.1. Thin Films by Pure Water–DNA Solutions

Tables 1 and 2 give values for DNA thin films from pure water–DNA solutions at varying DNA concentrations and spin rates. The standard deviation of a film divided by its average thickness was used to determine a quantitative value for uniformity weighted by the average so that thicker films will not be disadvantaged by a naturally higher standard deviation. Therefore, a lower Std/Ave corresponds to higher uniformity. We observed that pure water–DNA solutions created large sample-to-sample variation, which can be caused by uneven distribution of the solution onto silicon wafers and minor misalignment of the wafers, and relative humidity can be a significant factor in film formation by spin coating [36–38]. The two data sets in Tables 1 and 2 demonstrate that pure aqueous-based solutions of DNA produce inconsistent results particularly at the low RPM. However, we observe in the following sections that methanol and ethanol additions can greatly improve general uniformity and consistency of film formation; and that higher concentrations after dilution with alcohol experienced the greatest improvement in uniformity. This result can be explained by the viscosity change after dilution and faster drying, which affects the ability of DNA to coat evenly on the silicon surface.

Table 1. Average thickness (nm) and standard deviation (nm) from 5 × 5 measurements of a pure water–DNA sample set 1.

Pure Water–DNA Sample Set 1	0.4 wt.%	0.8 wt.%	1.2 wt.%
1k RPM	Ave: 28.84 Std: 6.35 Std/Ave: 22.04%	Ave: 184.0 Std: 14.08 Std/Ave: 7.66%	Ave: 216.5 Std: 5.52 Std/Ave: 2.56%
3k RPM	Ave: 15.88 Std: 0.24 Std/Ave: 1.50%	Ave: 76.88 Std: 1.86 Std/Ave: 2.42%	Ave: 67.71 Std: 2.43 Std/Ave: 3.59%
5k RPM	Ave: 10.86 Std: 0.1 Std/Ave: 1.50%	Ave: 50.85 Std: 2.0 Std/Ave: 3.93%	Ave: 55.84 Std: 1.53 Std/Ave: 2.74%

Table 2. Average thickness (nm) and standard deviation (nm) from 5 × 5 measurements of a pure water–DNA sample set 2.

Pure Water DNA Sample Set 2	0.4 wt.%	0.8 wt.%	1.2 wt.%
1k RPM	Ave: 38.6 Std: 5.05 Std/Ave: 13.1%	Ave: 202.4 Std: 47.57 Std/Ave: 23.5%	Ave: 412.7 Std: 103.0 Std/Ave: 24.96%
3k RPM	Ave: 31.7 Std: 4.04 Std/Ave: 12.8%	Ave: 104.9 Std: 59.7 Std/Ave: 57.2%	Ave: 212.2 Std: 71.5 Std/Ave: 33.6%
5k RPM	Ave: 19.14 Std: 3.90 Std/Ave: 20.3%	Ave: 49.7 Std: 4.51 Std/Ave: 9.07%	Ave: 57.9 Std: 25.6 Std/Ave: 29.1%

3.2. Thin Films by Water–Methanol DNA Solutions

Tables 3 and 4 demonstrate the effects of adding methanol to aqueous DNA solutions as the resulting solvent–DNA mixtures produce films consistent with expected results. Both show a highly consistent result in the high RPMs (3k and 5k). If the film thickness is greater than 80 nm, the film uniformity is visually monitored from the interference color patterns as shown in Figure 2. The inconsistencies of the pure water–DNA films demonstrate the poor ability of pure DNA solutions to make consistent results. Despite this, it is possible to have a film made from pure DNA solutions with good thickness and uniformity, but any perturbation from optimal conditions, such as small misalignments, inconsistent UV–ozone treatment, or other factors, will significantly affect the final film. Solvent-added solutions demonstrate an ability to overcome such perturbing.

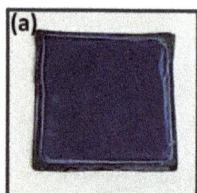 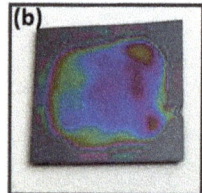

Figure 2. Pictures of two samples: (**a**) 1.2 wt.% + 25% volume of MeOH at 3k RPM with average thickness = 107 nm and Std/Ave = 2.01%; (**b**) 1.2 wt.% + 100% volume of MeOH at 1k RPM with average thickness = 327 nm and Std/Ave = 12%.

Table 3. Average thickness (nm) and standard deviation (nm) from 5 × 5 measurements of 0.5:2 MeOH/H$_2$O DNA solution. An amount of 2 mL of aqueous DNA with varying DNA concentration solutions diluted with 0.5 mL of methanol solvent.

0.5:2 (MeOH/H$_2$O)	0.4 wt.% + 25% Volume of MeOH	0.8 wt.% + 25% Volume of MeOH	1.2 wt.% + 25% Volume of MeOH
1k RPM	Ave: 54.58 Std: 5.28 Std/Ave: 9.67%	Ave: 251.82 Std: 12.42 Std/Ave: 4.9%	Ave: 349.60 Std: 88.58 Std/Ave: 25.3%
3k RPM	Ave: 16.27 Std: 0.16 Std/Ave: 0.98%	Ave: 76.20 Std: 1.90 Std/Ave: 2.49%	Ave: 106.77 Std: 2.15 Std/Ave: 2.01%
5k RPM	Ave: 11.6 Std: 0.14 Std/Ave: 1.2%	Ave: 49.23 Std: 1.35 Std/Ave: 2.74%	Ave: 76.19 Std: 0.98 Std/Ave: 1.29%

Table 4. Average thickness (nm) and standard deviation (nm) from 5 × 5 measurements of 1:1 MeOH/H$_2$O DNA solution. An amount of 2 mL of aqueous DNA with varying DNA concentrations diluted with 2 mL of methanol solvent.

1:1 (MeOH/H$_2$O)	0.4 wt.% + 100% Volume of MeOH	0.8 wt.% + 100% Volume of MeOH	1.2 wt.% + 100% Volume of MeOH
1k RPM	Ave: 38.56 Std: 5.05 Std/Ave: 13.1%	Ave: 230.87 Std: 21.36 Std/Ave: 9.25%	Ave: 256.45 Std: 42.72 Std/Ave: 16.7%
3k RPM	Ave: 12.49 Std: 0.84 Std/Ave: 6.72%	Ave: 49.68 Std: 0.87 Std/Ave: 1.75%	Ave: 93.24 Std: 1.57 Std/Ave: 1.68%
5k RPM	Ave: 9.28 Std: 0.15 Std/Ave: 1.62%	Ave: 31.51 Std: 1.5 Std/Ave: 4.76%	Ave: 62.14 Std: 2.20 Std/Ave: 3.54%

3.3. Thin Films by Water–Ethanol, Isopropanol, 1-Butanol, and Acetone DNA Solutions

Table 5 confirms observations from above discussion. We expect that the most significant effect of alcohol choice on the final film will be at the 1:1 ratio of solvent to aqueous DNA. Thus, focusing on that ratio, we may conclude simple relations between alcohol choice and optimal conditions for thin-film fabrication.

Table 5. Average thickness (nm) and standard deviation (nm) from 5 × 5 measurements of 1:1 EtOH/H$_2$O, 1:1 isopropanol/H$_2$O, and 1:1 acetone/H$_2$O DNA solutions. A total of 2 mL of aqueous DNA at 8 mg/mL (0.8 wt.%) diluted with 2 mL of ethanol, isopropanol, or acetone solvent.

1:1 (Solvent/DNA Solution)	0.8 wt.% + 100% Volume of Ethanol	0.8 wt.% + 100% Volume of Isopropanol	0.8 wt.% + 100% Volume of Acetone
1k RPM	Ave: 215.55 Std: 15.0 Std/Ave: 6.96%	Ave: 230.97 Std: 16.96 Std/Ave: 7.3%	Ave: 160.02 Std: 26.48 Std/Ave: 16.5%
3k RPM	Ave: 42.46 Std: 2.66 Std/Ave: 6.26%	Ave: 43.85 Std: 4.34 Std/Ave: 9.89%	Ave: 45.46 Std: 1.63 Std/Ave: 3.58%
5k RPM	Ave: 25.23 Std: 2.41 Std/Ave: 9.55%	Ave: 25.2 Std: 7.18 Std/Ave: 28.5%	Ave: 31.39 Std: 1.53 Std/Ave: 4.9%

We also note the importance of the time change between solution deposition and initiating the spin coating. Since spin coating is dependent on the quality of the initial solution, if there are bubbles within the deposited solution, then the time to remove the bubbles can create inconsistent results. Experiments with additions of acetone to DNA solution and allowing a waiting period between deposition and initial spin, we observed a decrease in the uniformity of the resulting film with only a 40 s wait. It is suspected this was due to the quicker drying rate of the acetone–DNA mixture. While this may be a more intuitive result of the experiment since a long-enough waiting period will result in a drop-casted film. The important observation is that even in a short time frame, the quality of the final film can be greatly affected.

Moreover, the problem of predicting uniformity even without a waiting is intractable. This is a result of the plateauing and in some cases a decline in the degree of uniformity as the RPM increases. For example, at 1000 RPM the films are the least uniform and exhibit a significant increase in uniformity at 3000 RPM (lowest ratio of standard deviation to average, Std/Ave). In most cases, at 5000 RPM the uniformity 'levels off' or experiences a slight decrease (represented by an uptick in Figures 3 and 4 Std/Ave). This trend is noticeable regardless of DNA concentration; thus, there are non-trivial optimal conditions that determine uniformity. That is, an increase in the RPM will not consistently produce films that are more uniform. Instead, for the solvents studied here, there may be an optimal uniformity that does not occur when the RPM is at its highest. The preceding observation is made based on the choice of defining a film's uniformity by the ratio of its standard deviation and its average.

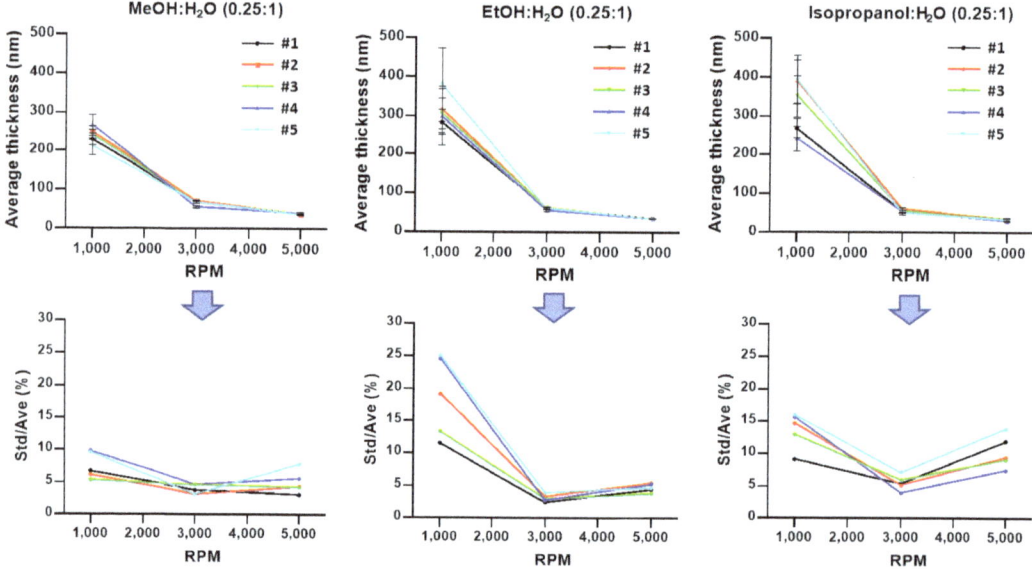

Figure 3. A total of 5 data sets of average thickness and standard deviation from 5 × 5 measurements of 0.25:1 MeOH/H$_2$O, 0.25:1 EtOH/H$_2$O, and 0.25:1 isopropanol/H$_2$O DNA solution films. An amount of 2 mL of aqueous DNA at 8 mg/mL (0.8 wt.%) diluted with 0.5 mL of methanol, ethanol, and isopropanol, respectively.

Solvent choice has little effect on final film thickness with a noted exception of solutions with 25% added isopropanol and were spun at 1000 RPM. The average thickness is greater than the average thickness produced by either 25% added methanol or ethanol (Figure 3). This appears to be a combination of the natural variability in film-to-film results of isopropanol solutions and the added variability of film-to-film results for highly concentrated solutions of DNA at a lower RPM. For the other series of data presented, in most cases thickness varies by less than 10 nanometers at higher 3000 and 5000 RPMs, suggesting that a high RPM dominates the spin-coating process even for concentrated solutions of DNA. Due to extra dilution, adding 100% solvent results in statistically thinner films at 1000 and 3000 RPMs when compared with an addition of only 25% solvent. At 5000 RPM this trend continues, however, the difference in thickness because of extra solvent dilution is significantly less when compared with thickness differences at 3000 RPM. As the RPM increases, the variability in thickness between films decreases, and thus it becomes significantly easier to predict the expected average thickness.

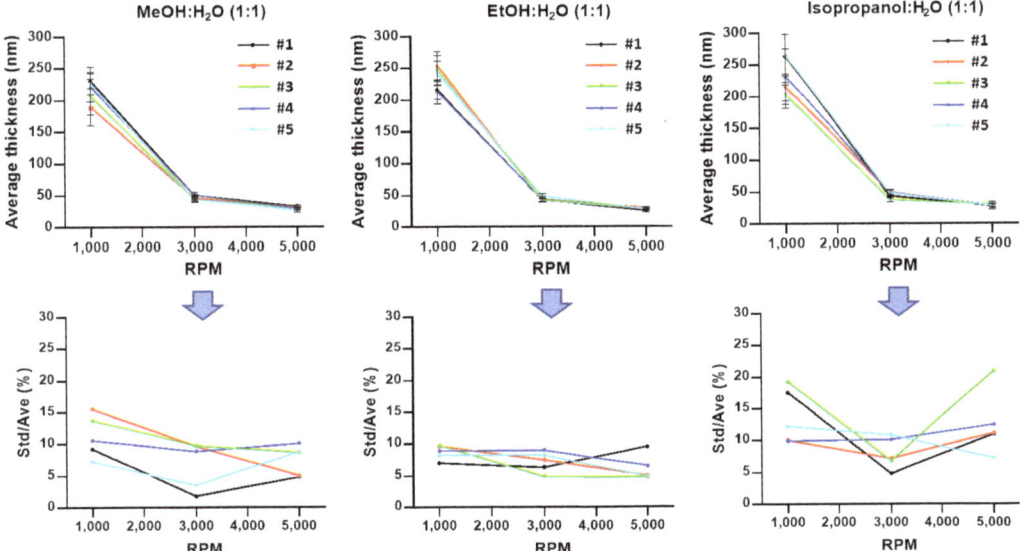

Figure 4. A total of 5 data sets of average thickness and standard deviation from 5 × 5 measurements of 1:1 MeOH/H_2O, 1:1 EtOH/H_2O, and 1:1 isopropanol/H_2O DNA solution films. An amount of 2 mL of aqueous DNA at 8 mg/mL (0.8 wt.%) diluted with 2 mL of methanol, ethanol, and isopropanol, respectively.

For 25% alcohol additions, the volume of the original solution, methanol was more consistently uniform than ethanol or isopropanol from a lower 1000 RPM to a higher 5000 RPM. At alcohol additions that were 100% of the original volume, ethanol was more consistently uniform than methanol, which was different from the trend observed at lower additions of methanol. However, in the case of 100% and 25% added solvents, there were individual films of methanol, ethanol, or isopropanol that appeared to be more uniform than the solvent that performs the best on average. This statistical phenomenon is well observed in methanol- and ethanol-added solutions at higher RPMs, where their performance was similar, and at higher spin rates despite the solvent addition, isopropanol produces the most inconsistent and non-uniform films.

Thus, in the case of uniformity, alcohol choice is important. It should also be noted that at 1:1 solvent/DNA ratio with 1-butanol, DNA will dissociate from the original solution, leading to an insufficient solution due to a clumping of large visible chains of DNA and immiscible solvents; this is a case where DNA–CTMA performs well due to the change in polarity. Upon examination of these tables alone, we cannot conclude why DNA performs best under these conditions, but it does motivate further research.

We illustrate uniformity in a visually intuitive manner via contour maps (Figures 5 and 6). The contour maps are constructed from selected data sets of 5 × 5 data points from the ellipsometry measurements of Figures 3 and 4; however, for the purpose of creating the contour maps, each point is modified by subtracting the average film thickness from the thickness at that point, and then the subsequent absolute value of the difference is divided by the average thickness, thus normalizing the points to the average film thickness, which is given by

$$Countour\ map\ point = \frac{\text{thickness at a point on film} - \text{average thickness of film}}{\text{average thickness of film}} \qquad (1)$$

The modified matrix of points is then mapped as a contour. The results illustrate how uniform the samples are among neighboring points, which provides a more robust understanding of uniformity and avoids complications with describing a sample uniformity using only standard deviation and thickness and has the advantage of utilizing complete information about the geometry of the films from the original data. For example, Figure 5 corresponds to thin films made by a one-to-one ratio of DNA solution-to-solvent addition (100% addition), and isopropanol at all RPMs exhibits random distributions of large variations in thickness occurring at the edges and center of the samples, whereas methanol- and ethanol-added solutions appear to have more centralized points of discontinuity and exhibit less significant continuity changes from point to point. Figure 6 demonstrates a similar pattern for 25% solvent additions, where again methanol and ethanol have centralized regions where thickness is significantly different from the average.

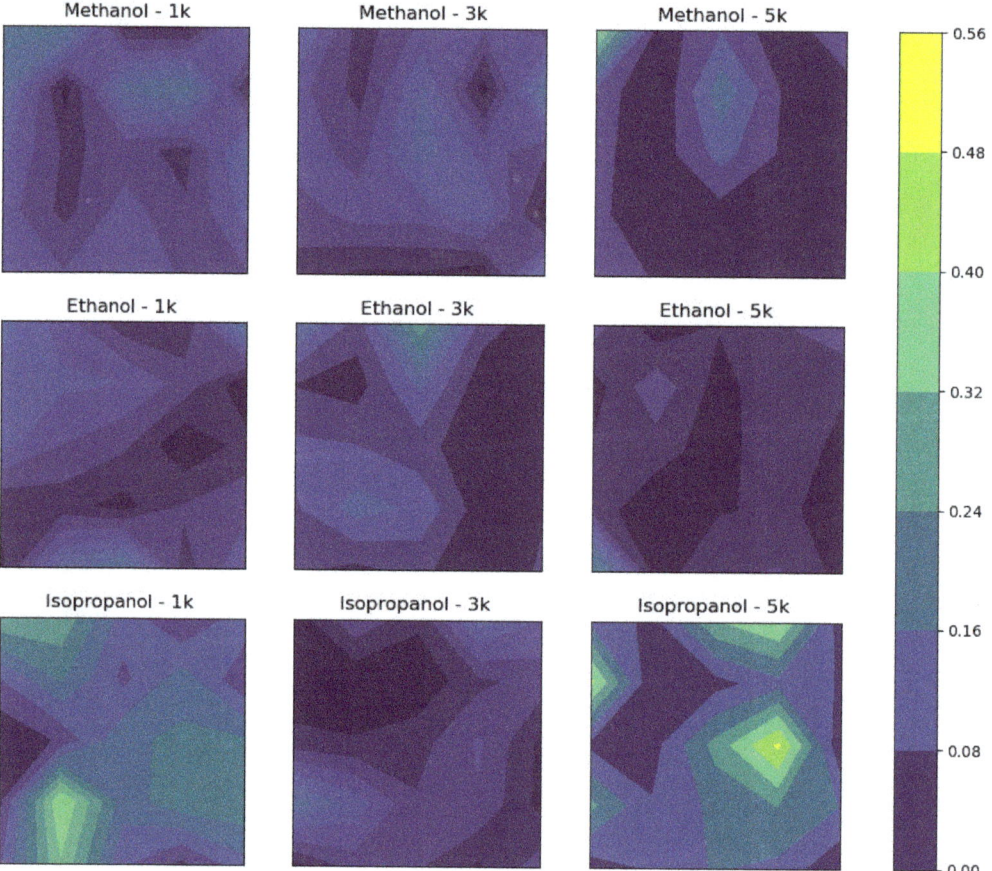

Figure 5. Contour plots constructed from the measurements made on films developed by 100% solvent additions at the specified RPM. A selected data set from Figure 4.

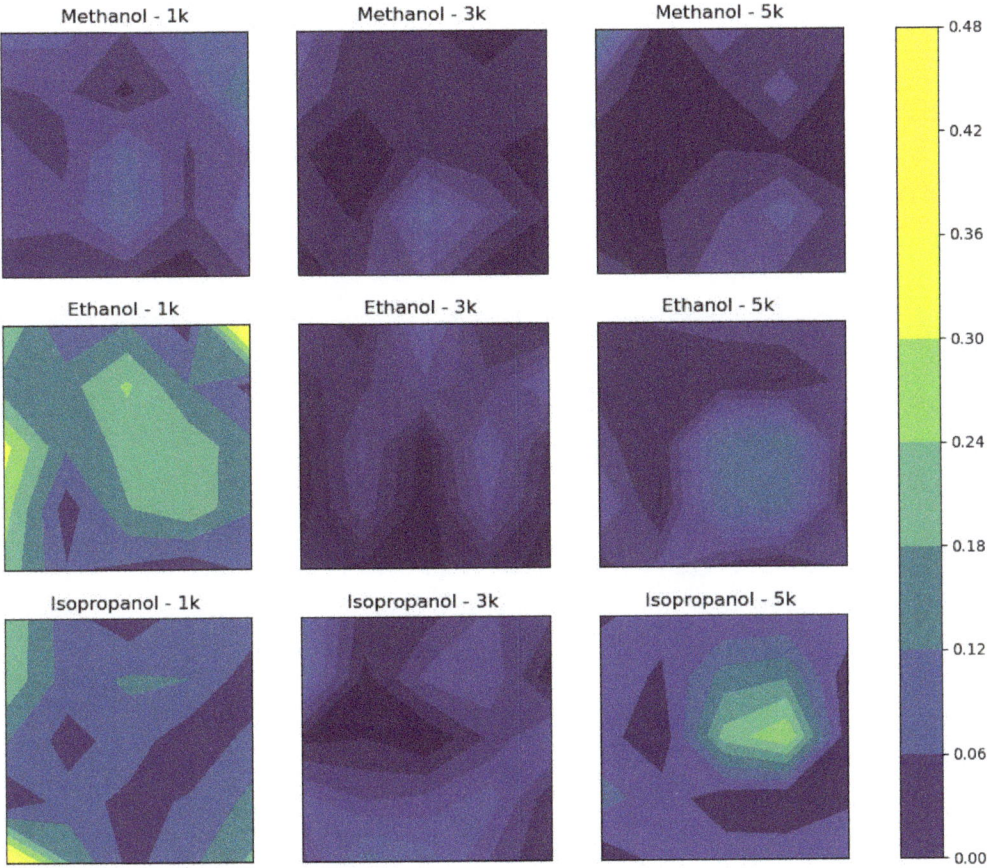

Figure 6. Contour plots constructed from the measurements made on films developed by 25% solvent additions at the specified RPM. A selected data set from Figure 3.

3.4. UV–VIS Measurements

UV–VIS analysis indicates that there is no significant absorbance peak shift in DNA regardless of the existence or choice of solvent (Figure 7). This is expected from previous research [20,30,39]; however, methanol, ethanol, and isopropanol exhibit different magnitudes in the peaks of DNA, and acetone solution is not graphed here since it exhibits a hard UV cutoff where the expected DNA peak is. Methanol and ethanol produce peaks of similar magnitude, while isopropanol produces a noticeably lower peak. This is also observed in cases where the DNA concentration is higher (data are not shown). We note that it is common for adding solvents to aqueous-based DNA solutions, which will lead to contraction in total volume, compared with the entire solution being purely aqueous, in which the contraction will lead to an increase in absorbance since there ought to be a denser spatial distribution of DNA in solvent-added solutions. However, DNA solutions interact differently with the alcohols, which may partially be explained by the polarity of the solvents, while all of the solvents (methanol, ethanol, isopropanol, acetone) are miscible in water. The polarity matches the magnitude of the curves, with methanol being the most polar and isopropanol being the least. This would similarly explain lab observations that when isopropanol is added to solutions with DNA, the DNA is visibly dissociated and coalesced in water and isopropanol mixture, but after magnetic stirring, the DNA aggregates become stable and are homogeneous within the solution by eye as shown in Figure 8. Due to the original coalescence, it is suspected that the DNA polymers may be greatly affected, and UV–VIS measurements indicate that the number of concentrations is lower despite volume contractions from adding non-aqueous solvents, thus resulting in solutions that are less dense in the number of DNA particles in the solutions.

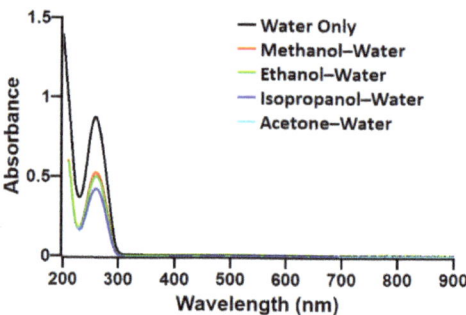

Figure 7. UV–VIS absorbance spectra of DNA solutions with/without solvents. Acetone–water DNA solution does not show the characteristic absorbance peak at 260 nm due to 330 nm UV cutoff of acetone. The optical path length of the cuvettes is 10 mm.

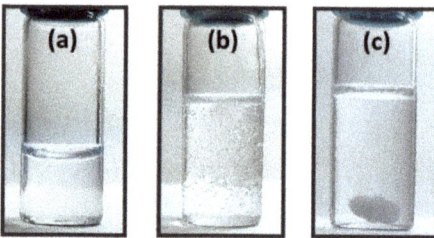

Figure 8. Pictures: (**a**) A highly homogeneous aqueous DNA solution. (**b**) The same volume of isopropanol is added into the sample of (**a**) with a vigorous quick shake. Bubbles in the entire solution and coalesced DNA at the top are present. (**c**) After 1 h of stirring of the sample of (**b**) with a magnetic stir bar, the DNA was dissolved homogeneously in isopropanol–water mixture.

4. Summary

Salmon DNA was dissolved in water, mixed, and later diluted with solvents of methanol, ethanol, isopropanol, acetone, and 1-butanol. The solutions were then used to make thin films by spin coating. The efficacy of the solvent additions to make uniform films was tested by varying DNA concentration, solvents, and spin rates, which are the major parameters of spin coating. Thickness measurements were made on a 5×5 grid of points on the substrate via focused-beam ellipsometry to determine an average and standard deviation, and subsequently quantify the uniformity. UV–VIS analysis was also performed on the solutions to demonstrate any changes in solutions after solvent addition.

Observations show that DNA aqueous solutions do not mix with 1-butanol, and thus are insufficient for thin-film fabrication. Acetone consistently produces the least uniform films given extra discontinuities on the surface. Of the alcohols, isopropanol is the least able to consistently produce a uniform thin film. Methanol and ethanol perform similarly; however, methanol is more consistent in producing uniform films regardless of spin rate. UV–VIS analysis indicates that the differences in performance may be a result of how concentrations are changed after solvent addition. This may also be a result of conformational changes to DNA structure as a result of alcohol addition, and due to these changes, the final film uniformity may be greatly affected by alcohol choice [40]. Greater addition of isopropanol may induce greater conformational changes that are ill suited for fabrication of thin films. Importantly, it was found the alcohol-added solutions are generally more capable of producing more uniform films, and optimal uniformity will not always be at the highest spin rate.

Author Contributions: Conceptualization, H.P.; methodology, J.R., M.N. and H.P.; formal analysis, J.R., M.N. and H.P.; investigation, J.R., M.N. and H.P.; data curation, J.R. and H.P.; writing—original draft preparation, J.R.; writing—review and editing, J.R. and H.P.; visualization, J.R. and H.P.; supervision, H.P.; project administration, H.P. All authors have read and agreed to the published version of the manuscript.

Funding: This research received no external funding.

Institutional Review Board Statement: Not applicable.

Informed Consent Statement: Not applicable.

Data Availability Statement: The data that support the findings of this study are available from the corresponding author upon reasonable request.

Acknowledgments: The authors acknowledge the kind support from Richard Selvaggi, a geriatrician in Commerce, Texas, who made a generous donation that helped fund the purchase of the optical ellipsometer used in this research.

Conflicts of Interest: The authors declare no conflict of interest.

References

1. Heckman, E.M.; Grote, J.G.; Yaney, P.P.; Hopkins, F.K. DNA-Based Nonlinear Photonic Materials. In Proceedings of the Optical Science and Technology, the SPIE 49th Annual Meeting, Denver, CO, USA, 2–6 August 2004; p. 47.
2. Grote, J.G.; Diggs, D.E.; Nelson, R.L.; Zetts, J.S.; Hopkins, F.K.; Ogata, N.; Hagen, J.A.; Heckman, E.; Yaney, P.P.; Stone, M.O.; et al. DNA Photonics [Deoxyribonucleic Acid]. *Mol. Cryst. Liq. Cryst.* **2005**, *426*, 3–17. [CrossRef]
3. Steckl, A.J. DNA—A new material for photonics? *Nat. Photonics* **2007**, *1*, 3–5. [CrossRef]
4. Ogata, N.; Yamaoka, K.; Yoshida, J. Progress of DNA biotronics and other applications. In Proceedings of the SPIE NanoScience + Engineering, San Diego, CA, USA, 1–5 August 2010; p. 776508.
5. Khazaeinezhad, R.; Hosseinzadeh Kassani, S.; Paulson, B.; Jeong, H.; Gwak, J.; Rotermund, F.; Yeom, D.-I.; Oh, K. Ultrafast nonlinear optical properties of thin-solid DNA film and their application as a saturable absorber in femtosecond mode-locked fiber laser. *Sci. Rep.* **2017**, *7*, 41480. [CrossRef]
6. Liang, L.; Fu, Y.; Wang, D.; Wei, Y.; Kobayashi, N.; Minari, T. DNA as Functional Material in Organic-Based Electronics. *Appl. Sci.* **2018**, *8*, 90. [CrossRef]
7. Sun, Q.; Chang, D.W.; Dai, L.; Grote, J.; Naik, R. Multilayer white polymer light-emitting diodes with deoxyribonucleic acid-cetyltrimetylammonium complex as a hole-transporting/electron-blocking layer. *Appl. Phys. Lett.* **2008**, *92*, 251108. [CrossRef]
8. Madhwal, D.; Rait, S.S.; Verma, A.; Kumar, A.; Bhatnagar, P.K.; Mathur, P.C.; Onoda, M. Increased luminance of MEH–PPV and PFO based PLEDs by using salmon DNA as an electron blocking layer. *J. Lumin.* **2010**, *130*, 331–333. [CrossRef]
9. Gomez, E.F.; Steckl, A.J. Improved Performance of OLEDs on Cellulose/Epoxy Substrate Using Adenine as a Hole Injection Layer. *ACS Photonics* **2015**, *2*, 439–445. [CrossRef]
10. Nakamura, K.; Ishikawa, T.; Nishioka, D.; Ushikubo, T.; Kobayashi, N. Color-tunable multilayer organic light emitting diode composed of DNA complex and tris(8-hydroxyquinolinato)aluminum. *Appl. Phys. Lett.* **2010**, *97*, 193301. [CrossRef]
11. Gomez, E.F.; Venkatraman, V.; Grote, J.G.; Steckl, A.J. DNA Bases Thymine and Adenine in Bio-Organic Light Emitting Diodes. *Sci. Rep.* **2015**, *4*, 7105. [CrossRef] [PubMed]
12. Wang, F.; Jin, S.; Sun, W.; Lin, J.; You, B.; Li, Y.; Zhang, B.; Hayat, T.; Alsaedi, A.; Tan, Z. Enhancing the Performance of Blue Quantum Dots Light-Emitting Diodes through Interface Engineering with Deoxyribonucleic Acid. *Adv. Opt. Mater.* **2018**, *6*, 1800578. [CrossRef]
13. Chopade, P.; Dugasani, S.R.; Jeon, S.; Jeong, J.-H.; Park, S.H. White light emission produced by CTMA-DNA nanolayers embedded with a mixture of organic light-emitting molecules. *RSC Adv.* **2019**, *9*, 31628–31635. [CrossRef]
14. Dagar, J.; Scarselli, M.; De Crescenzi, M.; Brown, T.M. Solar Cells Incorporating Water/Alcohol-Soluble Electron-Extracting DNA Nanolayers. *ACS Energy Lett.* **2016**, *1*, 510–515. [CrossRef]
15. Son, W.-H.; Reddy, M.S.P.; Choi, S.-Y. Hydrogenated amorphous silicon thin film solar cell with buffer layer of DNA-CTMA biopolymer. *Mod. Phys. Lett. B* **2014**, *28*, 1450107. [CrossRef]
16. Yusoff AR, B.M.; Kim, J.; Jang, J.; Nazeeruddin, M.K. New Horizons for Perovskite Solar Cells Employing DNA-CTMA as the Hole-Transporting Material. *ChemSusChem* **2016**, *9*, 1736–1742. [CrossRef]

17. Reddy, M.S.P.; Park, H.; Lee, J.-H. Residue-and-polymer-free graphene transfer: DNA-CTMA/graphene/GaN bio-hybrid photodiode for light-sensitive applications. *Opt. Mater.* **2018**, *76*, 302–307. [CrossRef]
18. Matsuo, Y.; Sugita, K.; Ikehata, S. Doping effect for ionic conductivity in DNA film. *Synth. Met.* **2005**, *154*, 13–16. [CrossRef]
19. Paulson, B.; Shin, I.; Jeong, H.; Kong, B.; Khazaeinezhad, R.; Dugasani, S.R.; Jung, W.; Joo, B.; Lee, H.-Y.; Park, S.; et al. Optical dispersion control in surfactant-free DNA thin films by vitamin B2 doping. *Sci. Rep.* **2018**, *8*, 9358. [CrossRef] [PubMed]
20. Jeong, H.; Oh, K. Uracil-doped DNA thin solid films: A new way to control optical dispersion of DNA film using a RNA constituent. *Opt. Express* **2019**, *27*, 36075. [CrossRef]
21. Yamahata, C.; Collard, D.; Takekawa, T.; Kumemura, M.; Hashiguchi, G.; Fujita, H. Humidity Dependence of Charge Transport through DNA Revealed by Silicon-Based Nanotweezers Manipulation. *Biophys. J.* **2008**, *94*, 63–70. [CrossRef] [PubMed]
22. Lee, H.; Lee, H.; Lee, J.E.; Lee, U.R.; Choi, D.H. Dielectric Function and Electronic Excitations of Functionalized DNA Thin Films. *Jpn. J. Appl. Phys.* **2010**, *49*, 061601. [CrossRef]
23. Johnson, L.E.; Latimer, L.N.; Benight, S.J.; Watanabe, Z.H.; Elder, D.L.; Robinson, B.H.; Bartsch, C.M.; Heckman, E.M.; Depotter, G.; Clays, K. Novel cationic dye and crosslinkable surfactant for DNA biophotonics. In Proceedings of the SPIE NanoScience + Engineering, San Diego, CA, USA, 12–16 August 2012; p. 84640D.
24. Heckman, E.M.; Hagen, J.A.; Yaney, P.P.; Grote, J.G.; Hopkins, F.K. Processing techniques for deoxyribonucleic acid; Biopolymer for photonics applications. *Appl. Phys. Lett.* **2005**, *87*, 211115. [CrossRef]
25. Hebda, E.; Jancia, M.; Kajzar, F.; Niziol, J.; Plelichowski, J.; Rau, I.; Tane, A. Optical Properties of Thin Films of DNA-CTMA and DNA-CTMA Doped with Nile Blue. *Mol. Cryst. Liq. Cryst.* **2012**, *556*, 309–316. [CrossRef]
26. Arasu, V.; Reddy Dugasani, S.; Son, J.; Gnapareddy, B.; Jeon, S.; Jeong, J.-H.; Ha Park, S. Thickness, morphology, and optoelectronic characteristics of pristine and surfactant-modified DNA thin films. *J. Phys. Appl. Phys.* **2017**, *50*, 415602. [CrossRef]
27. Samoc, A.; Galewski, Z.; Samoc, M.; Grote, J.G. Prism coupler and microscopic investigations of DNA films. In Proceedings of the NanoScience + Engineering, San Diego, CA, USA, 26–30 August 2007; p. 664607.
28. Samoc, A.; Miniewicz, A.; Samoc, M.; Grote, J.G. Refractive-index anisotropy and optical dispersion in films of deoxyribonucleic acid. *J. Appl. Polym. Sci.* **2007**, *105*, 236–245. [CrossRef]
29. Niziol, J.; Makyla-Juzak, K.; Marzec, M.M.; Ekiert, R.; Marzec, M.; Gondek, E. Thermal stability of the solid DNA as a novel optical material. *Opt. Mater.* **2017**, *66*, 344–350. [CrossRef]
30. Prajzler, V.; Jung, W.; Oh, K.; Cajzl, J.; Nekvindova, P. Optical properties of deoxyribonucleic acid thin layers deposited on an elastomer substrate. *Opt. Mater. Express* **2020**, *10*, 421. [CrossRef]
31. Pethrick, R.A.; Rankin, K.E. Criteria for uniform thin film formation for polymeric materials. *J. Mater. Sci. Mater. Electron.* **1999**, *10*, 141–144. [CrossRef]
32. Norrman, K.; Ghanbari-Siahkali, A.; Larsen, N.B. 6 Studies of spin-coated polymer films. *Annu. Rep. Sect. C Phys. Chem.* **2005**, *101*, 174. [CrossRef]
33. Mouhamad, Y.; Mokarian-Tabari, P.; Clarke, N.; Jones, R.A.L.; Geoghegan, M. Dynamics of polymer film formation during spin coating. *J. Appl. Phys.* **2014**, *116*, 123513. [CrossRef]
34. Na, J.Y.; Kang, B.; Sin, D.H.; Cho, K.; Park, Y.D. Understanding Solidification of Polythiophene Thin Films during Spin-Coating: Effects of Spin-Coating Time and Processing Additives. *Sci. Rep.* **2015**, *5*, 13288. [CrossRef]
35. Danglad-Flores, J.; Eickelmann, S.; Riegler, H. Deposition of polymer films by spin casting: A quantitative analysis. *Chem. Eng. Sci.* **2018**, *179*, 257–264. [CrossRef]
36. Jaczewska, J.; Budkowski, A.; Bernasik, A.; Raptis, I.; Raczkowska, J.; Goustouridis, D.; Rysz, J.; Sanopoulou, M. Humidity and solvent effects in spin-coated polythiophene–polystyrene blends. *J. Appl. Polym. Sci.* **2007**, *105*, 67–79. [CrossRef]
37. Plassmeyer, P.N.; Mitchson, G.; Woods, K.N.; Johnson, D.C.; Page, C.J. Impact of Relative Humidity during Spin-Deposition of Metal Oxide Thin Films from Aqueous Solution Precursors. *Chem. Mater.* **2017**, *29*, 2921–2926. [CrossRef]
38. Wang, J.; McGinty, C.; West, J.; Bryant, D.; Finnemeyer, V.; Reich, H.; Berry, S.; Clark, H.; Yaroshchuk, O.; Bos, P. Effects of humidity and surface on photoalignment of brilliant yellow. *Liq. Cryst.* **2017**, *44*, 863–872. [CrossRef]
39. Jeong, H.; Bjorn, P.; Hong, S.; Cheon, S.; Oh, K. Irreversible denaturation of DNA: A method to precisely control the optical and thermo-optic properties of DNA thin solid films. *Photonics Res.* **2018**, *6*, 918. [CrossRef]
40. Nakano, S.; Sugimoto, N. The structural stability and catalytic activity of DNA and RNA oligonucleotides in the presence of organic solvents. *Biophys. Rev.* **2016**, *8*, 11–23. [CrossRef]

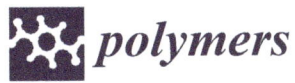 

Article

Strontium Aluminate-Based Long Afterglow PP Composites: Phosphorescence, Thermal, and Mechanical Characteristics

Anesh Manjaly Poulose [1,*], Arfat Anis [1,*], Hamid Shaikh [1], Abdullah Alhamidi [1], Nadavala Siva Kumar [2], Ahmed Yagoub Elnour [1] and Saeed M. Al-Zahrani [1]

[1] SABIC Polymer Research Center, Department of Chemical Engineering, King Saud University, Riyadh 11421, Saudi Arabia; hamshaikh@ksu.edu.sa (H.S.); AKFHK90@hotmail.com (A.A.); aelnour@ksu.edu.sa (A.Y.E.); szahrani@ksu.edu.sa (S.M.A.-Z.)

[2] Department of Chemical Engineering, King Saud University, Riyadh 11421, Saudi Arabia; snadavala@ksu.edu.sa

* Correspondence: apoulose@ksu.edu.sa (A.M.P.); aarfat@ksu.edu.sa (A.A.)

Abstract: A tremendous potential has been observed in the designing of long afterglow materials for sensing, bioimaging, and encryption applications. In this study, two different strontium aluminate-based luminescent materials; $SrAl_2O_4$: Eu, Dy (S_1), and $Sr_4Al_{14}O_{25}$: Eu, Dy (S_2) were melt-mixed with polypropylene (PP) matrix, and the phosphorescence properties were evaluated. After excitation at 320 nm, the PP/S_1 composite exhibited a green emission and the PP/S_2 generated a blue emission at 520 nm and 495 nm, respectively. The emission spectra intensity increased by increasing the content of these luminescent fillers. The attenuated total reflection-Fourier transform infrared (ATR-FTIR) experiments show that no chemical reaction occurred during the melt-mixing process. The differential scanning calorimetry (DSC) results revealed that the total crystallinity of the composites reduced by increasing the amount of the fillers; however, no changes in the temperature of melting (Tm) and crystallization (Tc) of PP were observed. Both fillers improved the impact strength of the composites, but the tensile strength (TS) and modulus (TM) decreased. Poly (ethylene glycol) dimethyl ether (P) plasticizer was used to improve the filler-matrix interaction and its dispersion; nevertheless, it adversely affected the intensity of the luminescence emissions.

Keywords: long afterglow PP composites; plasticizer; thermal; mechanical

Citation: Poulose, A.M.; Anis, A.; Shaikh, H.; Alhamidi, A.; Siva Kumar, N.; Elnour, A.Y.; Al-Zahrani, S.M. Strontium Aluminate-Based Long Afterglow PP Composites: Phosphorescence, Thermal, and Mechanical Characteristics. Polymers 2021, 13, 1373. https://doi.org/10.3390/polym13091373

Academic Editor: Swarup Roy

Received: 1 April 2021
Accepted: 21 April 2021
Published: 22 April 2021

Publisher's Note: MDPI stays neutral with regard to jurisdictional claims in published maps and institutional affiliations.

Copyright: © 2021 by the authors. Licensee MDPI, Basel, Switzerland. This article is an open access article distributed under the terms and conditions of the Creative Commons Attribution (CC BY) license (https://creativecommons.org/licenses/by/4.0/).

1. Introduction

Luminescent materials emit light, especially in the visible region. When a material continuously emits visible light for longer time (hours) after stopping the radiating source (visible, UV, X-ray, or gamma-ray radiation), a persistence of luminescence or phosphorescence is observed [1]. The first-generation phosphors were Cu or Mn-doped ZnS-based materials (green emission at 530 nm) [2,3]. These materials have been exploited in catalysts and optoelectronic devices. However, their functions are limited because of their low brightness and short afterglow time, and chemical instability in the presence of moisture and CO_2.

A new luminescence era started with the detection of rare-earth-doped phosphors. The first batch of modern luminescent materials was the rare-earth (R^{3+}) and Eu^{2+} doped alkaline-earth aluminates ($MAl_2O_4:Eu^{2+}$, R^{3+}; M = Ca, Sr, or Ba) [4–9]. The Eu^{2+} doped phosphor exhibits a bluish-green luminescence, and the glowing time can be enhanced by adding rare-earth ions, such as neodymium (Nd) or dysprosium (Dy) (e.g., $SrAl_xO_y$: Eu^{2+}, Dy^{3+}), or by adding Al_2O_3 [4,10]. These phosphors have attracted much attention because of their long phosphorescence, greater stability (i.e., moisture and photo-stability), and high quantum efficiency compared with sulfide-based phosphors [9,11,12]. These materials can obtain radiation energy from solar light, remaining photo-luminescent for long periods of time (12–20 h) [13,14]. The luminescent properties of these phosphors allowed their commercial acceptance because of the suitable usage in fluorescent lamps, glowing paints for highways buildings and airports, cathode ray tubes, plasma displays, textile, ceramic area, nighttime clocks, safety displays, among others. Significant growth in the use of these phosphors was observed in optoelectronics, telecommunications, optically active commercial products, and biomedical and way-finding systems [15–17].

Recently, various phases of strontium aluminates with rare-earth-doped were developed, such as $SrAl_2O_4:Eu^{2+}$, $Sr_2Al_6O_{11}:Eu^{2+}$, $Sr_4Al_{14}O_{25}:Eu^{2+}$, $SrAl_{12}O_{19}:Ce^{3+}$, $SrAl_{12}O_{19}:Pr^{3+}$, and $Sr_4Al_{14}O_{25}:Sm^{2+}$ [18–21]. The wavelength of the emitted visible light is decided by the crystalline phase structure of the resultant strontium aluminate [22,23]. Among the different $SrAl_xO_y: Eu^{2+}$, Dy^{3+} phosphors reported, $Sr_4Al_{14}O_{25}:$ Eu, Dy, and $SrAl_2O_4:$ Eu, Dy have exhibited the strongest potential for long phosphorescence and are commercially available [8,21,24]. Different synthesis techniques such as solid-state reactions [14,20,25], sol-gel method [11,24,26], combustion method [27], solvothermal method [28], chemical precipitation [29], microwave processing, and hydrothermal reaction [30] have been developed. These synthesis methods are often complex and require high temperatures for long-duration phosphorescence materials [31]. A mechanism for luminescence persistence was proposed for $SrAl_2O_4:Eu^{2+}$, Dy^{3+} and has been implied to explain the luminescence in several Dy^{3+} and Eu^{2+} co-doped silicates and aluminates. The mechanism is related to the thermally activated release of a hole from Eu^{2+} in its excited 5d state to the valence band, in which it is then trapped by Dy^{3+}. Luminescence is generated when the excited electron relaxes back to the ground state of Eu^{2+} [10,32]. The detailed mechanism of the process has been described [15]. However, $SrAl_xO_y: Eu^{2+}$, Dy^{3+} phosphors are not exempt from gradual luminescence decay due to their affinity towards moisture. Various encapsulation procedures have been described for gaining stable phosphor, such as Al_2O_3 [33], SrF_2 coating [34], phosphoric acid [35], organic ligands [36], amino-functionalized [37], among others, have been implied and reported. These processes are complicated, require elaborate equipment, and inversely affect the luminescence output. The easy and inexpensive method reported in the literature to prolong the afterglow properties of $SrAl_xO_y:$ Eu, Dy is the encapsulation within a polymer matrix, which acts as an insulator for moisture. This process enables the composites to exhibit better chemical stability, good physical properties and can be processed very easily [38–45]. In this study, two different strontium aluminate materials doped with Eu, Dy were incorporated in the selected poly(propylene) matrix, and the phosphorescent characteristics of the resultant composites were studied in detail. Moreover, a known plasticizer poly (ethylene glycol) dimethyl ether (P) was used to enhance the phosphor dispersion in the PP matrix and, also to evaluate its effect on the phosphorescence emission. The characterization studies on these composites provide valuable information on the fabrication of polymer-based luminescent films.

2. Materials and Methods

2.1. Materials

Poly(propylene) (TASNEE PP H4120) was provided by TASNEE with a density of 0.9 g/cm^3. It has a melt flow rate (MFR) of 12 g/10 min (ISO 1133). The strontium aluminate phosphors, $SrAl_2O_4:$ Eu, Dy (Mw = 209.11 g/mol) (S_1) and $Sr_4Al_{14}O_{25}:$ Eu, Dy (1139.55 g/mol) (S_2), were supplied by Sigma Aldrich. The plasticizer employed in this study was poly (ethylene glycol) dimethyl ether (P) purchased from Aldrich Company having a number average molecular weight of Mn 1000.

2.2. Methods

2.2.1. Preparation of the Composites

Different weight percentages of phosphors (1, 3, 5, and 10) were melt-mixed with the PP matrix in a Polylab QC (Brabender mixer) for a mixing period of 3 min at a temperature of 190–200 °C at 40 rpm. Thin films of 0.5 mm average thickness were made using COLLIN Press, Germany for the phosphorescence measurements.

2.2.2. Characterization of the Composites

Phosphorescence measurements

The phosphorescence tests were carried out in a Fluorescence Spectrophotometer (Agilent Technologies, Santa Clara, CA, USA) using a Xe ultraviolet (UV) lamp. The emission spectra were collected at the wavelength of excitation 320 nm.

Scanning Electron Microscope (SEM)

The morphological and elemental analyses were performed in a JEOL JSM-6360A, Japan SEM model, with energy-dispersive X-ray spectroscopy (EDS) facility. A thin cut surface of the composite was prepared for the analyses. The gold coating for these samples is performed in an auto fine coater (JFC/1600) for 30 s. The coating with gold was carried out to prevent the effect of charging and to improve the quality of the image.

Attenuated Total Reflection-Fourier Transform Infrared Spectroscopy (ATR-FTIR)

ATR-FTIR tests were performed in a Thermo-Scientific Nicolet iN10 FTIR model with germanium micro-tip attachment (400–4000 cm^{-1}).

Differential Scanning Calorimetry (DSC)

DSC tests were done in Shimadzu DSC-60A model. Approximately 6–10 mg of sample were taken in an aluminum pan and is heated from 30 °C to 220 °C at a ramp of 10 °C/min with 4 min holding time.

The percent crystallinity was evaluated as follows

$$X_c(\%) = \frac{\Delta H_m}{(1-\Phi)\Delta H_m^0} \times 100$$

where (Φ) is the filler weight fraction in the composites, (ΔH_m) is the melting enthalpy, and ($\Delta H^0{}_m$) is the melting enthalpy of 100% crystalline PP, and was reported as 207 J/g [46].

X-Ray Diffraction (XRD)

The crystalline studies were performed in a wide-angle XRD (Bruker D8 advance). The diffractometer was endowed with a wide-angle goniometer attached to a sealed-tube Cu-Kα radiation source (λ = 1.54056 Å). The scanning was done in the 2θ range of 5° to 50° at 5°/min in the reflection mode.

Thermo-gravimetric analysis (TGA)

TGA was done in a Shimadzu DTG-60H model. For the analysis, 10 ± 1.5 mg of the samples were maintained in an aluminum pan and is heated to a temperature of 600 °C (inert atmosphere) with a heating rate of 20 °C/min; and the loss in weight was monitored.

Mechanical properties

The standard tensile testing specimens (ASTM Type1, Dumb-bell shaped) were prepared using a DSM Xplore, Netherlands (12 cm^3, microinjection molding). The mold was maintained at room temperature and at a pressure of 6 bar. Tensile testing was performed in Hounsfield H100 KS model UTM (ASTM D638), and the mean of five test results was reported.

3. Results

3.1. DSC and ATR-FTIR Data

The DSC data for the PP and its composites are displayed in Tables 1 and 2. The presence of both S_1 and S_2 did not considerably affect the temperature of melting (T_m) and crystallization (T_c) of the composites, demonstrating that the PP was not interacting chemically with S_1 and S_2 and the mixing process was purely physical. Additionally, the FTIR spectra also support this observation as PP, PP/10S_1, PP/10S_2, and PP/10S_1/5P have similar FTIR spectra, in which no new peaks were observed nor either peaks diminished (Figure 1). This observation confirms the absence of chemical reactions between both S_1 and S_2 and PP. The sharp peaks at 2900 cm^{-1} were due to the asymmetrical CH$_2$ bending and the peaks at 1450 cm^{-1} and 1380 cm^{-1} were assigned to the symmetrical CH$_3$ bending and asymmetrical CH$_3$ bending, respectively, of PP [47].

Table 1. DSC data of PP, PP/S_1, and plasticized composites.

Material	T_c (°C)	T_m (°C)	ΔH_m (J/g)	X_c (%)
PP	121.9	164.6	87.0	42.0
PP/$1S_1$	122.6	165.7	83.8	40.5
PP/$3S_1$	122.7	164.5	83.2	40.2
PP/$5S_1$	122.5	164.1	83.3	40.2
PP/$10S_1$	122.9	164.2	83.8	40.5
PP/$10S_1$/2.5P	122.0	164.0	82.5	39.9
PP$10S_1$/5P	121.6	163.9	76.6	37.0

Table 2. DSC data of PP and PP/S_2 composites.

Material	T_c (°C)	T_m (°C)	ΔH_m (J/g)	X_c (%)
PP	121.9	164.6	87.0	42.0
PP/$1S_2$	120.9	163.5	74.9	36.2
PP/$3S_2$	120.0	164.4	71.4	34.5
PP/$5S_2$	120.1	164.6	68.6	33.1
PP/$10S_2$	120.7	164.9	66.9	32.3

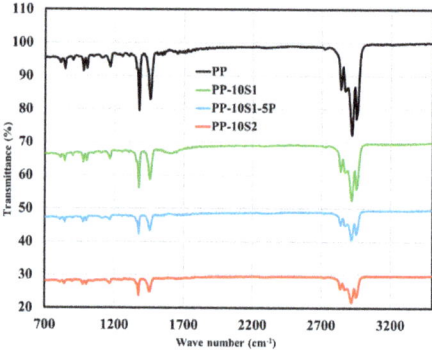

Figure 1. ATR-FTIR spectra of PP, PP/$10S_1$, PP/$10S_2$, and PP/$10S_1$/5P.

Conversely, the crystallinity of the composites decreased by increasing the S_1 and S_2 contents. The crystallinity decrease was more pronounced in S_2 because of the bulky chemical structure of S_2 compared with S_1, and in turn, restricts the PP chains mobility; thus, decreasing the crystallinity values [48]. The incorporation of the plasticizer in the composites led to a further reduction in the crystallinity percentage of the composites, as shown in Table 1.

3.2. X-ray Diffraction Studies of PP, PP/$10S_1$, PP/$10S_2$, and PP/$10S_1$/5P

Figure 2 illustrates the XRD patterns of neat PP, PP/$10S_1$, PP/$10S_2$, and PP/$10S_1$/5P. All composites show the characteristics diffraction peaks of α-PP, i.e., (110), (040), (130), and (111) [49]. Hence it was clear the absence of chemical reaction between PP and S_1, S_2 fillers, or the plasticizer. The (020) peak at 20°; (−211), (220), (211) peaks at 30°, and (031) peaks at 35° are the characteristics diffraction peaks of S_1 [39], and S_2 shows characteristics orthorhombic crystal structure with diffraction peaks at 25°, 27°, and 32° [50].

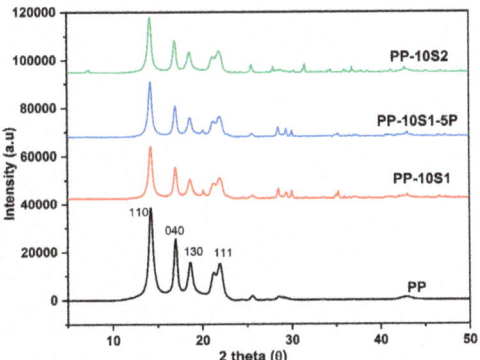

Figure 2. XRD pattern of PP, PP/10S$_1$, PP/10S$_1$/5P, and PP/10S$_2$.

3.3. Thermal Gravimetric Analysis (TGA)

The TGA results collected for PP, PP/S$_1$, PP/S$_2$, S$_1$, and S$_2$ under an inert atmosphere are shown in Figure 3A–C, respectively. For both S$_1$ and S$_2$ composites, the degradation process occurred in a single step and the composite with the highest filler loading have better thermal stability than that of neat PP. The better thermal stability of the composites is because of the fact that the inorganic filler particles (S$_1$ and S$_2$) can act as a barrier, slowing down the decomposition process of PP [43]. However, in low filler loading concentrations, the thermal stability decreased. Moreover, the residual weight left at the end of the TGA curve was proportional to the loading percentage of the S$_1$ and S$_2$ fillers. Both S$_1$ and S$_2$ are inorganic materials and are very much stable as seen from the TGA graph (Figure 3C). The weight loss is very much negligible for both the S$_1$ and S$_2$, and on comparing S$_1$ and S$_2$; S$_2$ is found to be slightly more stable than S$_1$. The S$_1$ and S$_2$ are found to be stable until 300 °C and the minor weight loss starts from that temperature, as shown in Figure 3C.

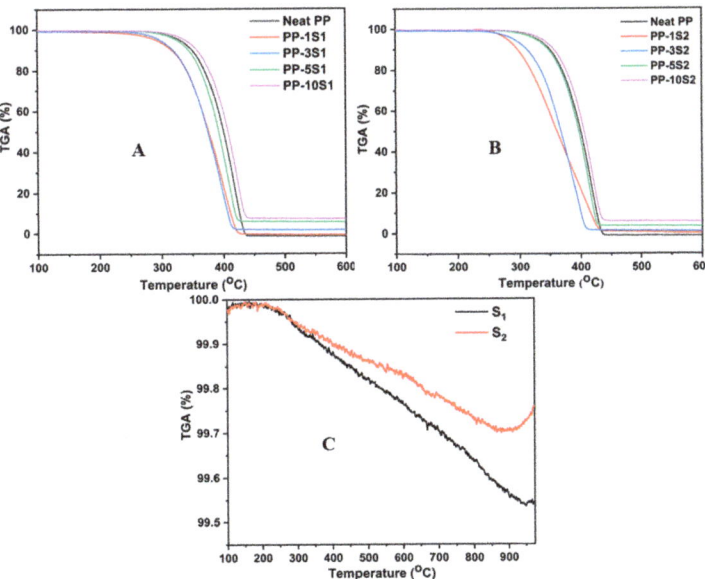

Figure 3. TGA curves of PP/S$_1$ (**A**), PP/S$_2$ composites (**B**), S$_1$ and S$_2$ (**C**).

3.4. Phosphorescence Emission

The phosphorescence emission spectra of PP/S_1 and PP/S_2 composites are presented in Figures 4 and 5, respectively. The emission spectra were collected at an excitation of 320 nm. As expected, the intensity of emission in the spectra of S_1 and S_2 were higher than that of PP composites because of the opacity and UV resistance of PP. In the PP/S_1 and PP/S_2 composites spectra, the emission intensity increased by increasing the percent loading of both S_1 and S_2. The PP/S_1 composite generated a green emission at 520 nm attributed to the electronic transition of europium divalent ion (Eu^{2+}) in the S_1 phosphors ($4f^65d^1$ to $4f^7$) [10,51]; the detailed mechanism of phosphorescence has already been described [52]. The PP/S_2 composites produced a blue emission at 495 nm and the emission intensity increased by increasing the S_2 content (Figure 5). The green (PP/S_1) and blue (PP/S_2) emissions in the dark are shown in Figure 6A,B, respectively.

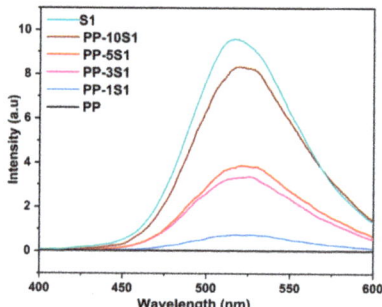

Figure 4. Phosphorescence emissions of S_1 and PP/S_1 composites (excitation wavelength: 320 nm; green emission at 520 nm).

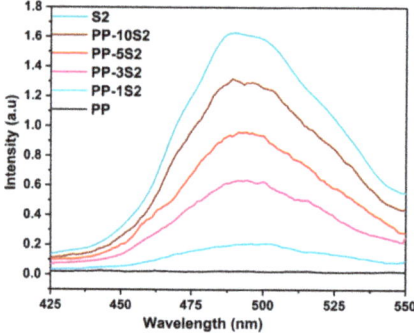

Figure 5. Phosphorescence emissions of S_2 and PP/S_2 composites (excitation wavelength: 320 nm; blue emission at 495 nm).

To investigate the effect of incorporation of plasticizer on the phosphorescence emission, 2.5 and 5 wt.% of plasticizers were added to composite with 10 wt.% of S_1; however, the plasticizer exhibited an adverse effect on the intensity of emissions. The incorporation of the plasticizer in the 10 wt.% S_1 composite decreases the overall phosphorescence emission intensity, as shown in Figure 7. The excitation process in S_1 and S_2 by absorbing UV light may get hindered in the presence of the plasticizer, which is more prone to degradation in UV light [53]. Because of the negative outcome of the plasticizer incorporation on the phosphorescence intensity of the PP/S_1 composites, they were not studied for the PP/S_2 composites.

Figure 6. Phosphorescent composites—(**A**) PP/10S$_1$ in the dark (green emission) and (**B**) PP/10S$_2$ in the dark (blue emission).

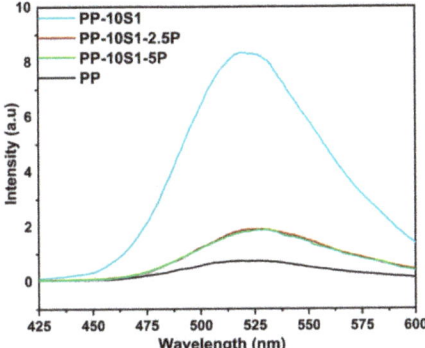

Figure 7. Phosphorescence emission of PP/10S$_1$ composites with 2.5 and 5 wt.% P plasticizers (excitation wavelength—320 nm; blue emission at 495 nm).

3.5. Mechanical Characteristics of PP/S$_1$ and PP/S$_2$ Composites

For practical application purposes, the composites must exhibit suitable physical properties. Therefore, the Izod impact strength and tensile data of the PP with PP/S$_1$ and PP/S$_2$ composites were evaluated and are illustrated in Figures 8 and 10, respectively.

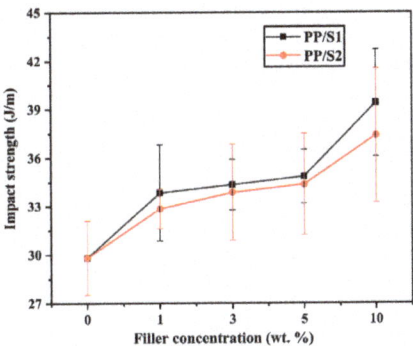

Figure 8. Impact strength of PP/S$_1$ and PP/S$_2$ composites.

The impact strength of PP/S$_1$ and PP/S$_2$ composites increased gradually by increasing the S$_1$ and S$_2$ contents as shown in Figure 8. The PP composites with 10 wt.% filler (S$_1$ and S$_2$) showed the highest notch impact strength, which is ~32% greater than that for PP.

The increase in the impact strength of the composites is because of the better interfacial adhesion among the PP and fillers (S_1 and S_2), which allows more efficient stress transfer. The distribution of fillers and the increased adhesion between the fillers and PP are visible in the SEM images of the composites with the highest S_1 and S_2 loading of 10 wt.% (Figure 9A,B).

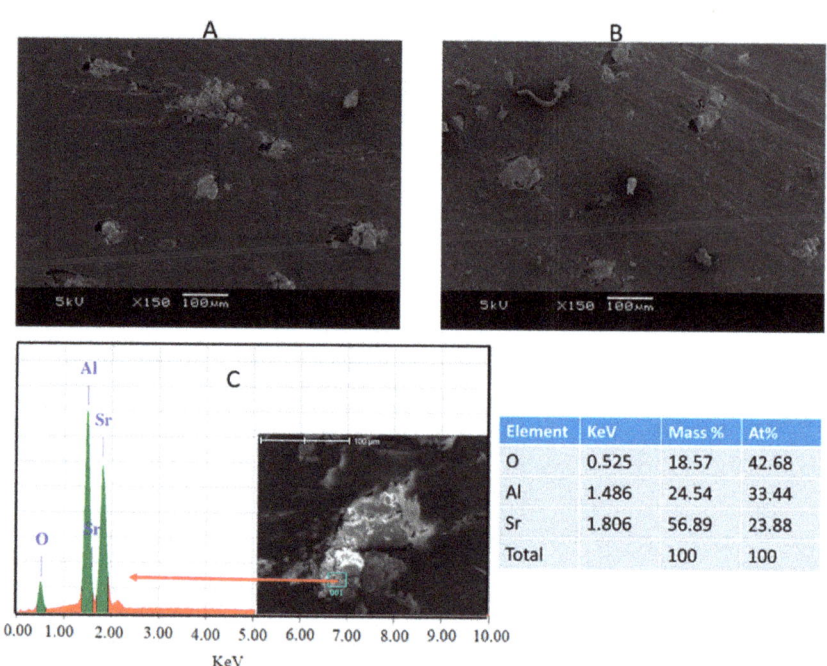

Figure 9. SEM images of PP/10S_1 (A) and PP/10S_2 (B) composites, and (C) SEM-EDS of PP/10S_1.

The tensile properties of PP and PP/S_1 and PP/S_2 composites are presented in Figure 10A,B, respectively. A gradual decrease in TS and tensile TM of the composites can be seen with the increase in the weight content of S_1 and S_2. In the highest filler loading (10 wt.% S_1 and S_2), the TS and TM decreased from 34.5 to 30.5 MPa and 1.1 to 0.94 GPa, respectively. The decrease in the tensile properties of PP/S_1 and PP/S_2 composites by increasing the filler loading can be due to the fact that the presence of inorganic fillers (S_1 and S_2) generally influences the elastic properties of PP because of their intrinsic stiffness and incapability to transfer the applied stress [37]. These observations are in agreement with the decrease in elongation at the yield values for these composites, as shown in Figure 10C. This is because of the decrease in PP ductility in the presence of S_1 and S_2 particles, which decreases the PP chain mobility. The inorganic-polymer composites often cause phase separation due to their incompatibility leading to a reduction in elongation at yield and break [54]. Additionally, the agglomeration of S_1 and S_2 filler, as shown in the SEM images (Figure 9A–C), adversely impacts the tensile modulus values.

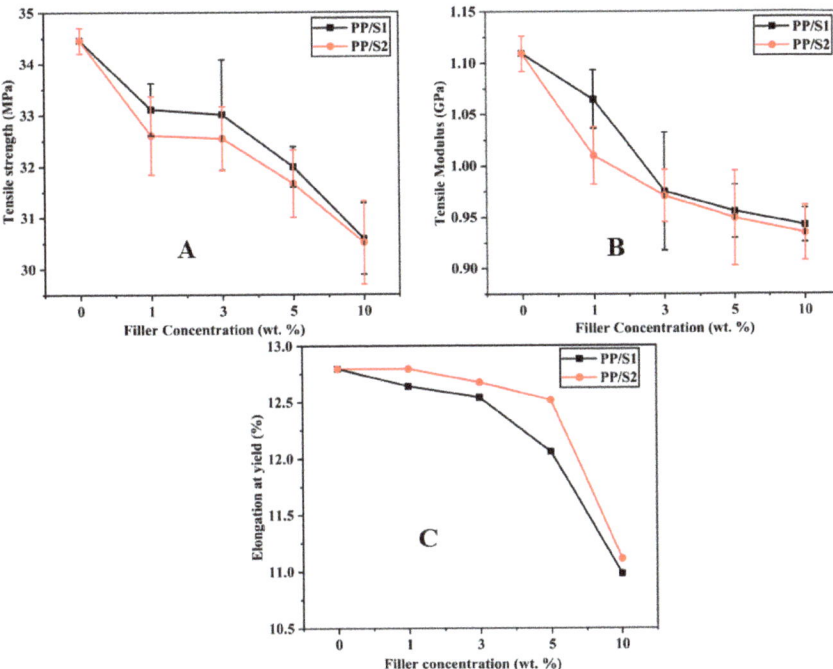

Figure 10. Tensile strength (**A**), tensile modulus (**B**), and elongation at yield (**C**) for PP/S_1 and PP/S_2 composites.

4. Conclusions

In this study, strontium aluminate-based phosphors (S_1 and S_2) were melt-mixed with PP matrix to achieve the long afterglow properties. A long-lasting, PP encapsulated S_1 and S_2 composites with long afterglow properties, which lasts for hours, were obtained. The ATR-FTIR spectra confirmed that the melt-blending process was physical. Moreover, the luminescence spectra of the composite have a major excitation peak at 320 nm and an emission peak at 520 nm (S_1; green) and 495 nm (S_2; blue), respectively. The thermal studies show that the Tc and Tm of PP were not affected by the S_1 and S_2 fillers. However, there was a significant decrease in the crystallinity of the composites with S_2 fillers, owing to the comparatively bulky chemical structure of these fillers. The impact strength of the resultant composites increased with the filler amount, but an adverse effect was witnessed on the TS and TM. These results demonstrated the satisfactory prospects for the formulation of phosphorescence films based on low-cost PP, which has great potential for applications in a new generation of light sources such as traffic signage and emergency signals.

Author Contributions: Conceptualization, A.M.P.; methodology, A.M.P.; investigation, A.A. (Abdullah Alhamidi) and A.Y.E.; writing draft preparation, A.M.P., A.A. (Arfat Anis) and N.S.K.; writing—review and editing, A.M.P., H.S., N.S.K., and A.A. (Arfat Anis); supervision, A.M.P. and S.M.A.-Z.; project administration, A.M.P. All authors have read and agreed to the published version of the manuscript.

Funding: This Project was funded by the National Plan for Science, Technology and Innovation (MAARIFAH), King Abdulaziz City for Science and Technology (13-ADV1044-02).

Institutional Review Board Statement: Not applicable.

Informed Consent Statement: Not Applicable.

Data Availability Statement: The data presented in this study are available on request from the corresponding author.

Acknowledgments: This Project was funded by the National Plan for Science, Technology and Innovation (MAARIFAH), King Abdulaziz City for Science and Technology, Kingdom of Saudi Arabia, Award Number (13-ADV1044-02).

Conflicts of Interest: The authors declare no conflict of interest.

References

1. Harvey, E.N. *A History of Luminescence from the Earliest Times Until 1900*; American Philosophical Society: Philadelphia, PA, USA, 1957. [CrossRef]
2. Liu, X.; Tian, B.; Yu, C.; Tu, B.; Zhao, D. Microwave-assisted solvothermal synthesis of radial ZnS nanoribbons. *Chem. Lett.* **2004**, *33*, 522–523. [CrossRef]
3. Wang, H.; Lu, X.; Zhao, Y.; Wang, C. Preparation and characterization of ZnS:Cu/PVA composite nanofibers via electrospinning. *Mater. Lett.* **2006**, *60*, 2480–2484. [CrossRef]
4. Matsuzawa, T.; Aoki, Y.; Takeuchi, N.; Murayama, Y. A new long phosphorescent phosphor with high brightness, $SrAl_2O_4$:Eu^{2+}, Dy^{3+}. *J. Electrochem. Soc.* **1996**, *143*, 2670–2673. [CrossRef]
5. Palilla, F.C.; Levine, A.K.; Tomkus, M.R. Fluorescent properties of alkaline Earth aluminates of the type MAl_2O_4 activated by divalent europium. *J. Electrochem. Soc.* **1968**, *115*, 642–644. [CrossRef]
6. Nakazawa, T.M.E. Traps in $SrAl_2O_4$: Eu^{2+} phosphor with rare-Earth ion doping. *J. Lumin.* **1997**, *72–74*, 236–237. [CrossRef]
7. Sakai, R.; Katsumata, T.; Komuro, S.; Morikawa, T. Effect of composition on the phosphorescence from $BaAl_2O_4$:Eu^{2+}, Dy^{3+} crystals. *J. Lumin.* **1999**, *85*, 149–154. [CrossRef]
8. Arellano-Tánori, O.; Meléndrez, R.; Pedroza-Montero, M.; Castañeda, B.; Chernov, V.; Yen, W.M.; Barboza-Flores, M. Persistent luminescence dosimetric properties of UV-irradiated $SrAl_2O_4$:Eu^{2+}, Dy^{3+} phosphor. *J. Lumin.* **2008**, *128*, 173–184. [CrossRef]
9. Smets, B.; Rutten, J.; Hoeks, G.; Verlijsdonk, J. $2SrO·3Al_2O_3$:Eu^{2+} and $1.29 (Ba, Ca) O, 6Al_2O_3$:$Eu^{2+}$: Two new blue-emitting phosphors. *J. Electrochem. Soc.* **1989**, *136*, 2119–2123. [CrossRef]
10. Nakazawa, E.; Murazaki, Y.; Saito, S. Mechanism of the persistent phosphorescence in $Sr_4Al_{14}O_{25}$: Eu and $SrAl_2O_4$: Eu codoped with rare Earth ions. *J. Appl. Phys.* **2006**, *100*, 113113. [CrossRef]
11. Zhang, P.; Xu, M.; Zheng, Z.; Liu, L.; Li, L. Synthesis and characterization of europium-doped $Sr_3Al_2O_6$ phosphors by sol–gel technique. *J. Sol Gel Sci. Technol.* **2007**, *43*, 59–64. [CrossRef]
12. Toh, K.; Nagata, S.; Tsuchiya, B.; Shikama, T. Luminescence characteristics of $Sr_4Al_{14}O_{25}$: Eu,Dy under proton irradiation. *Nucl. Instrum. Methods Phys. Res. B* **2006**, *249*, 209–212. [CrossRef]
13. Lu, Y.; Li, Y.; Xiong, Y.; Wang, D.; Yin, Q. $SrAl_2O_4$:Eu^{2+}, Dy^{3+} phosphors derived from a new sol–gel route. *Microelectron. J.* **2004**, *35*, 379–382. [CrossRef]
14. Katsumata, T.; Nabae, T.; Sasajima, K.; Komuro, S.; Morikawa, T. Effects of composition on the long phosphorescent $SrAl_2O_4$:Eu^{2+}, Dy^{3+} phosphor crystals. *J. Electrochem. Soc.* **1997**, *144*, L243–L245. [CrossRef]
15. Nance, J.; Sparks, T.D. From streetlights to phosphors: A review on the visibility of roadway markings. *Prog. Org. Coat.* **2020**, *148*, 105749. [CrossRef]
16. Tan, H.; Wang, T.; Shao, Y.; Yu, C.; Hu, L. Crucial breakthrough of functional persistent luminescence materials for biomedical and information technological applications. *Front. Chem.* **2019**, *7*, 387. [CrossRef]
17. Lin, Y.; Tang, Z.; Zhang, Z. Preparation of long-afterglow $Sr_4Al_{14}O_{25}$-based luminescent material and its optical properties. *Mater. Lett.* **2001**, *51*, 14–18. [CrossRef]
18. Zheng, R.; Xu, L.; Qin, W.; Chen, J.; Dong, B.; Zhang, L.; Song, H. Electrospinning preparation and photoluminescence properties of $SrAl_2O_4$:Ce^{3+} nanowires. *J. Mater. Sci.* **2011**, *46*, 7517–7524. [CrossRef]
19. Sakirzanovas, S.; Katelnikovas, A.; Dutczak, D.; Kareiva, A.; Jüstel, T. Synthesis and Sm^{2+}/Sm^{3+} doping effects on photoluminescence properties of $Sr_4Al_{14}O_{25}$. *J. Lumin.* **2011**, *131*, 2255–2262. [CrossRef]
20. Kim, S.J.; Won, H.I.; Hayk, N.; Won, C.W.; Jeon, D.Y.; Kirakosyan, A.G. Preparation and characterization of $Sr_4Al_2O_7$:Eu^{3+}, Eu^{2+} phosphors. *Mater. Sci. Eng. B* **2011**, *176*, 1521–1525. [CrossRef]
21. Chang, C.C.; Yang, C.Y.; Lu, C.H. Preparation and photoluminescence properties of $Sr_4Al_{14}O_{25}$:Eu^{2+} phosphors synthesized via the microemulsion route. *J. Mater. Sci. Mater. Electron.* **2013**, *24*, 1458–1462. [CrossRef]
22. Katsumata, T.; Sasajima, K.; Nabae, T.; Komuro, S.; Morikawa, T. Characteristics of strontium aluminate crystals used for long-duration phosphors. *J. Am. Ceram. Soc.* **1998**, *81*, 413–416. [CrossRef]
23. Suriyamurthy, N.; Panigrahi, B.S. Effects of non-stoichiometry and substitution on photoluminescence and afterglow luminescence of $Sr_4Al_{14}O_{25}$:Eu^{2+}, Dy^{3+} phosphor. *J. Lumin.* **2008**, *128*, 1809–1814. [CrossRef]
24. Peng, T.; Huajun, L.; Yang, H.; Yan, C. Synthesis of $SrAl_2O_4$: Eu, Dy phosphor nanometer powders by sol–gel processes and its optical properties. *Mater. Chem. Phys.* **2004**, *85*, 68–72. [CrossRef]

25. Yeşilay Kaya, S.; Karacaoglu, E.; Karasu, B. Effect of Al/Sr ratio on the luminescence properties of SrAl$_2$O$_4$:Eu^{2+}, Dy^{3+} phosphors. *Ceram. Int.* **2012**, *38*, 3701–3706. [CrossRef]
26. Chen, I.; Chen, T. Effect of host compositions on the afterglow properties of phosphorescent strontium aluminate phosphors derived from the sol–gel method. *J. Mater. Res.* **2001**, *16*, 1293–1300. [CrossRef]
27. Bem, D.B.; Swart, H.C.; Luyt, A.S.; Duvenhage, M.M.; Dejene, F.B. Characterization of luminescent and thermal properties of long afterglow SrAl$_x$O$_y$: Eu^{2+}, Dy^{3+} phosphor synthesized by combustion method. *Polym. Compos.* **2011**, *32*, 219–226. [CrossRef]
28. Xue, Z.; Deng, S.; Liu, Y. Synthesis and luminescence properties of SrAl$_2$O$_4$:Eu^{2+}, Dy^{3+} nanosheets. *Phys. B* **2012**, *407*, 3808–3812. [CrossRef]
29. Chang, C.; Yuan, Z.; Mao, D. Eu^{2+} activated long persistent strontium aluminate nano scaled phosphor prepared by precipitation method. *J. Alloys Compd.* **2006**, *415*, 220–224. [CrossRef]
30. Ravichandran, D.; Johnson, S.T.; Erdei, S.; Roy, R.; White, W.B. Crystal chemistry and luminescence of the Eu^{2+}-activated alkaline Earth aluminate phosphors. *Displays* **1999**, *19*, 197–203. [CrossRef]
31. Pathak, P.K.; Kurchania, R. Synthesis and thermoluminescence properties of SrAl$_2$O$_4$ (EU) phosphor irradiated with cobalt-60, 6 MV and 16 MV photon beams. *Radiat. Phys. Chem.* **2015**, *117*, 48–53. [CrossRef]
32. Dorenbos, P. Mechanism of persistent luminescence in Eu^{2+} and Dy^{3+} codoped aluminate and silicate compounds. *J. Electrochem. Soc.* **2005**, *152*, H107–H110. [CrossRef]
33. Lü, X.; Zhong, M.; Shu, W.; Yu, Q.; Xiong, X.; Wang, R. Alumina encapsulated SrAl$_2$O$_4$:Eu^{2+}, Dy^{3+} phosphors. *Powder Technol.* **2007**, *177*, 83–86. [CrossRef]
34. Guo, C.; Luan, L.; Huang, D.; Su, Q.; Lv, Y. Study on the stability of phosphor SrAl$_2$O$_4$:Eu^{2+}, Dy^{3+} in water and method to improve Its moisture resistance. *Mater. Chem. Phys.* **2007**, *106*, 268–272. [CrossRef]
35. Zhu, Y.; Zeng, J.; Li, W.; Xu, L.; Guan, Q.; Liu, Y. Encapsulation of strontium aluminate phosphors to enhance water resistance and luminescence. *Appl. Surf. Sci.* **2009**, *255*, 7580–7585. [CrossRef]
36. Wu, S.; Zhang, S.; Liu, Y.; Yang, J. The organic ligands coordinated long afterglow phosphor. *Mater. Lett.* **2007**, *61*, 3185–3188. [CrossRef]
37. Tian, S.; Wen, J.; Fan, H.; Chen, Y.; Yan, J.; Zhang, P. Sunlight-activated long persistent luminescent polyurethane incorporated with amino-functionalized SrAl$_2$O$_4$:Eu^{2+}, Dy^{3+} phosphor. *Polym. Int.* **2016**, *65*, 1238–1244. [CrossRef]
38. Bem, D.B.; Luyt, A.S.; Dejene, F.B.; Botha, J.R.; Swart, H.C. Structural, luminescent and thermal properties of blue SrAl$_2$O$_4$:Eu^{2+}, Dy^{3+} phosphor filled low-density polyethylene composites. *Phys. B* **2009**, *404*, 4504–4508. [CrossRef]
39. Bem, D.B.; Swart, H.C.; Luyt, A.S.; Coetzee, E.; Dejene, F.B. Properties of green SrAl$_2$O$_4$ phosphor in LDPE and PMMA polymers. *J. Appl. Polym. Sci.* **2010**, *117*, 2635–2640. [CrossRef]
40. Mishra, S.B.; Mishra, A.K.; Luyt, A.S.; Revaprasadu, N.; Hillie, K.T.; vdM Steyn, W.J.; Coetsee, E.; Swart, H.C. Ethylvinyl acetate copolymer-SrAl$_2$O$_4$: Eu,Dy and Sr$_4$Al$_{14}$O$_{25}$:Eu,Dy phosphor-based composites: Preparation and material properties. *J. Appl. Polym. Sci.* **2010**, *115*, 579–587. [CrossRef]
41. Mishra, S.B.; Mishra, A.K.; Revaprasadu, N.; Hillie, K.T.; Steyn, W.V.; Coetsee, E.; Swart, H.C. Strontium aluminate/polymer composites: Morphology, luminescent properties, and durability. *J. Appl. Polym. Sci.* **2009**, *112*, 3347–3354. [CrossRef]
42. Nance, J.; Sparks, T.D. Comparison of coatings for SrAl$_2$O$_4$:Eu^{2+}, Dy^{3+} powder in waterborne road striping paint under wet conditions. *Prog. Org. Coat.* **2020**, *144*, 105637. [CrossRef]
43. Wan, M.; Jiang, X.; Nie, J.; Cao, Q.; Zheng, W.; Dong, X.; Fan, Z.H.; Zhou, W. Phosphor powders-incorporated polylactic acid polymeric composite used as 3D printing filaments with green luminescence properties. *J. Appl. Polym. Sci.* **2020**, *137*, 48644. [CrossRef]
44. Piramidowicz, R.; Jusza, A.; Lipińska, L.; Gil, M.; Mergo, P. Re^{3+}: LaAlO$_3$ doped luminescent polymer composites. *Opt. Mater.* **2019**, *87*, 35–41. [CrossRef]
45. Barletta, M.; Puopolo, M.; Trovalusci, F.; Vesco, S. High-density polyethylene/SrAl$_2$O$_4$:Eu^{2+}, Dy^{3+} photoluminescent pigments: Material design, melt processing, and characterization. *Polym. Plast. Technol. Eng.* **2017**, *56*, 400–410. [CrossRef]
46. Paukkeri, R.; Lehtinen, A. Thermal behaviour of polypropylene fractions: 1. influence tacticity mol. weight crystallization melting behaviour. *Polymer* **1993**, *34*, 4075–4082. [CrossRef]
47. Poulose, A.M.; Elnour, A.Y.; Kumar, N.S.; Alhamidi, A.; George, J.; Al-Ghurabi, E.H.; Boumaza, M.; Al-Zahrani, S. Utilization of polyethylene terephthalate waste as a carbon filler in polypropylene matrix: Investigation of mechanical, rheological, and thermal properties. *J. Appl. Polym. Sci.* **2021**, *138*, 50292. [CrossRef]
48. Vidović, E.; Faraguna, F.; Jukić, A. Influence of inorganic fillers on PLA crystallinity and thermal properties. *J. Therm. Anal. Calorim.* **2017**, *127*, 371–380. [CrossRef]
49. Chiu, F.C.; Chu, P.H. Characterization of solution-mixed polypropylene/clay nanocomposites without compatibilizers. *J. Polym. Res.* **2006**, *13*, 73–78. [CrossRef]
50. Luitel, H.N.; Watari, T.; Torikai, T.; Yada, M. Luminescent properties of Cr^{3+} doped Sr$_4$Al$_{14}$O$_{25}$: Eu/Dy blue–green and red phosphor. *Opt. Mater.* **2009**, *31*, 1200–1204. [CrossRef]
51. Aitasalo, T.; Hölsä, J.; Jungner, H.; Lastusaari, M.; Niittykoski, J. Mechanisms of persistent luminescence in Eu^{2+}, Re^{3+} doped alkaline Earth aluminates. *J. Lumin.* **2001**, *94–95*, 59–63. [CrossRef]

52. Clabau, F.; Rocquefelte, X.; Jobic, S.; Deniard, P.; Whangbo, M.H.; Garcia, A.; Le Mercier, T. Mechanism of phosphorescence appropriate for the long-lasting phosphors Eu^{2+}-doped $SrAl_2O_4$ with codopants Dy^{3+} and B^{3+}. *Chem. Mater.* **2005**, *17*, 3904–3912. [CrossRef]
53. Das, I.; Gupta, S.K. Polyethylene glycol degradation by UV irradiation. *Indian J. Chem.* **2005**, *44a*, 1355–1358.
54. Mammeri, F.; Bourhis, E.L.; Rozes, L.; Sanchez, C. Mechanical properties of hybrid organic–inorganic materials. *J. Mater. Chem.* **2005**, *15*, 3787–3811. [CrossRef]

MDPI
St. Alban-Anlage 66
4052 Basel
Switzerland
Tel. +41 61 683 77 34
Fax +41 61 302 89 18
www.mdpi.com

Polymers Editorial Office
E-mail: polymers@mdpi.com
www.mdpi.com/journal/polymers

www.ingramcontent.com/pod-product-compliance
Lightning Source LLC
LaVergne TN
LVHW070706100526
838202LV00013B/1043